COMPUTATIONAL CHEMISTRY

COMPUTATIONAL CHEMISTRY

A Practical Guide for Applying Techniques to Real-World Problems

David C. Young
Cytoclonal Pharmaceutics Inc.

A JOHN WILEY & SONS, INC., PUBLICATION

New York • Chichester • Weinheim • Brisbane • Singapore • Toronto

The cover depicts a model compound for a new class of anti-cancer drugs under development of Cytoclonal Pharmaceutics.

This book is printed on acid-free paper. ∞

Published simultaneously in Canada.

For ordering and customer service, call 1-800-CALL-WILEY.

Library of Congress Cataloging-in-Publication Data:

Young, David C., 1964–
Computational chemistry: a practical guide for applying techniques to real world problems / David C. Young.
p. cm.
Includes index.
ISBN 0-471-33368-9 (cloth: alk paper)
1. Chemistry—Mathematics. 2 Chemistry—Data processing. 3. Chemistry—Computer simulation. I. Title.

QD39.3.M3 Y68 2001
540′.295—dc21 00-063341

Printed in the United States of America

10 9 8 7 6 5 4

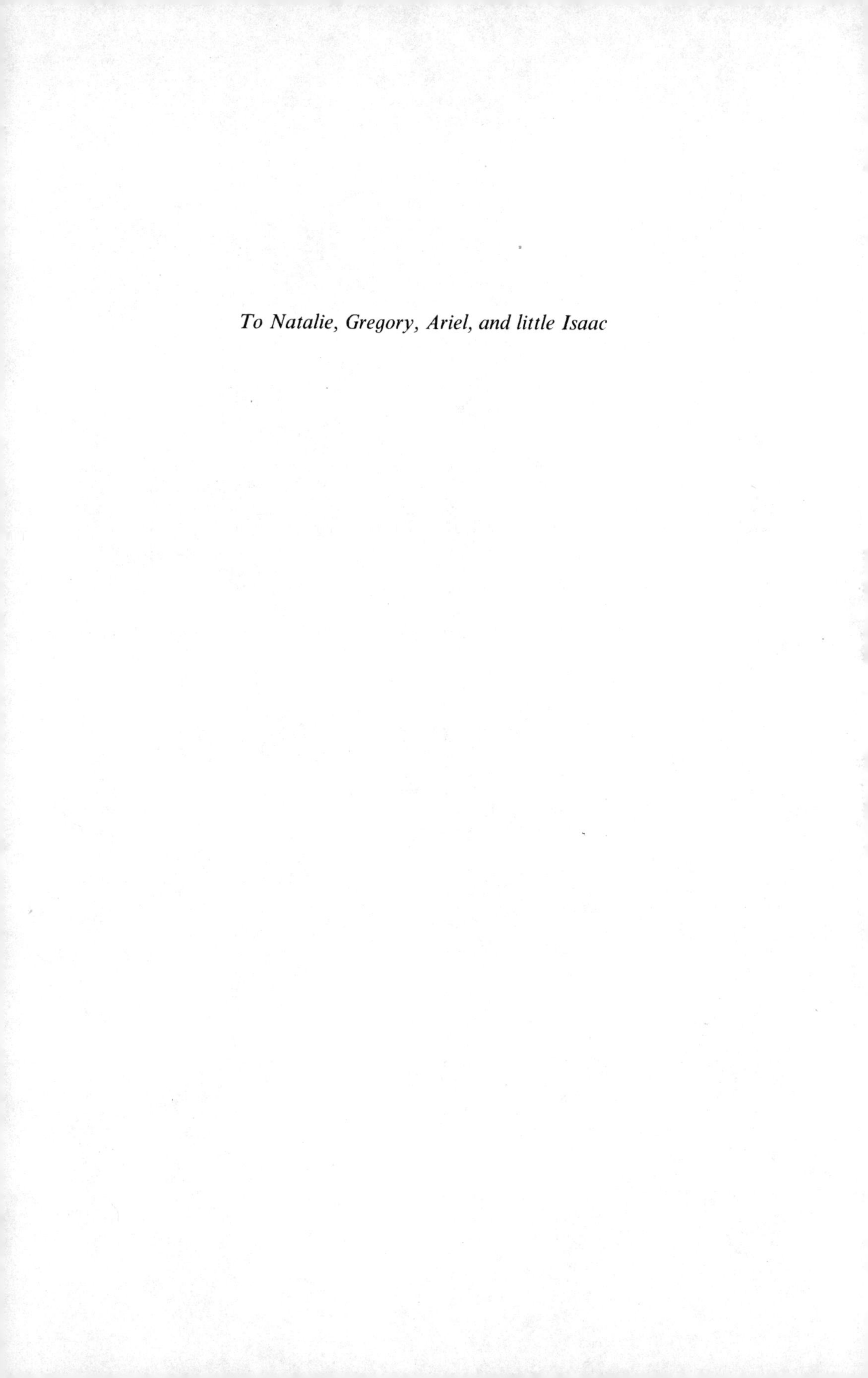

To Natalie, Gregory, Ariel, and little Isaac

CONTENTS

PREFACE xvii

ACKNOWLEDGMENTS xxi

SYMBOLS USED IN THIS BOOK xxiii

1. Introduction 1

1.1 Models, Approximations, and Reality 1
1.2 How Computational Chemistry Is Used 3
Bibliography 4

Part I. BASIC TOPICS 5

2. Fundamental Principles 7

2.1 Energy 7
2.2 Electrostatics 8
2.3 Atomic Units 9
2.4 Thermodynamics 9
2.5 Quantum Mechanics 10
2.6 Statistical Mechanics 12
Bibliography 16

3. *Ab initio* Methods 19

3.1 Hartree–Fock Approximation 19
3.2 Correlation 21
3.3 Møller–Plesset Perturbation Theory 22
3.4 Configuration Interaction 23
3.5 Multi-configurational Self-consistent Field 24
3.6 Multi-reference Configuration Interaction 25
3.7 Coupled Cluster 25
3.8 Quantum Monte Carlo Methods 26
3.9 Natural Orbitals 27
3.10 Conclusions 27
Bibliography 28

4. Semiempirical Methods **32**

4.1 Hückel 33
4.2 Extended Hückel 33
4.3 PPP 33
4.4 CNDO 34
4.5 MINDO 34
4.6 MNDO 34
4.7 INDO 35
4.8 ZINDO 35
4.9 SINDO1 35
4.10 PRDDO 36
4.11 AM1 36
4.12 PM3 37
4.13 PM3/TM 37
4.14 Fenske–Hall Method 37
4.15 TNDO 37
4.16 SAM1 38
4.17 Gaussian Theory 38
4.18 Recommendations 39
Bibliography 39

5. Density Functional Theory **42**

5.1 Basic Theory 42
5.2 Linear Scaling Techniques 43
5.3 Practical Considerations 45
5.4 Recommendations 46
Bibliography 46

6. Molecular Mechanics **49**

6.1 Basic Theory 49
6.2 Existing Force Fields 53
6.3 Practical Considerations 56
6.4 Recommendations 57
Bibliography 58

7. Molecular Dynamics and Monte Carlo Simulations **60**

7.1 Molecular Dynamics 60
7.2 Monte Carlo Simulations 62
7.3 Simulation of Molecules 63
7.4 Simulation of Liquids 64
7.5 Practical Considerations 64
Bibliography 65

8. Predicting Molecular Geometry **67**

8.1 Specifying Molecular Geometry 67
8.2 Building the Geometry 67
8.3 Coordinate Space for Optimization 68
8.4 Optimization Algorithm 70
8.5 Level of Theory 70
8.6 Recommendations 71
Bibliography 71

9. Constructing a Z-Matrix **73**

9.1 Z-Matrix for a Diatomic Molecule 73
9.2 Z-Matrix for a Polyatomic Molecule 73
9.3 Linear Molecules 74
9.4 Ring Systems 75
Bibliography 77

10. Using Existing Basis Sets **78**

10.1 Contraction Schemes 78
10.2 Notation 81
10.3 Treating Core Electrons 84
10.4 Common Basis Sets 85
10.5 Studies Comparing Results 89
Bibliography 90

11. Molecular Vibrations **92**

11.1 Harmonic Oscillator Approximation 92
11.2 Anharmonic Frequencies 94
11.3 Peak Intensities 95
11.4 Zero-point Energies and Thermodynamic Corrections 96
11.5 Recommendations 96
Bibliography 96

12. Population Analysis **99**

12.1 Mulliken Population Analysis 99
12.2 Löwdin Population Analysis 100
12.3 Natural Bond-Order Analysis 100
12.4 Atoms in Molecules 101
12.5 Electrostatic Charges 102
12.6 Charges from Structure Only 102
12.7 Recommendations 103
Bibliography 105

13. Other Chemical Properties 107

13.1 Methods for Computing Properties 107
13.2 Multipole Moments 110
13.3 Fermi Contact Density 110
13.4 Electronic Spatial Extent and Molecular Volume 111
13.5 Electron Affinity and Ionization Potential 111
13.6 Hyperfine Coupling 112
13.7 Dielectric Constant 112
13.8 Optical Activity 113
13.9 Biological Activity 113
13.10 Boiling Point and Melting Point 114
13.11 Surface Tension 114
13.12 Vapor Pressure 115
13.13 Solubility 115
13.14 Diffusivity 115
13.15 Visualization 115
13.16 Conclusions 121
Bibliography 121

14. The Importance of Symmetry 125

14.1 Wave Function Symmetry 127
14.2 Transition Structures 127
Bibliography 127

15. Efficient Use of Computer Resources 128

15.1 Time Complexity 128
15.2 Labor Cost 132
15.3 Parallel Computers 132
Bibliography 133

16. How to Conduct a Computational Research Project 135

16.1 What Do You Want to Know? How Accurately? Why? 135
16.2 How Accurate Do You Predict the Answer Will Be? 135
16.3 How Long Do You Predict the Research Will Take? 136
16.4 What Approximations Are Being Made? Which Are Significant? 136
Bibliography 142

Part II. ADVANCED TOPICS **145**

17. Finding Transition Structures **147**

17.1 Introduction 147
17.2 Molecular Mechanics Prediction 148
17.3 Level of Theory 149
17.4 Use of Symmetry 151
17.5 Optimization Algorithms 151
17.6 From Starting and Ending Structures 152
17.7 Reaction Coordinate Techniques 154
17.8 Relaxation Methods 155
17.9 Potential Surface Scans 155
17.10 Solvent Effects 155
17.11 Verifying That the Correct Geometry Was Obtained 155
17.12 Checklist of Methods for Finding Transition Structures 156
Bibliography 157

18. Reaction Coordinates **159**

18.1 Minimum Energy Path 159
18.2 Level of Theory 160
18.3 Least Motion Path 161
18.4 Relaxation Methods 161
18.5 Reaction Dynamics 162
18.6 Which Algorithm to Use 162
Bibliography 162

19. Reaction Rates **164**

19.1 Arrhenius Equation 164
19.2 Relative Rates 165
19.3 Hard-sphere Collision Theory 165
19.4 Transition State Theory 166
19.5 Variational Transition State Theory 166
19.6 Trajectory Calculations 167
19.7 Statistical Calculations 168
19.8 Electronic-state Crossings 169
19.9 Recommendations 169
Bibliography 170

20. Potential Energy Surfaces **173**

20.1 Properties of Potential Energy Surfaces 173
20.2 Computing Potential Energy Surfaces 175

20.3 Fitting PES Results to Analytic Equations 176
20.4 Fitting PES Results to Semiempirical Models 177
Bibliography 177

21. Conformation Searching **179**

21.1 Grid Searches 180
21.2 Monte Carlo Searches 182
21.3 Simulated Annealing 183
21.4 Genetic Algorithms 184
21.5 Distance-geometry Algorithms 185
21.6 The Fragment Approach 186
21.7 Chain-Growth 186
21.8 Rule-based Systems 186
21.9 Using Homology Modeling 187
21.10 Handling Ring Systems 189
21.11 Level of Theory 190
21.12 Recommended Search Algorithms 190
Bibliography 190

22. Fixing Self-Consistent Field Convergence Problems **193**

22.1 Possible Results of an SCF Procedure 193
22.2 How to Safely Change the SCF Procedure 194
22.3 What to Try First 195
Bibliography 196

23. QM/MM **198**

23.1 Nonautomated Procedures 198
23.2 Partitioning of Energy 198
23.3 Energy Subtraction 200
23.4 Self Consistent Method 201
23.5 Truncation of the QM Region 202
23.6 Region Partitioning 203
23.7 Optimization 203
23.8 Incorporating QM Terms in Force Fields 203
23.9 Recommendations 204
Bibliography 204

24. Solvation **206**

24.1 Physical Basis for Solvation Effects 206
24.2 Explicit Solvent Simulations 207
24.3 Analytic Equations 207
24.4 Group Additivity Methods 208

24.5 Continuum Methods 208
24.6 Recommendations 212
Bibliography 213

25. Electronic Excited States **216**

25.1 Spin States 216
25.2 CIS 216
25.3 Initial Guess 217
25.4 Block Diagonal Hamiltonians 218
25.5 Higher Roots of a CI 218
25.6 Neglecting a Basis Function 218
25.7 Imposing Orthogonality: DFT Techniques 218
25.8 Imposing Orthogonality: QMC Techniques 219
25.9 Path Integral Methods 219
25.10 Time-dependent Methods 219
25.11 Semiempirical Methods 220
25.12 State Averaging 220
25.13 Electronic Spectral Intensities 220
25.14 Recommendations 220
Bibliography 221

26. Size Consistency **223**

26.1 Correction Methods 224
26.2 Recommendations 225
Bibliography 226

27. Spin Contamination **227**

27.1 How Does Spin Contamination Affect Results? 227
27.2 Restricted Open-shell Calculations 228
27.3 Spin Projection Methods 229
27.4 Half-electron Approximation 229
27.5 Recommendations 230
Bibliography 230

28. Basis Set Customization **231**

28.1 What Basis Functions Do 231
28.2 Creating Basis Sets from Scratch 231
28.3 Combining Existing Basis Sets 232
28.4 Customizing a Basis Set 233
28.5 Basis Set Superposition Error 237
Bibliography 238

29. Force Field Customization 239

29.1 Potential Pitfalls 239
29.2 Original Parameterization 240
29.3 Adding New Parameters 240
Bibliography 241

30. Structure–Property Relationships 243

30.1 QSPR 243
30.2 QSAR 247
30.3 3D QSAR 247
30.4 Comparative QSAR 249
30.5 Recommendations 249
Bibliography 249

31. Computing NMR Chemical Shifts 252

31.1 *Ab initio* Methods 252
31.2 Semiempirical Methods 253
31.3 Empirical Methods 253
31.4 Recommendations 254
Bibliography 254

32. Nonlinear Optical Properties 256

32.1 Nonlinear Optical Properties 256
32.2 Computational Algorithms 257
32.3 Level of Theory 259
32.4 Recommendations 259
Bibliography 260

33. Relativistic Effects 261

33.1 Relativistic Terms in Quantum Mechanics 261
33.2 Extension of Nonrelativistic Computational Techniques 262
33.3 Core Potentials 262
33.4 Explicit Relativistic Calculations 263
33.5 Effects on Chemistry 263
33.6 Recommendations 264
Bibliography 264

34. Band Structures 266

34.1 Mathematical Description of Energy Bands 266
34.2 Computing Band Gaps 266
34.3 Computing Band Structures 268

34.4 Describing the Electronic Structure of Crystals 269
34.5 Computing Crystal Properties 270
34.6 Defect Calculations 271
Bibliography 271

35. Mesoscale Methods **273**

35.1 Brownian Dynamics 273
35.2 Dissipative Particle Dynamics 274
35.3 Dynamic Mean-field Density Functional Method 274
35.4 Nondynamic Methods 275
35.5 Validation of Results 275
35.6 Recommendations 275
Bibliography 276

36. Synthesis Route Prediction **277**

36.1 Synthesis Design Systems 277
36.2 Applications of Traditional Modeling Techniques 279
36.3 Recommendations 280
Bibliography 280

Part III. APPLICATIONS **281**

37. The Computational Chemist's View of the Periodic Table **283**

37.1 Organic Molecules 283
37.2 Main Group Inorganics, Noble Gases, and Alkali Metals 285
37.3 Transition Metals 286
37.4 Lanthanides and Actinides 289
Bibliography 290

38. Biomolecules **296**

38.1 Methods for Modeling Biomolecules 296
38.2 Site-specific Interactions 297
38.3 General Interactions 298
38.4 Recommendations 298
Bibliography 298

39. Simulating Liquids **302**

39.1 Level of Theory 302
39.2 Periodic Boundary Condition Simulations 303

39.3 Recommendations 305
Bibliography 305

40. Polymers **307**

40.1 Level of Theory 307
40.2 Simulation Construction 309
40.3 Properties 310
40.4 Recommendations 315
Bibliography 315

41. Solids and Surfaces **318**

41.1 Continuum Models 318
41.2 Clusters 318
41.3 Band Structures 319
41.4 Defect Calculations 319
41.5 Molecular Dynamics and Monte Carlo Methods 319
41.6 Amorphous Materials 319
41.7 Recommendations 319
Bibliography 320

Appendix. Software Packages **322**

A.1 Integrated Packages 322
A.2 *Ab initio* and DFT Software 332
A.3 Semiempirical Software 340
A.4 Molecular Mechanics/Molecular Dynamics/Monte Carlo Software 344
A.5 Graphics Packages 349
A.6 Special-purpose Programs 352
Bibliography 358

GLOSSARY **360**

Bibliography 370

INDEX **371**

Preface

At one time, computational chemistry techniques were used only by experts extremely experienced in using tools that were for the most part difficult to understand and apply. Today, advances in software have produced programs that are easily used by any chemist. Along with new software comes new literature on the subject. There are now books that describe the fundamental principles of computational chemistry at almost any level of detail. A number of books also exist that explain how to apply computational chemistry techniques to simple calculations appropriate for student assignments. There are, in addition, many detailed research papers on advanced topics that are intended to be read only by professional theorists.

The group that has the most difficulty finding appropriate literature are working chemists, not theorists. These are experienced researchers who know chemistry and now have computational tools available. These are people who want to use computational chemistry to address real world research problems and are bound to run into significant difficulties. This book is for those chemists.

We have chosen to cover a large number of topics, with an emphasis on when and how to apply computational techniques rather than focusing on theory. Each chapter gives a clear description with just the amount of technical depth typically necessary to be able to apply the techniques to computational problems. When possible, the chapter ends with a list of steps to be taken for difficult cases.

There are many good books describing the fundamental theory on which computational chemistry is built. The description of that theory as given here in the first few chapters is very minimal. We have chosen to include just enough theory to explain the terminology used in later chapters.

The core of this book is the description of the many computation techniques available and when to use them. Prioritizing which techniques work better or worse for various types of problems is a double-edged sword. This is certainly the type of information that is of use in solving practical problems, but there is no rigorous mathematical way to prove which techniques work better than others. Even though this prioritization cannot be proven, it is better to have an approximate idea of what works best than to have no idea at all. These suggestions are obtained from a compilation of information based on lessons from our own experience, those of colleagues, and a large body of literature covering chemistry from organic to inorganic, from polymers to drug design. Unfortunately, making generalizations from such a broad range of applications means that there are bound to be exceptions to many of the general rules of thumb given here.

The reader is advised to start with this book and to then delve further into the computational literature pertaining to his or her specific work. It is impossible to reference all relevant works in a book such as this. The bibliography included at the end of each chapter primarily lists textbooks and review articles. These are some of the best sources from which to begin a serious search of the literature. It is always advisable to run several tests to determine which techniques work best for a given project.

The section on applications examines the same techniques from the standpoint of the type of chemical system. A number of techniques applicable to biomolecular work are mentioned, but not covered at the level of detail presented throughout the rest of the book. Likewise, we only provide an introduction to the techniques applicable to modeling polymers, liquids, and solids. Again, our aim was to not repeat in unnecessary detail information contained elsewhere in the book, but to only include the basic concepts needed for an understanding of the subjects involved.

We have supplied brief reviews on the merits of a number of software packages in the appendix. Some of these were included due to their widespread use. Others were included based on their established usefulness for a particular type of problem discussed in the text. Many other good programs are available, but space constraints forced us to select a sampling only. The description of the advantages and limitations of each software package is again a generalization for which there are bound to be exceptions. The researcher is advised to carefully consider the research task at hard and what program will work best in addressing it. Both software vendors and colleagues doing similar work can provide useful suggestions.

Although there are now many problems that can be addressed by occasional users of computational tools, a large number of problems exist that only career computational chemists, with the time and expertise, can effectively solve. Some of the readers of this book will undoubtedly decide to forego using computational chemistry, thus avoiding months of unproductive work that they cannot afford. Such a decision in and of itself is a valuable reason for doing a bit of reading rather than blindly attempting a difficult problem.

This book was designed to aid in research, rather than as a primary text on the subject. However, students may find some sections helpful. Advanced undergraduate students and graduate students will find the basic topics and applications useful. Beginners are advised to first become familiar with the use of computational chemistry software before delving into the advanced topics section. It may even be best to come back to this book when problems arise during computations. Some of the information in the advanced topics section is not expected to be needed until postgraduate work.

The availability of easily used graphic user interfaces makes computational chemistry a tool that can now be used readily and casually. Results may be obtained often with a minimum amount of work. However, if the methods used are not carefully chosen for the project at hand, these results may not in any way reflect reality. We hope that this book will help chemists solve the real-world problems they face.

DAVID C. YOUNG

Acknowledgments

This book grew out of a collection of technical-support web pages. Those pages were also posted to the computational chemistry list server maintained by the Ohio Supercomputer Center. Many useful comments came from the subscribers of that list. In addition, thanks go to Dr. James F. Harrison at Michigan State University for providing advice born of experience.

The decision to undertake this project was prompted by Barbara Goldman at John Wiley & Sons, who was willing to believe in a first-time author. Her suggestions greatly improved the quality of the finished text. Darla Henderson and Jill Roter were also very helpful in bringing the project to completion and making the existence of bureaucracy transparent.

Thanks go to Dr. Michael McKee at Auburn University and the Alabama Research and Education Network, both of which allowed software to be tested on their computers. Thanks are also due the Nichols Research Corporation and Computer Sciences Corporation and particularly Scott von Laven and David Ivey for being so tolerant of employees engaged in such job-related extra-curricular activities.

A special acknowledgment also needs to be made to my family, who have now decided that Daddy will always be involved in some sort of big project so they might as well learn to live with it. My 14-year-old son observed that the computer intended for creating this book's illustrations was the best game-playing machine in the neighborhood and took full advantage of it. Our third child was born half-way through this book's writing. Much time was spent at 2:00 A.M. with a bottle in one hand and a review article in the other.

Symbols Used in This Book

Note: A few symbols are duplicated. Although this is at times confusing, it does reflect common usage in the literature. Thus, it is an important notation for the reader to understand. Acronyms are defined in the glossary at the end of the book.

$\langle \rangle$	expectation value
Å	Angstroms
∇^2	Laplacian operator
α	a constant, or polarizability
β	a constant, or hyperpolarizability
χ	susceptibility tensor, or Flory–Huggins parameter
ε_0	vacuum permitivity constant
ε_s	relative permitivity
ϕ	electrostatic potential
Γ	a point in phase space, or a point in k-space
γ	overlap between orbitals, or second hyperpolarizability
$\hat{H}$	Hamiltonian operator
κ	dielectric constant
ν	frequency of light
ρ	electron density, also called the charge density
ρ	density of states
σ	surface tension
θ	bond angle
Ψ	wave function
φ	an orbital
ζ	exponent of a basis function
A	number of active space orbitals, preexponential factor, a constant, surface area, or a point in k-space
a	a constant
amu	atomic mass units
B	a constant
C	molecular orbital coefficient, contraction coefficient, or a constant
C_0	weight of the HF reference determinant in the CI
C_p	heat capacity
c	a constant
D	a derminant, bond dissociation energy, or number of degrees of freedom

d	a descriptor
E	energy, or electric field
E_a	activation energy
eV	electron volts
F	force
$f(\ )$	correlation function
G	Gibbs free energy
$g(r)$	radial distribution function
$\hat{H}$	Hamiltonian operator or matrix
$H^{(1)}$	first-order transition matrix
J	Joules
K	Kelvin, or a point in k-space
k	a constant
k_B	Boltzmann constant
kg	kilograms
k_x, k_y, k_z	coordinates in k-space
L	length of the side of a periodic box
l	bond length
M	number of atoms, number of angles
m	mass
N	number of molecules, particles, orbitals, basis functions, or bonds
n	number of cycles in the periodicity
$O(\ )$	time complexity
P	polarization
Q	partition function
q	charge
R	ideal gas constant
$R(\ )$	radial function
r	distance between two particles, or reaction rate
S	total spin
s	spin
T	temperature, or CPU time
T_g	glass transition temperature
V	volume
$w(\ )$	probability used for a weighted average
X	a point in k-space
Y	a point in k-space
Y_{lm}	angular function
x, y, z	Cartesian coordinates

COMPUTATIONAL CHEMISTRY

1 Introduction

Anyone can do calculations nowadays.
Anyone can also operate a scalpel.
That doesn't mean all our medical problems are solved.

—Karl Irikura

Recent years have witnessed an increase in the number of people using computational chemistry. Many of these newcomers are part-time theoreticians who work on other aspects of chemistry the rest of the time. This increase has been facilitated by the development of computer software that is increasingly easy to use. It is now so easy to do computational chemistry that calculations can be performed with no knowledge of the underlying principles. As a result, many people do not understand even the most basic concepts involved in a calculation. Their work, as a result, is largely unfocused and often third-rate.

The term *theoretical chemistry* may be defined as the mathematical description of chemistry. The term *computational chemistry* is generally used when a mathematical method is sufficiently well developed that it can be automated for implementation on a computer. Note that the words "exact" and "perfect" do not appear in these definitions. Very few aspects of chemistry can be computed exactly, but almost every aspect of chemistry has been described in a qualitative or approximately quantitative computational scheme. The biggest mistake a computational chemist can make is to assume that any computed number is exact. However, just as not all spectra are perfectly resolved, often a qualitative or approximate computation can give useful insight into chemistry if the researcher understands what it does and does not predict.

Most chemists want to avoid the paper-and-pencil type of work that theoretical chemistry in its truest form entails. However, keep in mind that it is precisely for this kind of painstaking and exacting research that many Nobel prizes have been awarded. This book will focus almost exclusively on the knowledge needed to effectively use existing computer software for molecular modeling.

1.1 MODELS, APPROXIMATIONS, AND REALITY

By the end of their college career, most chemistry students have noticed that the information being disseminated in their third- and fourth-year chemistry classes-level seems to conflict with what was taught in introductory courses.

The course instructors or professors have not tried to intentionally deceive their students. Most individuals cannot grasp the full depth and detail of any chemical concept the first time that it is presented to them. It has been found that most people learn complex subjects best when first given a basic description of the concepts and then left to develop a more detailed understanding over time. Despite the best efforts of educators, a few misconceptions are at times possibly introduced in the attempt to simplify complex material for freshmen students. The part of this process that perpetuates any confusion is the fact that texts and instructors alike often do not acknowledge the simplifications being presented.

The scientific method is taught starting in elementary school. The first step in the scientific method is to form a hypothesis. A hypothesis is just an educated guess or logical conclusion from known facts. It is then compared against all available data and its details developed. If the hypothesis is found to be consistent with known facts, it is called a theory and usually published. The characteristics most theories have in common are that they explain observed phenomena, predict the results of future experiments, and can be presented in mathematical form. When a theory is found to be always correct for many years, it is eventually referred to as a scientific law. However useful this process is, we often use constructs that do not fit in the scientific method scheme as it is typically described.

One of the most commonly used constructs is a model. A model is a simple way of describing and predicting scientific results, which is known to be an incorrect or incomplete description. Models might be simple mathematical descriptions or completely nonmathematical. Models are very useful because they allow us to predict and understand phenomena without the work of performing the complex mathematical manipulations dictated by a rigorous theory. Experienced researchers continue to use models that were taught to them in high school and freshmen chemistry courses. However, they also realize that there will always be exceptions to the rules of these models.

A very useful model is the Lewis dot structure description of chemical bonding. It is not a complete description of the molecules involved since it does not contain the kinetic energies of the particles or Coulombic interactions between the electrons and nuclei. The theory of quantum mechanics, which accounts correctly for these factors, does predict that only two electrons can have the same spatial distribution (one of α spin and one of β spin). The Lewis model accounts for this pairing and for the number of energy levels likely to be occupied in the electronic ground state. This results in the Lewis model being able to predict chemical bonding patterns and give an indication of the strength of the bonds (single bonds, double bonds, etc.). However, none of the quantum mechanics equations are used in applying this technique. An example of a quantitative model would be Troutan's rule for predicting the boiling points of normal liquids. Group additivity methods would be another example.

Approximations are another construct that is often encountered in chemistry. Even though a theory may give a rigorous mathematical description of chemical phenomena, the mathematical difficulties might be so great that it is just not feasible to solve a problem exactly. If a quantitative result is desired, the best technique is often to do only part of the work. One approximation is to completely leave out part of the calculation. Another approximation is to use an average rather than an exact mathematical description. Some other common approximation methods are variations, perturbations, simplified functions, and fitting parameters to reproduce experimental results.

Quantum mechanics gives a mathematical description of the behavior of electrons that has never been found to be wrong. However, the quantum mechanical equations have never been solved exactly for any chemical system other than the hydrogen atom. Thus, the entire field of computational chemistry is built around approximate solutions. Some of these solutions are very crude and others are expected to be more accurate than any experiment that has yet been conducted. There are several implications of this situation. First, computational chemists require a knowledge of each approximation being used and how accurate the results are expected to be. Second, obtaining very accurate results requires extremely powerful computers. Third, if the equations can be solved analytically, much of the work now done on supercomputers could be performed faster and more accurately on a PC.

This discussion may well leave one wondering what role reality plays in computation chemistry. Only some things are known exactly. For example, the quantum mechanical description of the hydrogen atom matches the observed spectrum as accurately as any experiment ever done. If an approximation is used, one must ask how accurate an answer should be. Computations of the energetics of molecules and reactions often attempt to attain what is called *chemical accuracy*, meaning an error of less than about 1 kcal/mol. This is sufficient to describe van der Waals interactions, the weakest interaction considered to affect most chemistry. Most chemists have no use for answers more accurate than this.

A chemist must realize that theories, models, and approximations are powerful tools for understanding and achieving research goals. The price of having such powerful tools is that not all of them are perfect. This may not be an ideal situation, but it is the best that the scientific community has to offer. Chemists are advised to develop an understanding of the nature of computational chemistry approximations and what results can be trusted with any given degree of accuracy.

1.2 HOW COMPUTATIONAL CHEMISTRY IS USED

Computational chemistry is used in a number of different ways. One particularly important way is to model a molecular system prior to synthesizing that molecule in the laboratory. Although computational models may not be perfect, they are often good enough to rule out 90% of possible compounds as being unsuitable for their intended use. This is very useful information because synthesizing a single compound could require months of labor and raw materials, and generate toxic waste.

A second use of computational chemistry is in understanding a problem more completely. There are some properties of a molecule that can be obtained computationally more easily than by experimental means. There are also insights into molecular bonding, which can be obtained from the results of computations, that cannot be obtained from any experimental method. Thus, many experimental chemists are now using computational modeling to gain additional understanding of the compounds being examined in the laboratory.

As computational chemistry has become easier to use, professional computational chemists have shifted their attention to more difficult modeling problems. No matter how easy computational chemistry becomes, there will always be problems so difficult that only an expert in the field can tackle them.

BIBLIOGRAPHY

For more discussion of the application of the scientific method to chemistry see

R. M. Hazen, J. Trefil, *Science Matters* Anchor, New York (1991).

L. Pauling, *General Chemistry* 13, Dover, New York (1970).

A historical perspective is given in

G. G. Hall, *Chem. Soc. Rev.* **2**, 21 (1973).

PART I
Basic Topics

2 Fundamental Principles

This chapter is in no way meant to impart a thorough understanding of the theoretical principles on which computational techniques are based. There are many texts available on these subjects, a selection of which are listed in the bibliography. This book assumes that the reader is a chemist and has already taken introductory courses outlining these fundamental principles. This chapter presents the notation and terminology that will be used in the rest of the book. It will also serve as a reminder of a few key points of the theory upon which computation chemistry is based.

2.1 ENERGY

Energy is one of the most useful concepts in science. The analysis of energetics can predict what molecular processes are likely to occur, or able to occur. All computational chemistry techniques define energy such that the system with the lowest energy is the most stable. Thus, finding the shape of a molecule corresponds to finding the shape with the lowest energy.

The amount of energy in a system is often broken down into kinetic energy and potential energy. The kinetic energy may be further separated into vibrational, translational and rotational motion. A distinction is also made between the kinetic energy due to nuclear motion versus that due to electron motion. The potential energy may be expressed purely as Coulomb's law, or it might be broken down into energies of bond stretching, bond bending, conformational energy, hydrogen bonds, and so on.

Chemical processes, such as bond stretching or reactions, can be divided into adiabatic and diabatic processes. Adiabatic processes are those in which the system does not change state throughout the process. Diabatic, or nonadiabatic, processes are those in which a change in the electronic state is part of the process. Diabatic processes usually follow the lowest energy path, changing state as necessary.

In formulating a mathematical representation of molecules, it is necessary to define a reference system that is defined as having zero energy. This zero of energy is different from one approximation to the next. For *ab initio* or density functional theory (DFT) methods, which model all the electrons in a system, zero energy corresponds to having all nuclei and electrons at an infinite distance from one another. Most semiempirical methods use a valence energy that corresponds to having the valence electrons removed and the resulting ions at an infinite distance. A few molecular mechanics methods use chemical standard states as zero energy, but most use a strainless molecule as zero energy. For some molecular mechanics methods, the zero of energy is completely arbitrary.

Even within a particular approximation, total energy values relative to the method's zero of energy are often very inaccurate. It is quite common to find that this inaccuracy is almost always the result of systematic error. As such, the most accurate values are often relative energies obtained by subtracting total energies from separate calculations. This is why the difference in energy between conformers and bond dissociation energies can be computed extremely accurately.

2.2 ELECTROSTATICS

Electrostatics is the study of interactions between charged objects. Electrostatics alone will not described molecular systems, but it is very important to the understanding of interactions of electrons, which is described by a wave function or electron density. The central pillar of electrostatics is Coulombs law, which is the mathematical description of how like charges repel and unlike charges attract. The Coulombs law equations for energy and the force of interaction between two particles with charges q_1 and q_2 at a distance r_{12} are

$$E = \frac{q_1 q_2}{r_{12}} \tag{2.1}$$

$$F = \frac{q_1 q_2}{r_{12}^2} \tag{2.2}$$

Note that these equations do not contain the constants that are typically included in introductory texts, such as the vacuum permitivity constant. Theoreticians, and thus software developers, work with a system of units called atomic units. Within this unit system, many of the fundamental constants are defined as having a value of 1. Atomic units will be used throughout this book unless otherwise specified.

Another very useful function from electrostatics is the electrostatic potential ϕ. The electrostatic potential is a function that is defined at every point in three-dimensional real space. If a charged particle is added to a system, without disturbing the system, the energy of placing it at any point in space is the electrostatic potential times the charge on the particle. The requirement that there is no movement of existing charges (polarization of electron density) is sometimes described by stating that the electrostatic potential is the energy of placing an infinitesimal point charge in the system. The application of electrostatic potentials to chemical systems will be discussed further in Chapter 13.

The statement of Coulombs law above assumes that the charges are sepa-

TABLE 2.1 Conversion Factors for Atomic Units

Property	Atomic Unit	Conversion
Length	Bohr	1 Bohr = 0.529177249 Å
Weight	atomic mass unit (amu)	1 amu = $1.6605402 \times 10^{-27}$ kg
Charge	electron charge	1 electron = 1.602188×10^{-19} Coulombs
Energy	Hartree	1 Hartree = 27.2116 eV
Charge separation	Bohr electron	1 Bohr electron = 2.541765 Debye

rated by a vacuum. If the charges are separated by some continuum medium, this interaction will be modified by the inclusion of a dielectric constant for that medium. For the description of molecules, it is correct to assume that the nuclei and electrons are in a vacuum. However, dielectric effects are often included in the description of solvent effects as described in Chapter 24.

The Poisson equation relates the electrostatic potential ϕ to the charge density ρ. The Poisson equation is

$$\nabla^2\phi = -\rho \tag{2.3}$$

This may be solved numerically or within some analytic approximation. The Poisson equation is used for obtaining the electrostatic properties of molecules.

2.3 ATOMIC UNITS

The system of atomic units was developed to simplify mathematical equations by setting many fundamental constants equal to 1. This is a means for theorists to save on pencil lead and thus possible errors. It also reduces the amount of computer time necessary to perform chemical computations, which can be considerable. The third advantage is that any changes in the measured values of physical constants do not affect the theoretical results. Some theorists work entirely in atomic units, but many researchers convert the theoretical results into more familiar unit systems. Table 2.1 gives some conversion factors for atomic units.

2.4 THERMODYNAMICS

Thermodynamics is one of the most well-developed mathematical descriptions of chemistry. It is the field of thermodynamics that defines many of the concepts of energy, free energy and entropy. This is covered in physical chemistry text books.

Thermodynamics is no longer a subject for an extensive amount of research. The reasons for this are two-fold: the completeness of existing or previous work and the general inability to provide detailed insight into chemical processes. Very often, any thermodynamic treatment is left for trivial pen-and-paper work since many aspects of chemistry are so accurately described with very simple mathematical expressions.

Computational results can be related to thermodynamics. The result of computations might be internal energies, free energies, and so on, depending on the computation done. Likewise, it is possible to compute various contributions to the entropy. One frustration is that computational software does not always make it obvious which energy is being listed due to the differences in terminology between computational chemistry and thermodynamics. Some of these differences will be noted at the appropriate point in this book.

2.5 QUANTUM MECHANICS

Quantum mechanics (QM) is the correct mathematical description of the behavior of electrons and thus of chemistry. In theory, QM can predict any property of an individual atom or molecule exactly. In practice, the QM equations have only been solved exactly for one electron systems. A myriad collection of methods has been developed for approximating the solution for multiple electron systems. These approximations can be very useful, but this requires an amount of sophistication on the part of the researcher to know when each approximation is valid and how accurate the results are likely to be. A significant portion of this book addresses these questions.

Two equivalent formulations of QM were devised by Schrödinger and Heisenberg. Here, we will present only the Schrödinger form since it is the basis for nearly all computational chemistry methods. The Schrödinger equation is

$$\hat{H}\Psi = E\Psi \tag{2.4}$$

where $\hat{H}$ is the Hamiltonian operator, Ψ a wave function, and E the energy. In the language of mathematics, an equation of this form is called an eigen equation. Ψ is then called the eigenfunction and E an eigenvalue. The operator and eigenfunction can be a matrix and vector, respectively, but this is not always the case.

The wave function Ψ is a function of the electron and nuclear positions. As the name implies, this is the description of an electron as a wave. This is a probabilistic description of electron behavior. As such, it can describe the probability of electrons being in certain locations, but it cannot predict exactly where electrons are located. The wave function is also called a probability amplitude because it is the square of the wave function that yields probabilities. This is the only rigorously correct meaning of a wave function. In order to obtain a physically relevant solution of the Schrödinger equation, the wave function must be continuous, single-valued, normalizable, and antisymmetric with respect to the interchange of electrons.

The Hamiltonian operator $\hat{H}$ is, in general,

$$\hat{H} = -\sum_{i}^{\text{particles}} \frac{\nabla_i^2}{2m_i} + \sum_{i<j}^{\text{particles}} \sum \frac{q_i q_j}{r_{ij}} \tag{2.5}$$

$$\nabla_i^2 = \frac{\partial^2}{\partial x_i^2} + \frac{\partial^2}{\partial y_i^2} + \frac{\partial^2}{\partial z_i^2} \tag{2.6}$$

where ∇_i^2 is the Laplacian operator acting on particle i. Particles are both electrons and nuclei. The symbols m_i and q_i are the mass and charge of particle i, and r_{ij} is the distance between particles. The first term gives the kinetic energy of the particle within a wave formulation. The second term is the energy due to Coulombic attraction or repulsion of particles. This formulation is the time-independent, nonrelativistic Schrödinger equation. Additional terms can appear in the Hamiltonian when relativity or interactions with electromagnetic radiation or fields are taken into account.

In currently available software, the Hamiltonian above is nearly never used. The problem can be simplified by separating the nuclear and electron motions. This is called the Born–Oppenheimer approximation. The Hamiltonian for a molecule with stationary nuclei is

$$\hat{H} = -\sum_{i}^{\text{electrons}} \frac{\nabla_i^2}{2} - \sum_{i}^{\text{nuclei}} \sum_{j}^{\text{electrons}} \frac{Z_i}{r_{ij}} + \sum_{i<j}^{\text{electrons}} \sum \frac{1}{r_{ij}} \tag{2.7}$$

Here, the first term is the kinetic energy of the electrons only. The second term is the attraction of electrons to nuclei. The third term is the repulsion between electrons. The repulsion between nuclei is added onto the energy at the end of the calculation. The motion of nuclei can be described by considering this entire formulation to be a potential energy surface on which nuclei move.

Once a wave function has been determined, any property of the individual molecule can be determined. This is done by taking the expectation value of the operator for that property, denoted with angled brackets $\langle\ \rangle$. For example, the energy is the expectation value of the Hamiltonian operator given by

$$\langle E \rangle = \int \Psi^* \hat{H} \Psi \tag{2.8}$$

For an exact solution, this is the same as the energy predicted by the Schrödinger equation. For an approximate wave function, this gives an approximation of the energy, which is the basis for many of the techniques described in subsequent chapters. This is called variational energy because it is always greater than or equal to the exact energy. By substituting different operators, it is possible to obtain different observable properties, such as the dipole moment or electron density. Properties other than the energy are not variational, because only the Hamiltonian is used to obtain the wave function in the widely used computational chemistry methods.

Another way of obtaining molecular properties is to use the Hellmann–Feynman theorem. This theorem states that the derivative of energy with respect to some property P is given by

$$\frac{dE}{dP} = \left\langle \frac{\partial \hat{H}}{\partial P} \right\rangle \tag{2.9}$$

This relationship is often used for computing electrostatic properties. Not all approximation methods obey the Hellmann–Feynman theorem. Only variational methods obey the Hellmann–Feynman theorem. Some of the variational methods that will be discussed in this book are denoted HF, MCSCF, CI, and CC.

2.6 STATISTICAL MECHANICS

Statistical mechanics is the mathematical means to calculate the thermodynamic properties of bulk materials from a molecular description of the materials. Much of statistical mechanics is still at the paper-and-pencil stage of theory. Since quantum mechanicians cannot exactly solve the Schrödinger equation yet, statistical mechanicians do not really have even a starting point for a truly rigorous treatment. In spite of this limitation, some very useful results for bulk materials can be obtained.

Statistical mechanics computations are often tacked onto the end of *ab initio* vibrational frequency calculations for gas-phase properties at low pressure. For condensed-phase properties, often molecular dynamics or Monte Carlo calculations are necessary in order to obtain statistical data. The following are the principles that make this possible.

Consider a quantity of some liquid, say, a drop of water, that is composed of N individual molecules. To describe the geometry of this system if we assume the molecules are rigid, each molecule must be described by six numbers: three to give its position and three to describe its rotational orientation. This $6N$-dimensional space is called phase space. Dynamical calculations must additionally maintain a list of velocities.

An individual point in phase space, denoted by Γ, corresponds to a particular geometry of all the molecules in the system. There are many points in this phase space that will never occur in any real system, such as configurations with two atoms in the same place. In order to describe a real system, it is necessary to determine what configurations could occur and the probability of their occurrence.

The probability of a configuration occurring is a function of the energy of that configuration. This energy is the sum of the potential energy from intermolecular attractive or repulsive forces and the kinetic energy due to molecular motion. For an ideal gas, only the kinetic energy needs to be considered. For a molecular gas, the kinetic energy is composed of translational, rotational, and vibrational motion. For a monatomic ideal gas, the energy is due to the translational motion only. For simplicity of discussion, we will refer to the energy of the system or molecule without differentiating the type of energy.

There is a difference between the energy of the system, composed of all molecules, and the energy of the individual molecules. There is an amount of energy in the entire system that is measurable as the temperature of the system. However, not all molecules will have the same energy. Individual molecules will have more or less energy, depending on their motion and interaction with other molecules. There is some probability of finding molecules with any given energy. This probability depends on the temperature T of the system. The function that gives the ratio of the number of molecules, N_i, with various energies, E_i, to the number of molecules in state j is the Boltzmann distribution, which is expressed as

$$\frac{N_i}{N_j} = e^{-(E_i - E_j)/k_B T} \tag{2.10}$$

where k_B is the Boltzmann constant, 1.38066×10^{-23} J/K.

Equation (2.10) is valid if there are an equal number of ways to put the system in both energy states. Very often, there are more states available with higher energies due to there being an increasing number of degenerate states. When this occurs, the percentage of molecules in each state is determined by multiplying the equation above by the number of states available. Thus, there is often a higher probability of finding high-energy molecules at higher temperatures as shown in Figure 2.1. Note that the ground state may be a very poor approximation to the average.

When some property of a system is measured experimentally, the result is an average for all of the molecules with their respective energies. This observed quantity is a statistical average, called a weighted average. It corresponds to the result obtained by determining that property for every possible energy state of the system, $A(\Gamma)$, and multiplying by the probability of finding the system in that energy state, $w(\Gamma)$. This weighted average must be normalized by a partition function Q, where

$$\langle A \rangle = \frac{\sum w(\Gamma) A(\Gamma)}{Q} \tag{2.11}$$

$$Q = \sum w(\Gamma) \tag{2.12}$$

This technique for finding a weighted average is used for ideal gas properties and quantum mechanical systems with quantized energy levels. It is not a convenient way to design computer simulations for real gas or condensed-phase

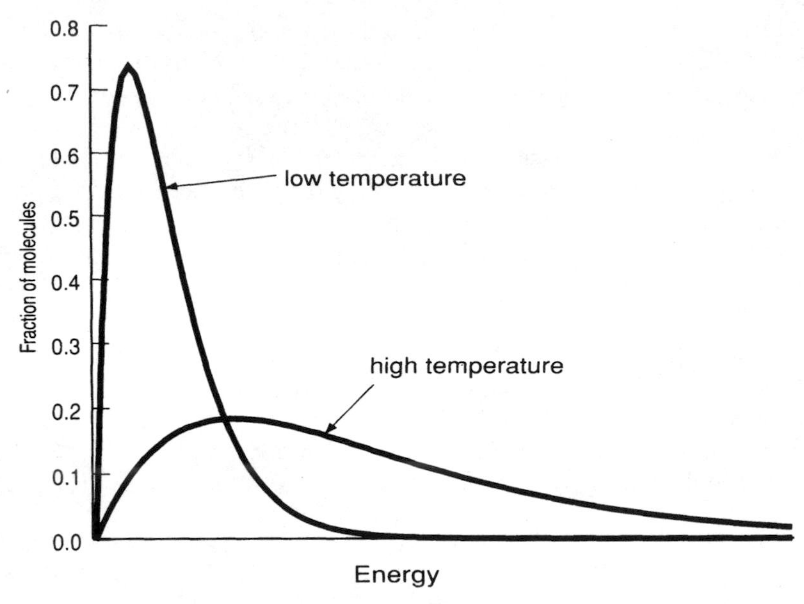

FIGURE 2.1 Fraction of molecules that will be found at various energies above the ground-state energy for two different temperatures.

systems, because determining every possible energy state is by no means a trivial task. However, a result can be obtained from a reasonable sampling of states. This results in values having a statistical uncertainty σ that is related to the number of states sampled M by

$$\sigma \alpha \frac{1}{\sqrt{M}} \tag{2.13}$$

There could also be systematic errors that are not indicated by this relationship.

Another way of formulating this problem is to use derivatives of the partition function without a weight function. This is done with the following relationships:

$$U = k_B T^2 \left(\frac{\partial \ln Q}{\partial T} \right)_V \tag{2.14}$$

$$A = -k_B T \ln Q \tag{2.15}$$

$$P = k_B T \left(\frac{\partial \ln Q}{\partial V} \right)_T \tag{2.16}$$

$$C_V = 2k_B T \left(\frac{\partial \ln Q}{\partial T} \right)_V + k_B T^2 \left(\frac{\partial^2 \ln Q}{\partial T^2} \right)_V \tag{2.17}$$

$$H = k_B T^2 \left(\frac{\partial \ln Q}{\partial T} \right)_V + k_B T V \left(\frac{\partial \ln Q}{\partial V} \right)_T \tag{2.18}$$

$$S = k_B T \left(\frac{\partial \ln Q}{\partial T} \right)_V + k_B \ln Q \tag{2.19}$$

$$G = k_B T V \left(\frac{\partial \ln Q}{\partial V} \right)_T - k_B T \ln Q \tag{2.20}$$

Other thermodynamic functions can be computed from these quantities. This is still not an ideal way to compute properties due to the necessity of accounting for all energy states of the system in order to obtain Q.

It is hardest to obtain precise values for the enthalpic values A, S, and G because they depend more heavily on high-energy states, which the system achieves infrequently. These functions depend on the actual value of Q, not just its derivatives.

There are several other, equivalent ways to obtain a statistical average. One of these is to use a time average. In this formulation, a calculation is designed to simulate the motion of molecules. At every step in the simulation, the property is computed for one molecule and averaged over all the time steps equally. This is equivalent to the weighted average because the molecule will be in more probable energy states a larger percentage of the time. The accuracy of this result depends on the number of time steps and the ability of the simulation to correctly describe how the real system will behave.

Another averaging technique is an ensemble average. Simulations often include thousands of molecules. A value can be averaged by including the result for every molecule in the simulation. This corresponds to the concept of an ensemble of molecules and is thus called an ensemble average. It is often most efficient to combine time averages and ensemble averages, thus averaging all molecules over many time steps.

Another type of property to examine is the geometric orientation of molecules. A set of Cartesian coordinates will describe a point in phase space, but it does not convey the statistical tendency of molecules to orient in a certain way. This statistical description of geometry is given by a radial distribution function, also called a pair distribution function. This is the function that gives the probability of finding atoms various distances apart. The radial distribution function gives an indication of phase behavior as shown in Figure 2.2. More detail can be obtained by using atom-specific radial distribution functions, such as the probability of finding a hydrogen atom various distances from an oxygen atom.

The connections between simulation and thermodynamics can be carried further. Simulations can be set up to be constant volume, pressure, temperature, and so on. Some of the most sophisticated simulations are those involving multiple phases or phase changes. These techniques are discussed further in Chapter 7.

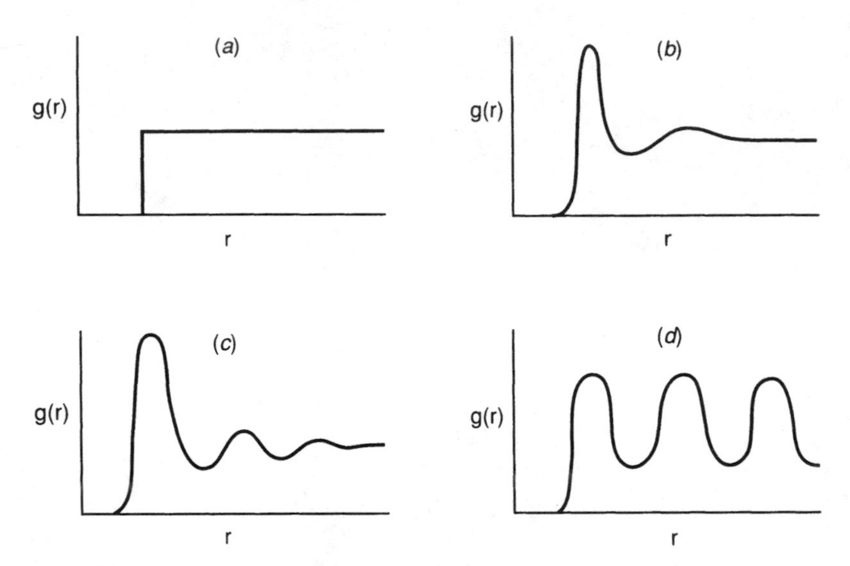

FIGURE 2.2 Radial distribution functions for (*a*) a hard sphere fluid, (*b*) a real gas, (*c*) a liquid, (*d*) a crystal.

BIBLIOGRAPHY

Sources covering fundamental principles are

T. P. Straatsma, J. A. McCammon, *Ann. Rev. Phys. Chem.* **43**, 407 (1992).

D. Halliday, R. Resnick, *Fundamentals of Physics* John Wiley & Sons, New York (1988).

R. P. Feynman, R. B. Leighton, M. Sands, *The Feynman Lectures on Physics* Addison-Wesley, Reading (1963).

A review of the implications of electrostatics is

G. Náray-Szabó, G. G. Ferency, *Chem. Rev.* **95**, 829 (1995).

Some texts covering thermodynamics are

I. N. Levine, *Physical Chemistry* McGraw-Hill, New York (1995).

J. W. Whalen, *Molecular Thermodynamics: A Statistical Approach* John Wiley & Sons, NY (1991).

P. W. Atkins, *Physical Chemistry* W. H. Freeman, New York (1990).

T. L. Hill, *Thermodynamics of Small Systems* Dover, New York (1964).

Some quantum mechanics texts are

J. E. House, *Fundamentals of Quantum Mechanics* Academic Press, San Diego (1998).

P. W. Atkins, R. S. Friedman, *Molecular Quantum Mechanics* Oxford, Oxford (1997).

J. Simons, J. Nichols, *Quantum Mechanics in Chemistry* Oxford, Oxford (1997).

R. H. Landau, *Quantum Mechanics II* John Wiley & Sons, New York (1996).

A. Szabo, N. S. Ostlund, *Modern Quantum Chemistry* Dover, New York (1996).

W. Greiner, *Quantum Mechanics An Introduction* Springer, Berlin (1994).

J. P. Lowe, *Quantum Chemistry* Academic Press, San Diego (1993).

P. W. Atkins, *Quanta* Oxford, Oxford (1991).

I. N. Levine, *Quantum Chemistry* Prentice Hall, Englewood Cliffs (1991).

R. E. Christoffersen, *Basic Principles and Techniques of Molecular Quantum Mechanics* Springer-Verlag, New York (1989).

D. A. McQuarrie *Quantum Chemistry* University Science Books, Mill Valley, CA (1983).

D. Bohm, *Quantum Theory* Dover, New York (1979).

C. Cohen-Tannoudji, B. Diu, F. Laloë, *Quantum Mechanics* Wiley-Interscience, New York (1977).

R. McWeeny, B. T. Sutcliffe, *Methods of Molecular Quantum Mechanics* Academic Press, London (1976).

E. Merzbacher, *Quantum Mechanics* John Wiley & Sons, New York (1970).

L. Pauling, E. B. Wilson, Jr., *Introduction to Quantum Mechanics With Applications to Chemistry* Dover, New York (1963).

P. A. M. Dirac, *The Principles of Quantum Mechanics* Oxford, Oxford (1958).

Some overviews of statistical mechanics are

A. Maczek, *Statistical Thermodynamics* Oxford, Oxford (1998).

J. Goodisman, *Statistical Mechanics for Chemists* John Wiley & Sons, New York (1997).

R. E. Wilde, S. Singh, *Statistical Mechanics: Fundamentals and Modern Applications* John Wiley & Sons, New York (1997).

D. Frenkel, B. Smit, *Understanding Molecular Simulations* Academic Press, San Diego (1996).

C. Garrod, *Statistical Mechanics and Thermodynamics* Oxford, Oxford (1995).

G. Jolicard, *Ann. Rev. Phys. Chem.* **46**, 83 (1995).

M. P. Allen, D. J. Tildesley, *Computer Simulation of Liquids* Oxford, Oxford (1987).

D. G. Chandler, *Introduction to Modern Statistical Mechanics* Oxford, Oxford (1987).

T. L. Hill, *An Introduction to Statistical Thermodynamics* Dover, New York (1986).

G. Sperber, *Adv. Quantum Chem.* **11**, 411 (1978).

D. A. McQuarrie, *Statistical Mechanics* Harper Collins, New York (1976).

W. Forst, *Chemn. Rev.* **71**, 339 (1971).

Adv. Chem. Phys. **vol. 11** (1967).

E. Schrödinger, *Statistical Thermodynamics* Dover, New York (1952).

Implications of the Born-Oppenheimer approximation are reviewed in

P. M. Kozlowski, L. Adamowicz, *Chem. Rev.* **93**, 2007 (1993).

Computation of free energies is reviewed in

D. A. Pearlman, B. G. Rao, *Encycl. Comput. Chem* **2**, 1036 (1998).

T. P. Straatsma, *Encycl. Comput. Chem.* **2**, 1083 (1998).

T. P. Straatsma, *Rev. Comput. Chem.* **9**, 81 (1996).

Path integral formulations of statistical mechanics are reviewed in

B. J. Berne, D. Thirumalai, *Ann. Rev. Phys. Chem.* **37**, 401 (1986).

Simulation and prediction of phase changes is reviewed in

G. H. Fredrickson, *Ann. Rev. Phys. Chem.* **39**, 149 (1988).

A. D. J. Haymet, *Ann. Rev. Phys. Chem.* **38**, 89 (1987).

T. Kihara, *Adv. Chem. Phys.* **1**, 267 (1958).

3 *Ab initio* Methods

The term *ab initio* is Latin for "from the beginning." This name is given to computations that are derived directly from theoretical principles with no inclusion of experimental data. This is an approximate quantum mechanical calculation. The approximations made are usually mathematical approximations, such as using a simpler functional form for a function or finding an approximate solution to a differential equation.

3.1 HARTREE–FOCK APPROXIMATION

The most common type of *ab initio* calculation is called a Hartree–Fock calculation (abbreviated HF), in which the primary approximation is the central field approximation. This means that the Coulombic electron–electron repulsion is taken into account by integrating the repulsion term. This gives the average effect of the repulsion, but not the explicit repulsion interaction. This is a variational calculation, meaning that the approximate energies calculated are all equal to or greater than the exact energy. The energies are calculated in units called Hartrees (1 Hartree = 27.2116 eV). Because of the central field approximation, the energies from HF calculations are always greater than the exact energy and tend to a limiting value called the Hartree–Fock limit as the basis set is improved.

One of the advantages of this method is that it breaks the many-electron Schrödinger equation into many simpler one-electron equations. Each one-electron equation is solved to yield a single-electron wave function, called an orbital, and an energy, called an orbital energy. The orbital describes the behavior of an electron in the net field of all the other electrons.

The second approximation in HF calculations is due to the fact that the wave function must be described by some mathematical function, which is known exactly for only a few one-electron systems. The functions used most often are linear combinations of Gaussian-type orbitals $\exp(-ar^2)$, abbreviated GTO. The wave function is formed from linear combinations of atomic orbitals or, stated more correctly, from linear combinations of basis functions. Because of this approximation, most HF calculations give a computed energy greater than the Hartree–Fock limit. The exact set of basis functions used is often specified by an abbreviation, such as STO−3G or 6−311++g**. Basis sets are discussed further in Chapters 10 and 28.

The Gaussian functions are multiplied by an angular function in order to give the orbital the symmetry of a *s*, *p*, *d*, and so on. A constant angular term yields *s* symmetry. Angular terms of x, y, z give *p* symmetry. Angular terms of xy, xz, yz, x^2-y^2, $4z^2-2x^2-2y^2$ yield *d* symmetry. This pattern can be continued for the other orbitals.

These orbitals are then combined into a determinant. This is done to satisfy two requirements of quantum mechanics. One is that the electrons must be indistinguishable. By having a linear combination of orbitals in which each electron appears in each orbital, it is only possible to say that an electron was put in a particular orbital but not which electron it is. The second requirement is that the wave function for fermions (an electron is a fermion) must be antisymmetric with respect to interchanging two particles. Thus, if electron 1 and electron 2 are switched, the sign of the total wave function must change and only the sign can change. This is satisfied by a determinant because switching two electrons is equivalent to interchanging two columns of the determinant, which changes its sign.

The functions put into the determinant do not need to be individual GTO functions, called Gaussian primitives. They can be a weighted sum of basis functions on the same atom or different atoms. Sums of functions on the same atom are often used to make the calculation run faster, as discussed in Chapter 10. Sums of basis functions on different atoms are used to give the orbital a particular symmetry. For example, a water molecule with C_{2v} symmetry will have orbitals that transform as A_1, A_2, B_1, B_2, which are the irreducible representations of the C_{2v} point group. The resulting orbitals that use functions from multiple atoms are called molecular orbitals. This is done to make the calculation run much faster. Any overlap integral over orbitals of different symmetry does not need to be computed because it is zero by symmetry.

The steps in a Hartree–Fock calculation start with an initial guess for the orbital coefficients, usually using a semiempirical method. This function is used to calculate an energy and a new set of orbital coefficients, which can then be used to obtain a new set, and so on. This procedure continues iteratively until the energies and orbital coefficients remain constant from one iteration to the next. This is called having the calculation converge. There is no guarantee the calculation will converge. In cases where it does not, some technical expertise is required to fix the problem, as discussed in Chapter 22. This iterative procedure is called a self-consistent field procedure (SCF). Some researchers refer to these as SCF calculations to distinguish them from the earlier method created by Hartree, but HF is used more widely.

A variation on the HF procedure is the way that orbitals are constructed to reflect paired or unpaired electrons. If the molecule has a singlet spin, then the same orbital spatial function can be used for both the α and β spin electrons in each pair. This is called the restricted Hartree–Fock method (RHF).

There are two techniques for constructing HF wave functions of molecules with unpaired electrons. One technique is to use two completely separate sets of orbitals for the α and β electrons. This is called an unrestricted Hartree–Fock wave function (UHF). This means that paired electrons will not have the same spatial distribution. This introduces an error into the calculation, called spin contamination. Spin contamination might introduce an insignificant error or the error could be large enough to make the results unusable depending on the chemical system involved. Spin contamination is discussed in more detail in Chapter 27. UHF calculation are popular because they are easy to implement and run fairly efficiently.

Another way of constructing wave functions for open-shell molecules is the restricted open shell Hartree–Fock method (ROHF). In this method, the paired electrons share the same spatial orbital; thus, there is no spin contamination. The ROHF technique is more difficult to implement than UHF and may require slightly more CPU time to execute. ROHF is primarily used for cases where spin contamination is large using UHF.

For singlet spin molecules at the equilibrium geometry, RHF and UHF wave functions are almost always identical. RHF wave functions are used for singlets because the calculation takes less CPU time. In a few rare cases, a singlet molecule has biradical resonance structures and UHF will give a better description of the molecule (i.e., ozone).

The RHF scheme results in forcing electrons to remain paired. This means that the calculation will fail to reflect cases where the electrons should uncouple. For example, a series of RHF calculations for H_2 with successively longer bond lengths will show that H_2 dissociates into H^+ and H^-, rather than two H atoms. This limitation must be considered whenever processes involving pairing and unpairing of electrons are modeled. This is responsible for certain systematic errors in HF results, such as activation energies that are too high, bond lengths slightly too short, vibrational frequencies too high, and dipole moments and atomic charges that are too large. UHF wave functions usually dissociate correctly.

There are a number of other technical details associated with HF and other *ab initio* methods that are discussed in other chapters. Basis sets and basis set superposition error are discussed in more detail in Chapters 10 and 28. For open-shell systems, additional issues exist: spin polarization, symmetry breaking, and spin contamination. These are discussed in Chapter 27. Size–consistency and size–extensivity are discussed in Chapter 26.

3.2 CORRELATION

One of the limitations of HF calculations is that they do not include electron correlation. This means that HF takes into account the average affect of electron repulsion, but not the explicit electron–electron interaction. Within HF theory the probability of finding an electron at some location around an atom is determined by the distance from the nucleus but not the distance to the other electrons as shown in Figure 3.1. This is not physically true, but it is the consequence of the central field approximation, which defines the HF method.

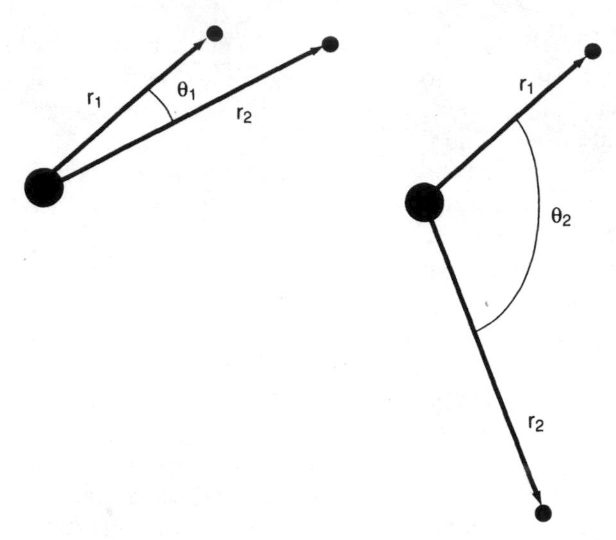

FIGURE 3.1 Two arrangements of electrons around the nucleus of an atom having the same probability within HF theory, but not in correlated calculations.

A number of types of calculations begin with a HF calculation and then correct for correlation. Some of these methods are Møller–Plesset perturbation theory (MPn, where n is the order of correction), the generalized valence bond (GVB) method, multi-configurational self-consistent field (MCSCF), configuration interaction (CI), and coupled cluster theory (CC). As a group, these methods are referred to as correlated calculations.

Correlation is important for many different reasons. Including correlation generally improves the accuracy of computed energies and molecular geometries. For organic molecules, correlation is an extra correction for very-high-accuracy work, but is not generally needed to obtain quantitative results. One exception to this rule are compounds exhibiting Jahn–Teller distortions, which often require correlation to give quantitatively correct results. An extreme case is transition metal systems, which often require correlation in order to obtain results that are qualitatively correct.

3.3 MØLLER–PLESSET PERTURBATION THEORY

Correlation can be added as a perturbation from the Hartree–Fock wave function. This is called Møller–Plesset perturbation theory. In mapping the HF wave function onto a perturbation theory formulation, HF becomes a first-order perturbation. Thus, a minimal amount of correlation is added by using the second-order MP2 method. Third-order (MP3) and fourth-order (MP4) calculations are also common. The accuracy of an MP4 calculation is roughly equivalent to the accuracy of a CISD calculation. MP5 and higher calculations are seldom done due to the high computational cost (N^{10} time complexity or worse).

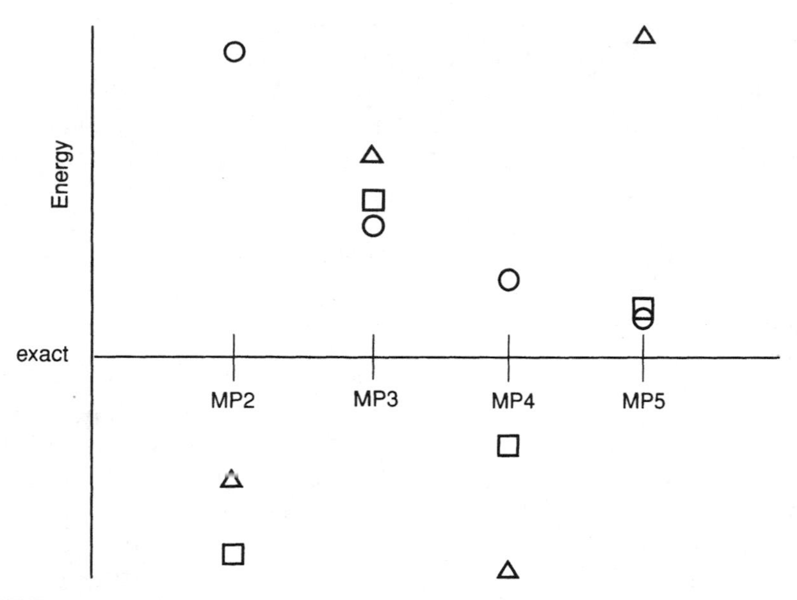

FIGURE 3.2 Possible results of increasing the order of Møller–Plesset calculations. The circles show monotonic convergence. The squares show oscillating convergence. The triangles show a diverging series.

Møller–Plesset calculations are not variational. In fact, it is not uncommon to find MP2 calculations that give total energies below the exact total energy. Depending on the nature of the chemical system, there seem to be two patterns in using successively higher orders of perturbation theory. For some systems, the energies become successively lower and closer to the total energy in going from MP2 to MP3, to MP4, and so on, as shown in Figure 3.2. For other systems, MP2 will have an energy lower than the exact energy, MP3 will be higher, MP4 will be lower, and so on, with each having an error that is lower in magnitude but opposite in sign. If the assumption of a small perturbation is not valid, the MPn energies may diverge as shown in Figure 3.2. This might happen if the single determinant reference is a poor qualitative description of the system. One advantage of Møller–Plesset is that it is size extensive.

There is also a local MP2 (LMP2) method. LMP2 calculations require less CPU time than MP2 calculations. LMP2 is also less susceptible to basis set superposition error. The price of these improvements is that about 98% of the MP2 energy correction is recovered by LMP2.

3.4 CONFIGURATION INTERACTION

A configuration interaction wave function is a multiple-determinant wave function. This is constructed by starting with the HF wave function and making new determinants by promoting electrons from the occupied to unoccupied orbitals. Configuration interaction calculations can be very accurate, but the cost in CPU time is very high (N^8 time complexity or worse).

Configuration interaction calculations are classified by the number of excitations used to make each determinant. If only one electron has been moved for each determinant, it is called a configuration interaction single-excitation (CIS) calculation. CIS calculations give an approximation to the excited states of the molecule, but do not change the ground-state energy. Single-and double-excitation (CISD) calculations yield a ground-state energy that has been corrected for correlation. Triple-excitation (CISDT) and quadruple-excitation (CISDTQ) calculations are done only when very-high-accuracy results are desired.

The configuration interaction calculation with all possible excitations is called a full CI. The full CI calculation using an infinitely large basis set will give an exact quantum mechanical result. However, full CI calculations are very rarely done due to the immense amount of computer power required.

CI results can vary a little bit from one software program to another for open-shell molecules. This is because of the HF reference state being used. Some programs, such as Gaussian, use a UHF reference state. Other programs, such as MOLPRO and MOLCAS, use a ROHF reference state. The difference in results is generally fairly small and becomes smaller with higher-order calculations. In the limit of a full CI, there is no difference.

3.5 MULTI-CONFIGURATIONAL SELF-CONSISTENT FIELD

MCSCF calculations also use multiple determinants. However, in an MCSCF calculation the orbitals are optimized for use with the multiple-determinant wave function. These calculations can often give the most accurate results for a given amount of CPU time. Compared to a CI calculation, an MCSCF gives more of the correlation energy with fewer configurations. However, CI calculations usually give more correlation energy in total because so many more configurations are included.

It is particularly desirable to use MCSCF or MRCI if the HF wave function yield a poor qualitative description of the system. This can be determined by examining the weight of the HF reference determinant in a single-reference CI calculation. If the HF determinant weight is less than about 0.9, then it is a poor description of the system, indicating the need for either a multiple-reference calculation or triple and quadruple excitations in a single-reference calculation.

Unfortunately, these methods require more technical sophistication on the part of the user. This is because there is no completely automated way to choose which configurations are in the calculation (called the active space). The user must determine which molecular orbitals to use. In choosing which orbitals to include, the user should ensure that the bonding and corresponding antibonding orbitals are correlated. The orbitals that will yield the most correlation energy can be determined by running an unrestricted or correlated calculation, then using the "virtual" natural orbitals with the highest occupation numbers along with their corresponding "occupied" orbitals. If the orbitals are chosen poorly, as is almost always the case without manual intervention, the results will not only fail to improve, but also they may actually become less accurate with the addition of more orbitals.

An MCSCF calculation in which all combinations of the active space orbitals are included is called a complete active space self-consistent field (CASSCF) calculation. This type of calculation is popular because it gives the maximum correlation in the valence region. The smallest MCSCF calculations are two-configuration SCF (TCSCF) calculations. The generalized valence bond (GVB) method is a small MCSCF including a pair of orbitals for each molecular bond.

3.6 MULTI-REFERENCE CONFIGURATION INTERACTION

It is possible to construct a CI wave function starting with an MCSCF calculation rather than starting with a HF wave function. This starting wave function is called the reference state. These calculations are called multi-reference configuration interaction (MRCI) calculations. There are more CI determinants in this type of calculation than in a conventional CI. This type of calculation can be very costly in terms of computing resources, but can give an optimal amount of correlation for some problems.

The notation for denoting this type of calculation is sometimes more specific. For example, the acronym MCSCF+1+2 means that the calculation is a MRCI calculation with single and double CI excitations out of an MCSCF reference space. Likewise, CASSCF+1+2 and GVB+1+2 calculations are possible.

3.7 COUPLED CLUSTER

Coupled cluster calculations are similar to configuration interaction calculations in that the wave function is a linear combination of many determinants. However, the means for choosing the determinants in a coupled cluster calculation is more complex than the choice of determinants in a CI. Like CI, there are various orders of the CC expansion, called CCSD, CCSDT, and so on. A calculation denoted CCSD(T) is one in which the triple excitations are included perturbatively rather than exactly.

Coupled cluster calculations give variational energies as long as the excitations are included successively. Thus, CCSD is variational, but CCD is not. CCD still tends to be a bit more accurate than CID.

The accuracy of these two methods is very similar. The advantage of doing coupled cluster calculations is that it is a size extensive method (see chapter 26). Often, coupled-cluster results are a bit more accurate than the equivalent size configuration interaction calculation results. When all possible configurations are included, a full coupled-cluster calculation is equivalent to a full CI calculation.

Quadratic configuration interaction calculations (QCI) use an algorithm that is a combination of the CI and CC algorithms. Thus, a QCISD calculation is an approximation to a CCSD calculation. These calculations are popular since they often give an optimal amount of correlation for high-accuracy calculations on organic molecules while using less CPU time than coupled cluster calculations. Most popular is the single- and double-excitation calculation, QCISD. Sometimes, triple excitations are included as well, QCISD(T). The T in parentheses indicates that the triple excitations are included perturbatively.

There is a variation on the coupled cluster method known as the symmetry adapted cluster (SAC) method. This is also a size consistent method. For excited states, a CI out of this space, called a SAC-CI, is done. This improves the accuracy of electronic excited-state energies.

Another technique, called Brueckner doubles, uses orbitals optimized to make single excitation contributions zero and then includes double excitations. This is essentially equivalent to CCSD in terms of both accuracy and CPU time.

3.8 QUANTUM MONTE CARLO METHODS

A method that avoids making the HF mistakes in the first place is called quantum Monte Carlo (QMC). There are several types of QMC: variational, diffusion, and Greens function Monte Carlo calculations. These methods work with an explicitly correlated wave function. This is a wave function that has a function of the electron–electron distance (a generalization of the original work by Hylleraas).

The wave function for a variational QMC calculation might take the functional form

$$\Psi = D_\alpha D_\beta \prod_{i<j} f(r_{ij}) \tag{3.1}$$

where D_α and D_β are determinants of α and β spin electrons. The use of two determinants in this way does not introduce any additional error as long as there are not any spin terms in the Hamiltonian (i.e., spin coupling). The $f(r_{ij})$ term is the term that accounts for electron correlation. This correlation function could include three body terms, denoted $f(r_i, r_j, r_{ij})$. The general shape of the correlation function is known, but there is not yet a consensus on the best mathematical function to use and new functions are still being proposed.

The diffusion and Greens function Monte Carlo methods use numerical wave functions. In this case, care must be taken to ensure that the wave function has the nodal properties of an antisymmetric function. Often, nodal surfaces from HF wave functions are used. In the most sophisticated calculations, the nodal surfaces are "relaxed," meaning that they are allowed to shift to optimize the wave function.

The integrals over the wave function are evaluated numerically using a Monte Carlo integration scheme. These calculations can be extremely time-consuming, but they are probably the most accurate methods known today. Because these calculations scale as N^3 and are extremely accurate, it is possible they could become important in the future if they can be made faster. At this current stage of development, most of the researchers using quantum Monte Carlo calculations are those writing their own computer codes and inventing the methods contained therein.

3.9 NATURAL ORBITALS

Once an energy and wave function has been found, it is often necessary to compute other properties of the molecule. For the multiple-determinant methods (MCSCF, CI, CC, MRCI), this is done most efficiently using natural orbitals. Natural orbitals are the eigenfunctions of the first-order reduced density matrix. The details of density matrix theory are beyond the scope of this text. Suffice it to say that once the natural orbitals have been found, the information in the wave function has been compressed from a many-determinant function down to a set of orbitals and occupation numbers. The occupation numbers are the number of electrons in each natural orbital. This is a real number, which is close to one or two electrons for those considered to be occupied orbitals. Many of the higher-energy orbitals have a small amount of electron occupation, which is roughly analogous to the excited-configuration weights in the wave function, but not mathematically equivalent.

Properties can be computed by finding the expectation value of the property operator with the natural orbitals weighted by the occupation number of each orbital. This is a much faster way to compute properties than trying to use the expectation value of a multiple-determinant wave function. Natural orbitals are not equivalent to HF or Kohn–Sham orbitals, although the same symmetry properties are present.

3.10 CONCLUSIONS

In general, *ab initio* calculations give very good qualitative results and can yield increasingly accurate quantitative results as the molecules in question become smaller. The advantage of *ab initio* methods is that they eventually converge to the exact solution once all the approximations are made sufficiently small in magnitude. In general, the relative accuracy of results is

$$\text{HF} \ll \text{MP2} < \text{CISD} \cong \text{MP4} \cong \text{CCSD} < \text{CCSD(T)} < \text{CCSDT} < \text{Full CI}$$

However, this convergence is not monotonic. Sometimes, the smallest calculation gives a very accurate result for a given property. There are four sources of error in *ab initio* calculations:

1. The Born–Oppenheimer approximation
2. The use of an incomplete basis set
3. Incomplete correlation
4. The omission of relativistic effects

The disadvantage of *ab initio* methods is that they are expensive. These methods often take enormous amounts of computer CPU time, memory, and disk space. The HF method scales as N^4, where N is the number of basis functions. This means that a calculation twice as big takes 16 times as long (2^4) to complete. Correlated calculations often scale much worse than this. In practice, extremely accurate solutions are only obtainable when the molecule contains a dozen electrons or less. However, results with an accuracy rivaling that of many experimental techniques can be obtained for moderate-size organic molecules. The minimally correlated methods, such as MP2 and GVB, are often used when correlation is important to the description of large molecules.

BIBLIOGRAPHY

Some textbooks discussing *ab initio* methods are

F. Jensen, *Introduction to Computational Chemistry* John Wiley & Sons, New York (1999).

T. Veszprémi, M. Fehér, *Quantum Chemistry; Fundamentals to Applications* Kluwer, Dordrecht (1999).

D. B. Cook, *Handbook of Computational Quantum Chemistry* Oxford, Oxford (1998).

P. W. Atkins, R. S. Friedman, *Molecular Quantum Mechanics* Oxford, Oxford (1997).

J. Simons, J. Nichols, *Quantum Mechanics in Chemistry* Oxford, Oxford (1997).

A. R. Leach, *Molecular Modelling Principles and Applications* Longman, Essex (1996).

P. Fulde, *Electron Correlations in Molecules and Solids Third Enlarged Edition* Springer, Berlin (1995).

B. L. Hammond, W. A. Lester, Jr., P. J. Reynolds, *Monte Carlo Methods in Ab Initio Quantum Chemistry* World Scientific, Singapore (1994).

I. N. Levine, *Quantum Chemistry* Prentice Hall, Englewood Cliffs (1991).

T. Clark, *A Handbook of Computational Chemistry* John Wiley & Sons, New York (1985).

S. Wilson, *Electron Correlation in Molecules* Clarendon, Oxford (1984).

D. A. McQuarrie, *Quantum Chemistry* University Science Books, Mill Valley (1983).

A. C. Hurley, *Electron Correlation in Small Molecules* Academic Press, London (1976).

Articles reviewing *ab initio* methods in general, or topics common to all methods

E. R. Davidson, *Encycl. Comput. Chem.* **3**, 1811 (1998).

S. Profeta, Jr., *Kirk-Othmer Encyclopedia of Chemical Technology Supplement* 315, J. I. Kroschwitz Ed. John Wiley & Sons, New York (1998).

S. Shaik, *Encycl. Comput. Chem.* **5**, 3143 (1998).

S. Sabo-Etienne, B. Chaudret, *Chem. Rev.* **98**, 2077 (1998).

J. Gerratt, D. L. Cooper, P. B. Karadakov, M. Raimond, *Chem. Soc. Rev.* **26**, 87 (1997).

Adv. Chem. Phys. vol **93** (1996).

J. Cioslowski, *Rev. Comput. Chem.* **4**, 1 (1993).

R. A. Friesner, *Ann. Rev. Phys. Chem.* **42**, 341 (1991).

D. L. Cooper, J. Gerratt, M. Raimondi, *Chem. Rev.* **91**, 929 (1991).

Adv. Chem. Phys. vol. **69** (1987).

G. Marc, W. G. McMillan, *Adv. Chem. Phys.* **58**, 209 (1985).

G. A. Gallup, R. L. Vance, J. R. Collins, J. M. Norbeck, *Adv. Quantum Chem.* **16**, 229 (1982).

H. F. Schaefer, III, *Ann. Rev. Phys. Chem.* **27**, 261 (1976).

E. R. Davidson, *Adv. Quantum Chem.* **6**, 235 (1972).

H. A. Bent, *Chem. Rev.* **61**, 275 (1961).

Sources comparing the accuracy of results are

F. Jensen, *Introduction to Computational Chemistry* John Wiley & Sons, New York (1999).

H. Partridge, *Encycl. Comput. Chem* **1**, 581 (1998).

J. B. Foresman, Æ. Frisch, *Exploring Chemistry with Electronic Structure Methods* Gaussian, Pittsburgh (1996).

K. Raghavachari, J. B. Anderson, *J. Phys. Chem.* 100, 12960 (1996).

R. J. Bartlett, *Modern Electronic Structure Theory Part 2* D. R. Yarkony, Ed., 1047, World Scientific, Singapore (1995).

T. J. Lee, G. E. Scuseria, *Quantum Mechanical Electronic Structure Calculations with Chemical Accuracy* S. R. Langhoff, Ed., 47, Kluwer, Dordrecht (1995).

W. J. Hehre, *Practical Strategies for Electronic Structure Calculations* Wavefunction, Irvine (1995).

H. F. Schaefer, III, J. R. Thomas, Y. Yamaguchi, B. J. DeLeeuw, G. Vacek, *Modern Electronic Structure Theory Part 1* D. R. Yarkony, Ed., 3, World Scientific, Singapore (1995).

C. W. Bauschlicker, Jr., S. R. Langhoff, P. R. Taylor, *Adv. Chem. Phys.* **77**, 103 (1990).

W. J. Hehre, L. Radom, P. v.R. Schleyer, J. A. Pople, *Ab Initio Molecular Orbital Theory* John Wiley & Sons, New York (1986).

V. Kvasnička, V. Laurinc, S. Biskupič, M. Haring, *Adv. Chem. Phys.* **52**, 181 (1983).

J. A. Pople, *Applications of Electronic Structure Theory* H. F. Schaefer, III, Ed., 1, Plenum, New York (1977).

Reviews of techniques for anion calculations are

J. Kalcher, A. F. Sax, *Chem Rev.* **94**, 2291 (1994).

J. Simons, M. Gutowski, *Chem. Rev.* **91**, 669 (1991).

J. Simons, K. D. Jordan, *Chem. Rev.* **87**, 535 (1987).

J. Simons, *Ann. Rev. Phys. Chem.* **28**, 15 (1977).

A review of bonding descriptions within *ab initio* methods is

W. A. Goddard, III, L. B. Harding, *Ann. Rev. Phys. Chem.* **29**, 363 (1978).

A discussion of choosing the CASSCF reference state is in

F. Bernardi, A. Bottini, J. J. W. McDougall, M. A. Robb, H. B. Schlegel, *Far. Symp. Chem. Soc.* **19**, 137 (1984).

Reviews of correlated methods are

K. Andersson, *Encycl. Comput. Chem.* **1**, 460 (1998).

P. Cársky, *Encycl. Comput. Chem.* **1**, 485 (1998).

H. Partridge, *Encycl. Comput. Chem.* **1**, 581 (1998).

J. Gauss, *Encycl. Comput. Chem.* **1**, 615 (1998).

D. Cremer, *Encycl. Comput. Chem.* **3**, 1706 (1998).

B. Jeziorski, R. Moszynski, K. Szalewiz, *Chem. Rev.* **94**, 1887 (1994).

S. Saebø, P. Pulay, *Ann. Rev. Phys. Chem.* **44**, 213 (1993).

K. Raghavachari, *Ann. Rev. Phys. Chem.* **42**, 615 (1991).

R. Ahlriches, P. Scharf, *Adv. Chem. Phys.* **67**, 501 (1987).

S. A. Kucharski, R. J. Bartlett, *Adv. Quantum Chem.* **18**, 207 (1986).

M. R. Hoffmann, H. F. Schaefer, III, *Adv. Quantum Chem.* **18**, 207 (1986).

I. Shavitt, *Advanced Theories and Computational Approaches to the Electronic Structure of Molecules* C. E. Dykstra, Ed., 185, Reidel, Dordrecht (1984).

J. Olsen, D. L. Yeager, P. Jørgensen, *Adv. Chem. Phys.* **54**, 1 (1983).

R. J. Bartlett, *Ann. Rev. Phys. Chem.* **32**, 359 (1981).

A. C. Wahl, G. Das, *Methods of Electronic Structure Theory* H. F. Schaefer III, 51, Plenum, New York (1977).

I. Shavitt, *Methods of Electronic Structure Theory* H. F. Schaefer III, 189, Plenum, New York (1977).

K. F. Freed, *Ann. Rev. Phys. Chem.* **22**, 313 (1971).

Adv. Chem. Phys. vol. **14** (1969).

R. K. Nesbet, *Adv. Chem. Phys.* **9**, 321 (1965).

P.-O. Löwdin, *Adv. Chem. Phys.* **2**, 207 (1959).

Reviews of coupled cluster methods are

T. D. Crawford, H. F. Schaefer III, *Rev. Comput. Chem.* **14**, 33 (2000).

I. Hubac, J. Masik, P. Mach, J. Urban, P. Babinec, *Computational Chemistry Reviews of Current Trends Volume 3* 1, J. Leszczynski, Ed., World Scientific, Singapore (1999).

A review of path integral techniques is

H. Grinberg, J. Maranon, *Adv. Quantum Chem.* **22**, 9 (1991).

A review of thermochemical results from the Hartree-Fock method is

A. C. Hurley, *Adv. Quantum Chem.* **7**, 315 (1973).

A review of the time dependent Hartree-Fock method is

P. Jørgensen, *Ann. Rev. Phys. Chem.* **26**, 359 (1975).

Reviews of quantum Monte Carlo methods are

J. B. Anderson, *Rev. Comput. Chem.* **13**, 132 (1999).

W. A. Lester, Jr., R. N. Barett, *Encycl. Comput. Chem.* **3**, 1735 (1998).

R. N. Burnett, W. A. Lester, Jr., *Computational Chemistry Reviews of Current Trends, Volume 2* J. Leszczynski, Ed., 125, World Scientific, Singapore (1997).

J. B. Anderson, *Quantum Mechanical Electronic Structure Calculations with Chemical Accuracy* S. R. Langhoff, Ed., 1, Kluwer, Dordrecht (1995).

Reviews of results for Van der Waals clusters are

G. Chalasinski, M. M. Szcześniak, *Chem. Rev.* **94**, 1723 (1994).

P. Hobza, H. L. Selzle, E. W. Schlag, *Chem. Rev.* **94**, 1767 (1994).

4 Semiempirical Methods

Semiempirical calculations are set up with the same general structure as a HF calculation in that they have a Hamiltonian and a wave function. Within this framework, certain pieces of information are approximated or completely omitted. Usually, the core electrons are not included in the calculation and only a minimal basis set is used. Also, some of the two-electron integrals are omitted. In order to correct for the errors introduced by omitting part of the calculation, the method is parameterized. Parameters to estimate the omitted values are obtained by fitting the results to experimental data or *ab initio* calculations. Often, these parameters replace some of the integrals that are excluded.

The advantage of semiempirical calculations is that they are much faster than *ab initio* calculations. The disadvantage of semiempirical calculations is that the results can be erratic and fewer properties can be predicted reliably. If the molecule being computed is similar to molecules in the database used to parameterize the method, then the results may be very good. If the molecule being computed is significantly different from anything in the parameterization set, the answers may be very poor. For example, the carbon atoms in cyclopropane and cubane have considerably different bond angles from those in most other compounds; thus, these molecules may not be predicted well unless they were included in the parameterization. However, semiempirical methods are not as sensitive to the parameterization set as are molecular mechanics calculations.

Semiempirical methods are parameterized to reproduce various results. Most often, geometry and energy (usually the heat of formation) are used. Some researchers have extended this by including dipole moments, heats of reaction, and ionization potentials in the parameterization set. A few methods have been parameterized to reproduce a specific property, such as electronic spectra or NMR chemical shifts. Semiempirical calculations can be used to compute properties other than those in the parameterization set.

Many semiempirical methods compute energies as heats of formation. The researcher should not add zero-point corrections to these energies because the thermodynamic corrections are implicit in the parameterization.

CIS calculations from the semiempirical wave function can be used for computing electronic excited states. Some software packages allow CI calculations other than CIS to be performed from the semiempirical reference space. This is a good technique for modeling compounds that are not described properly by a single-determinant wave function (see Chapter 26). Semiempirical CI calculations do not generally improve the accuracy of results since they include correlation twice: once by CI and once by parameterization.

Semiempirical calculations have been very successful in the description of organic chemistry, where there are only a few elements used extensively and the molecules are of moderate size. Some semiempirical methods have been devised specifically for the description of inorganic chemistry as well. The following are some of the most commonly used semiempirical methods.

4.1 HÜCKEL

The Hückel method and is one of the earliest and simplest semiempirical methods. A Hückel calculation models only the π valence electrons in a planar conjugated hydrocarbon. A parameter is used to describe the interaction between bonded atoms. There are no second atom affects. Hückel calculations do reflect orbital symmetry and qualitatively predict orbital coefficients. Hückel calculations can give crude quantitative information or qualitative insight into conjugated compounds, but are seldom used today. The primary use of Hückel calculations now is as a class exercise because it is a calculation that can be done by hand.

4.2 EXTENDED HÜCKEL

An extended Hückel calculation is a simple means for modeling the valence orbitals based on the orbital overlaps and experimental electron affinities and ionization potentials. In some of the physics literature, this is referred to as a tight binding calculation. Orbital overlaps can be obtained from a simplified single STO representation based on the atomic radius. The advantage of extended Hückel calculations over Hückel calculations is that they model all the valence orbitals.

The primary reason for interest in extended Hückel today is because the method is general enough to use for all the elements in the periodic table. This is not an extremely accurate or sophisticated method; however, it is still used for inorganic modeling due to the scarcity of full periodic table methods with reasonable CPU time requirements. Another current use is for computing band structures, which are extremely computation-intensive calculations. Because of this, extended Hückel is often the method of choice for band structure calculations. It is also a very convenient way to view orbital symmetry. It is known to be fairly poor at predicting molecular geometries.

4.3 PPP

The Pariser–Parr–Pople (PPP) method is an extension of the Hückel method that allows heteroatoms other than hydrogen. It is still occasionally used when very minimal amounts of electronic effects are required. For example, PPP-based terms have been incorporated in molecular mechanics calculations to describe aromaticity. This method is also popular for developing simple parameterized analytic expressions for molecular properties.

4.4 CNDO

The complete neglect of differential overlap (CNDO) method is the simplest of the neglect of differential overlap (NDO) methods. This method models valence orbitals only using a minimal basis set of Slater type orbitals. The CNDO method has proven useful for some hydrocarbon results but little else. CNDO is still sometimes used to generate the initial guess for *ab initio* calculations on hydrocarbons.

Practically all CNDO calculations are actually performed using the CNDO/2 method, which is an improved parameterization over the original CNDO/1 method. There is a CNDO/S method that is parameterized to reproduce electronic spectra. The CNDO/S method does yield improved prediction of excitation energies, but at the expense of the poorer prediction of molecular geometry. There have also been extensions of the CNDO/2 method to include elements with occupied *d* orbitals. These techniques have not seen widespread use due to the limited accuracy of results.

4.5 MINDO

There are three modified intermediate neglect of differential overlap (MINDO) methods: MINDO/1, MINDO/2, and MINDO/3. The MINDO/3 method is by far the most reliable of these. This method has yielded qualitative results for organic molecules. However its use today has been superseded by that of more accurate methods such as Austin model 1 (AM1) and parameterization method 3 (PM3). MINDO/3 is still sometimes used to obtain an initial guess for *ab initio* calculations.

4.6 MNDO

The modified neglect of diatomic overlap (MNDO) method has been found to give reasonable qualitative results for many organic systems. It has been incorporated into several popular semiempirical programs as well as the MNDO program. Today, it is still used, but the more accurate AM1 and PM3 methods have surpassed it in popularity.

There are a some known cases where MNDO gives qualitatively or quantitatively incorrect results. Computed electronic excitation energies are underestimated. Activation barriers tend to be too high. The correct conformer is not always computed to be lowest in energy. Barriers to bond rotation are often computed to be too small. Hypervalent compounds and sterically crowded molecules are computed to be too unstable. Four-membered rings are predicted to be too stable. Oxygenated functional groups on aromatic rings are predicted to be out-of-plane. The peroxide bond is too short by about 0.17 Å. The ether C–O–C bond angle is too large by about 9°. Bond lengths between electronegative elements are too short. Hydrogen bonds are too weak and long.

A variation on MNDO is MNDO/d. This is an equivalent formulation including *d* orbitals. This improves predicted geometry of hypervalent molecules. This method is sometimes used for modeling transition metal systems, but its accuracy is highly dependent on the individual system being studied. There is also a MNDOC method that includes electron correlation.

4.7 INDO

The intermediate neglect of differential overlap (INDO) method was at one time used for organic systems. Today, it has been superseded by more accurate methods. INDO is still sometimes used as an initial guess for *ab initio* calculations.

4.8 ZINDO

The Zerner's INDO method (ZINDO) is also called spectroscopic INDO (INDO/S). This is a reparameterization of the INDO method specifically for the purpose of reproducing electronic spectra results. This method has been found to be useful for predicting electronic spectra. ZINDO is also used for modeling transition metal systems since it is one of the few methods parameterized for metals. It predicts UV transitions well, with the exception of metals with unpaired electrons. However, its use is generally limited to the type of results for which it was parameterized. ZINDO often gives poor results when used for geometry optimization.

4.9 SINDO1

The symmetrically orthogonalized intermediate neglect of differential overlap method (SINDO1) is both a semiempirical method and a computer program incorporating that method. It is another variation on INDO. SINDO1 is designed for the prediction of the binding energies and geometries of the 1st and 2nd row elements as well as the 3rd row transition metals. Some of the parameters were taken directly from experimental or *ab initio* results, whereas the rest were parameterized to reproduce geometry and heats of formation. The method was originally designed for modeling ground states of organic molecules. More recently, it has been extended to predict photochemistry and transition metal results.

4.10 PRDDO

The PRDDO (partial retention of diatomic differential overlap) method is an attempt to get the optimal ratio of accuracy to CPU time. It has been parameterized for the periodic elements through Br, including the 3rd row transition metals. It was parameterized to reproduce *ab initio* results. PRDDO has been used primarily for inorganic compounds, organometallics, solid-state calculations, and polymer modeling. This method has seen less use than other methods of similar accuracy mostly due to the fact that it has not been incorporated into the most widely used semiempirical software.

There are several variations of this method. The PRDDO/M method is parameterized to reproduce electrostatic potentials. The PRDDO/M/FCP method uses frozen core potentials. PRDDO/M/NQ uses an approximation called "not quite orthogonal orbitals" in order to give efficient calculations on very large molecules. The results of these methods are fairly good overall, although bond lengths involving alkali metals tend to be somewhat in error.

4.11 AM1

The Austin Model 1 (AM1) method is still popular for modeling organic compounds. AM1 generally predicts the heats of formation (ΔH_f) more accurately than MNDO, although a few exceptions involving Br atoms have been documented. Depending on the nature of the system and information desired, either AM1 or PM3 will often give the most accurate results obtainable for organic molecules with semiempirical methods.

There are some known strengths and limitations in the results obtained from these methods. Activation energies are improved over MNDO. AM1 tends to predict results for aluminum better than PM3. It tends to poorly predict nitrogen paramidalization. AM1 tends to give O–Si–O bonds that are not bent enough. There are some known limitations to AM1 energies, such as predicting rotational barriers to be one-third the actual barrier and predicting five-membered rings to be too stable. The predicted heat of formation tends to be inaccurate for molecules with a large amount of charge localization. Geometries involving phosphorus are predicted poorly. There are systematic errors in alkyl group energies predicting them to be too stable. Nitro groups are too positive in energy. The peroxide bond is too short by about 0.17 Å. Hydrogen bonds are predicted to have the correct strength, but often the wrong orientation. On average, AM1 predicts energies and geometries better than MNDO, but not as well as PM3. Computed bond enthalpies are consistently low.

4.12 PM3

Parameterization method 3 (PM3) uses nearly the same equations as the AM1 method along with an improved set of parameters. The PM3 method is also currently extremely popular for organic systems. It is more accurate than AM1 for hydrogen bond angles, but AM1 is more accurate for hydrogen bond energies. The PM3 and AM1 methods are also more popular than other semiempirical methods due to the availability of algorithms for including solvation effects in these calculations.

There are also some known strengths and limitations of PM3. Overall heats of formation are more accurate than with MNDO or AM1. Hypervalent molecules are also predicted more accurately. PM3 tends to predict that the barrier to rotation around the C–N bond in peptides is too low. Bonds between Si and the halide atoms are too short. PM3 also tends to predict incorrect electronic states for germanium compounds. It tends to predict sp^3 nitrogen as always being pyramidal. Some spurious minima are predicted. Proton affinities are not accurate. Some polycyclic rings are not flat. The predicted charge on nitrogen is incorrect. Nonbonded distances are too short. Hydrogen bonds are too short by about 0.1 Å, but the orientation is usually correct. On average, PM3 predicts energies and bond lengths more accurately than AM1 or MNDO.

4.13 PM3/TM

PM3/TM is an extension of the PM3 method to include *d* orbitals for use with transition metals. Unlike the case with many other semiempirical methods, PM3/TM's parameterization is based solely on reproducing geometries from X-ray diffraction results. Results with PM3/TM can be either reasonable or not depending on the coordination of the metal center. Certain transition metals tend to prefer a specific hybridization for which it works well.

4.14 FENSKE–HALL

The Fenske–Hall method is a modification of crystal field theory. This is done by using a population analysis scheme, then replacing orbital interactions with point charge interactions. This has been designed for the description of inorganic metal-ligand systems. There are both parameterized and unparameterized forms of this method.

4.15 TNDO

The typed neglect of differential overlap (TNDO) method is a semiempirical method parameterized specifically to reproduce NMR chemical shifts. This parameterization goes one step further than other semiempiricals in that the method must distinguish between atoms of the same element but different hybridizations. For example, different parameters are used to describe an sp^2 carbon than are used for an sp^3 carbon. There are two versions of this method: TNDO/1 and TNDO/2. The prediction of NMR chemical shifts is discussed in Chapter 31.

4.16 SAM1

Semi-*ab initio* method 1 (SAM1) is different from the rest of the methods just discussed. It still neglects some of the integrals included in HF calculations, but includes more than other semiempirical methods, including *d* orbitals. Thus, the amount of CPU time for SAM1 calculations is more than for other semiempiricals but still significantly less than for a minimal basis set HF calculation. The method uses a parameterization to estimate the correlation effects. For organic molecules too large for correlated *ab initio* calculations, this is a reasonable way to incorporate correlation effects. Results tend to be slightly more accurate than with AM1 or PM3, but with the price of an increased amount of CPU time necessary. Vibrational frequencies computed with SAM1 are significantly more accurate than with other semiempiricals.

4.17 GAUSSIAN THEORY

The Gaussian methods (G1, G2, and G3) are also unique types of computations. These methods arose from the observation that certain *ab initio* methods tended to show a systematic error for predicting the energies of the ground states of organic molecules. This observation resulted in a correction equation that uses the energies from several different *ab initio* calculations in order to extrapolate to a very-high-accuracy result. All the calculations that go into this extrapolation are *ab initio* methods. However, the extrapolation equation itself is an empirically defined equation parameterized to reproduce results from a test set of molecules as accurately as possible. The extrapolation to complete correlation is based on the number of electrons times an empirically determined constant. For this reason, these methods show the same strengths and weaknesses as other semiempirical methods. The accuracy can be extremely good for systems similar to those for which they were parameterized, the ground state of organic molecules. However, for other systems, such as transition structures or clusters, these methods often are less accurate than some less computationally intensive *ab initio* methods. J. A. Pople once referred to this as a "slightly empirical theory."

The G1 method is seldom used since G2 yields an improved accuracy of results. G2 has proven to be a very accurate way to model small organic molecules, but gives poor accuracy when applied to chlorofluorocarbons. At the time this book was written, the G3 method had just been published. The initial results from G3 show some improvement in accuracy especially for chlorofluorocarbons.

There have been a number of variations on the G2 method proposed. The G2(MP2) and G2(MP2,SVP) methods are designed to require less CPU time, with a slight loss of accuracy. The G2(B3LYP/MP2/CC) method uses an amount of CPU similar to G2(MP2) with slightly better results. Some variations designed for improved accuracy over G2 are G2(COMPLETE), G2(BD), G2(CCSD), and G2M(RCC). Based on the improved accuracy of results, it is likely that the future will see more publications using G3, G2(COMPLETE), and G2M(RCC). The complete basis set method discussed in Chapter 10 is similar in application to Gaussian theory, but significantly different in the theoretical derivation. These issues are discussed in more detail in the references listed at the end of this chapter.

4.18 RECOMMENDATIONS

Semiempirical methods can provide results accurate enough to be useful, particularly for organic molecules with computation requirements low enough to make them convenient on PC or Macintosh computers. These methods are generally good for predicting molecular geometry and energetics. Semiempirical methods can be used for predicting vibrational modes and transition structures, but do so less reliably than *ab initio* methods. Semiempirical calculations generally give poor results for van der Waals and dispersion intermolecular forces, due to the lack of diffuse basis functions.

BIBLIOGRAPHY

Text books discussing semiempirical methods are

F. Jensen, *Introduction to Computational Chemistry* John Wiley & Sons, New York (1999).

P. W. Atkins, R. S. Friedman, *Molecular Quantum Mechanics* Oxford, Oxford (1997).

J. Simons, J. Nichols, *Quantum Mechanics in Chemistry* Oxford, Oxford (1997).

A. R. Leach, *Molecular Modelling Principles and Applications* Longman, Essex (1996).

I. N. Levine, *Quantum Chemistry* Prentice Hall, Englewood Cliffs (1991).

T. Clark, *A Handbook of Computational Chemistry* John Wiley & Sons, New York (1985).

J. Sadlej, *Semi-Empirical Methods of Quantum Chemistry* Ellis Harwood, Chichester (1985).

D. A. McQuarrie, *Quantum Chemistry* University Science Books, Mill Valley CA (1983).

C. A. Coulson, B. O'Leary, R. B. Mallion, *Hückel Theory for Organic Chemists* Academic Press, London (1978).

K. Yates, *Hückel Molecular Orbital Theory* Academic Press, New York (1978).

Semiempirical Methods of Electronic Structure Calculation G. A. Segal, Ed., Plenum, New York (1977).

J. N. Murrell, A. J. Harget, *Semi-empirical self-consistent-field molecular orbital theory of molecules* John Wiley & Sons, New York (1972).

J. A. Pople, D. L. Beveridge *Approximate Molecular Orbital Theory* McGraw-Hill, New York (1970).

M. Karplus, R. N. Porter, *Atoms & Molecules: An Introduction For Students of Physical Chemistry* W. A. Benjamine, Menlo Park, (1970).

Review articles over semiempirical methods in general are

K. Jug, F. Neumann, *Encycl Comput. Chem.* **1**, 507 (1998).

J. J. P. Stewart, *Encycl. Comput. Chem.* **2**, 1513 (1998).

J. J. P. Stewart, *Encycl. Comput. Chem.* **3**, 2000 (1998).

S. Profeta, Jr., *Kirk-Othmer Encyclopedia of Chemical Technology Supplement* 315 J. I. Kroschwitz Ed., John Wiley & Sons, New York (1998).

W. Thiel, *Adv. Chem. Phys.* **93**, 703 (1996).

M. C. Zerner, *Rev Comput. Chem.* **2**, 313 (1991).

J. J. P. Stewart, *J. Comput-Aided Mol. Design* **4**, 1 (1990).

J. J. P. Stewart, *Rev. Comput. Chem.* **1**, 45 (1990).

W. Thiel, *Tetrahedron* **44**, 7393 (1988).

P. Durand, J.-P. Malrieu, *Adv. Chem. Phys.* **67**, 321 (1987).

M. J. S. Dewar, *J. Phys. Chem.* **89**, 2145 (1985).

M. Simonetta, A. Gavezzotti, *Bonding Forces* J. D. Dunitz, P. Hemmerich, R. H. Holm, J. A. Ibers, C. K. Jorgensen, J. B. Neilands, D. Reinen, R. J. P. Williams, Eds., Springer-Verlag, Berlin (1976).

B. J. Nicholson, *Adv. Chem. Phys.* **18**, 249 (1970).

I. Fischer-Hjalmars, *Adv. Quantum Chem.* **2**, 25 (1965).

Adv. Quantum Chem. vol **1**, (1964).

The accuracy of results is compared in

F. Jensen, *Introduction to Computational Chemistry* John Wiley & Sons, New York (1999).

W. J. Hehre, *Practical Strategies for Electronic Structure Calculations* Wavefunction, Irvine (1995).

T. Clark, *A Handbook of Computational Chemistry* John Wiley & Sons, New York (1985).

J. Sadlej, *Semi-Empirical Methods of Quantum Chemistry* Ellis Harwood, Chichester (1985).

Semiempirical Methods of Electronic Structure Calculation G. A. Segal, Ed., Plenum, New York (1977).

J. N. Murrell, A. J. Harget, *Semi-empirical self-consistent-field molecular orbital theory of molecules* John Wiley & Sons, New York (1972).

J. A. Pople, D. L. Beveridge *Approximate Molecular Orbital Theory* McGraw-Hill, New York (1970).

Adv. Quantum Chem. vol **1**, (1964).

A review of AM1 is

A. J. Holder, *Encycl. Comput. Chem.* **1**, 8 (1998).

CNDO/S is presented in

H. M. Chang, H. H. Jaffé, C. A. Masmanides, *J. Phys Chem* **79**, 1118 (1975).

Reviews of Gaussian theory and other extrapolation methods are

J. C. Corchado, D. G. Truhlar, *ACS Symp. Ser.* **712** in press.

L. A. Curtiss, K. Raghavachari, *Encycl. Comput. Chem.* **2**, 1104 (1998).

M. R. A. Blomberg, P. E. M. Siegbahn, *ACS Symp. Ser.* **677**, 197 (1998).

J. M. L. Martin, *ACS Symp. Ser.* **677**, 212 (1998).

G. A. Petersson, *ACS Symp. Ser.* **677**, 237 (1998).

A review of MINDO/3 is

D. F. v. Lewis, *Chem. Rev.* **86**, 1111 (1986).

Reviews of MNDO are

W. Thiel, *Encycl. Comput. Chem.* **3**, 1599 (1998).

W. Thiel, *Encycl. Comput. Chem* **3**, 1604 (1998).

A review of PM3 is

J. J. P. Stewart, *Encycl. Comput. Chem* **3**, 2080 (1998).

A review of PRDDO is

D. S. Marynick, *Encycl. Comput. Chem.* **3**, 2153 (1998).

A review of SAM1 is

A. J. Holder, *Encycl. Comput. Chem.* **4**, 2542 (1998).

A review of SINDO1 is

K. Jug, T. Bredow, *Encycl. Comput. Chem.* **4**, 2599 (1998).

5 Density Functional Theory

Density functional theory (DFT) has become very popular in recent years. This is justified based on the pragmatic observation that it is less computationally intensive than other methods with similar accuracy. This theory has been developed more recently than other *ab initio* methods. Because of this, there are classes of problems not yet explored with this theory, making it all the more crucial to test the accuracy of the method before applying it to unknown systems.

5.1 BASIC THEORY

The premise behind DFT is that the energy of a molecule can be determined from the electron density instead of a wave function. This theory originated with a theorem by Hoenburg and Kohn that stated this was possible. The original theorem applied only to finding the ground-state electronic energy of a molecule. A practical application of this theory was developed by Kohn and Sham who formulated a method similar in structure to the Hartree–Fock method.

In this formulation, the electron density is expressed as a linear combination of basis functions similar in mathematical form to HF orbitals. A determinant is then formed from these functions, called Kohn–Sham orbitals. It is the electron density from this determinant of orbitals that is used to compute the energy. This procedure is necessary because Fermion systems can only have electron densities that arise from an antisymmetric wave function. There has been some debate over the interpretation of Kohn–Sham orbitals. It is certain that they are not mathematically equivalent to either HF orbitals or natural orbitals from correlated calculations. However, Kohn–Sham orbitals do describe the behavior of electrons in a molecule, just as the other orbitals mentioned do. DFT orbital eigenvalues do not match the energies obtained from photoelectron spectroscopy experiments as well as HF orbital energies do. The questions still being debated are how to assign similarities and how to physically interpret the differences.

A density functional is then used to obtain the energy for the electron density. A functional is a function of a function, in this case, the electron density. The exact density functional is not known. Therefore, there is a whole list of different functionals that may have advantages or disadvantages. Some of these functionals were developed from fundamental quantum mechanics and some were developed by parameterizing functions to best reproduce experimental results. Thus, there are in essence *ab initio* and semiempirical versions of DFT. DFT tends to be classified either as an *ab initio* method or in a class by itself.

The advantage of using electron density is that the integrals for Coulomb repulsion need be done only over the electron density, which is a three-dimensional function, thus scaling as N^3. Furthermore, at least some electron correlation can be included in the calculation. This results in faster calculations than HF calculations (which scale as N^4) and computations that are a bit more accurate as well. The better DFT functionals give results with an accuracy similar to that of an MP2 calculation.

Density functionals can be broken down into several classes. The simplest is called the Xα method. This type of calculation includes electron exchange but not correlation. It was introduced by J. C. Slater, who in attempting to make an approximation to Hartree–Fock unwittingly discovered the simplest form of DFT. The Xα method is similar in accuracy to HF and sometimes better.

The simplest approximation to the complete problem is one based only on the electron density, called a local density approximation (LDA). For high-spin systems, this is called the local spin density approximation (LSDA). LDA calculations have been widely used for band structure calculations. Their performance is less impressive for molecular calculations, where both qualitative and quantitative errors are encountered. For example, bonds tend to be too short and too strong. In recent years, LDA, LSDA, and VWN (the Vosko, Wilks, and Nusair functional) have become synonymous in the literature.

A more complex set of functionals utilizes the electron density and its gradient. These are called gradient-corrected methods. There are also hybrid methods that combine functionals from other methods with pieces of a Hartree–Fock calculation, usually the exchange integrals.

In general, gradient-corrected or hybrid calculations give the most accurate results. However, there are a few cases where Xα and LDA do quite well. LDA is known to give less accurate geometries and predicts binding energies significantly too large. The current generation of hybrid functionals are a bit more accurate than the present gradient-corrected techniques. Some of the more widely used functionals are listed in Table 5.1.

5.2 LINEAR SCALING TECHNIQUES

One recent development in DFT is the advent of linear scaling algorithms. These algorithms replace the Coulomb terms for distant regions of the molecule with multipole expansions. This results in a method with a time complexity of N for sufficiently large molecules. The most common linear scaling techniques are the fast multipole method (FMM) and the continuous fast multipole method (CFMM).

DFT is generally faster than Hartree–Fock for systems with more than 10–

TABLE 5.1 Density Functionals

Acronyms	Name	Type
Xα	X alpha	Exchange only
HFS	Hartree–Fock Slater	HF with LDA exchange
VWN	Vosko, Wilks, and Nusair	LDA
BLYP	Becke correlation functional with Lee, Yang, Parr exchange	Gradient-corrected
B3LYP, Becke3LYP	Becke 3 term with Lee, Yang, Parr exchange	Hybrid
PW91	Perdue and Wang 1991	Gradient-corrected
G96	Gill 1996	Exchange
P86	Perdew 1986	Gradient-corrected
B96	Becke 1996	Gradient-corrected
B3P86	Becke exchange, Perdew correlation	Hybrid
B3PW91	Becke exchange, Perdew and Wang correlation	Hybrid

15 nonhydrogen atoms, depending on the numeric integral accuracy and basis set. Linear scaling algorithms do not become advantageous until the number of heavy atoms exceeds 30 or more, depending on the general shape of the molecule.

The linear scaling DFT methods can be the fastest *ab initio* method for large molecules. However, there has been a lot of misleading literature in this field. The literature is ripe with graphs indicating that linear scaling methods take an order of magnitude less CPU time than conventional algorithms for some test systems, such as *n*-alkanes or graphite sheets. However, calculations with commercial software often indicate speedups of only a few percent or perhaps a slightly slower calculation. There are a number of reasons for these inconsistencies.

The first factor to note is that most software packages designed for efficient operation use integral accuracy cutoffs with *ab initio* calculations. This means that integrals involving distant atoms are not included in the calculation if they are estimated to have a negligible contribution to the final energy, usually less than 0.00001 Hartrees or one-hundredth the energy of a van der Waals interaction. In the literature, many of the graphs showing linear scaling DFT performance compare it to an algorithm that does not use integral accuracy cutoffs. Cases where the calculation runs faster without the linear scaling method are due to the integral accuracy cutoffs being more time-efficient than the linear scaling method.

The second consideration is the geometry of the molecule. The multipole estimation methods are only valid for describing interactions between distant regions of the molecule. The same is true of integral accuracy cutoffs. Because of this, it is common to find that the calculated CPU time can vary between different conformers. Linear systems can be modeled most efficiently and folded, globular, or planar systems less efficiently. In our test calculations on the $C_{40}H_{82}$ *n*-alkane, the energy calculation on a folded conformation took four times as much CPU time as the calculation on the linear conformation.

The bottom line is that linear scaling methods can use less CPU time than conventional methods, but the speedup is not as great as is indicated by some of the literature. We ran test calculations on a C_{40} *n*-alkane in various conformations and a C_{40} graphite sheet with two software packages. These calculations showed that linear scaling methods required 60–80% of the amount of CPU time required for the conventional calculation. It is possible to obtain better performance than this by manually setting the multipole order used by the algorithm, but researchers have advised extreme caution about doing this because it can affect the accuracy of results.

5.3 PRACTICAL CONSIDERATIONS

As mentioned above, DFT calculations must use a basis set. This raises the question of whether DFT-optimized or typical HF-optimized basis sets should be used. Studies using DFT-optimized basis sets have shown little or no improvement over the use of a similar-size conventional basis sets. Most DFT calculations today are being done with HF-optimized GTO basis sets. The accuracy of results tends to degrade significantly with the use of very small basis sets. For accuracy considerations, the smallest basis set used is generally 6–31G* or the equivalent. Interestingly, there is only a small increase in accuracy obtained by using very large basis sets. This is probably due to the fact that the density functional is limiting accuracy more than the basis set limitations.

Since DFT calculations use numerical integrals, calculations using GTO basis sets are no faster than those using other types of basis sets. It is reasonable to expect that STO basis sets or numeric basis sets (e.g., cubic splines) would be more accurate due to the correct representation of the nuclear cusp and exponential decay at long distances. The fact that so many DFT studies use GTO basis sets is not a reflection of accuracy or computation time advantages. It is because there were a large number of programs written for GTO HF calculations. HF programs can be easily turned into DFT programs, so it is very common to find programs that do both. There are programs that use cubic spline basis sets (e.g., the dMol and Spartan programs) and STO basis sets (e.g., ADF).

The accuracy of results from DFT calculations can be poor to fairly good, depending on the choice of basis set and density functional. The choice of density functional is made more difficult because creating new functionals is still an active area of research. At the time of this book's publication, the B3LYP hybrid functional (also called Becke3LYP) was the most widely used for molecular calculations by a fairly large margin. This is due to the accuracy of the B3LYP results obtained for a large range of compounds, particularly organic molecules. However, it would not be surprising if this functional's dominance changed within a few years. Table 5.1 lists a number of commonly used functionals.

Due to the newness of DFT, its performance is not completely known and continues to change with the development of new functionals. The bibliography at the end of this chapter includes references for studies comparing the accuracy of results. At the present time, DFT results have been very good for organic molecules, particularly those with closed shells. Results have not been so encouraging for heavy elements, highly charged systems, or systems known to be very sensitive to electron correlation. Also, the functionals listed in Table 5.1 do not perform well for problems dominated by dispersion forces.

5.4 RECOMMENDATIONS

Given the fact that DFT is newer than the other *ab initio* methods, it is quite likely that conventional wisdom over which technique works best will shift with the creation of new techniques in the not too distant future. DFT's recent heavy usage has been due to the often optimal accuracy versus CPU time. At the time of this book's publication, the B3LYP method with basis sets of 6–31G* or larger is the method of choice for many organic molecule calculations. Unfortunately, there is no systematic way to improve DFT calculations, thus making them unusable for very-high-accuracy work. Researchers are advised to look for relevant literature and run test calculations before using these methods.

BIBLIOGRAPHY

Introductory discussions are in

F. Jensen, *Introduction to Computational Chemistry* John Wiley & Sons, New York (1999).

J. Simons, J. Nichols, *Quantum Mechanics in Chemistry* Oxford, Oxford (1997).

J. B. Foresman, Æ. Frisch, *Exploring Chemistry with Electronic Structure Methods* Gaussian, Pittsburgh (1996).

A. R. Leach, *Molecular Modelling Principles and Applications* Longman, Essex (1996).

I. R. Levine, *Quantum Chemistry* Prentice Hall, Englewood Cliffs (1991).

Books on DFT are

W. Koch, M. C. Holthausen, *A Chemist's Guide to Density Functional Theory* Wiley-VCH, Weinheim (2000).

Electronic Density Functional Theory Recent Progress and New Directions J. F. Dobson, G. Vignale, M. P. Das, Eds., Plenum, New York (1998).

Density Functional Methods in Chemistry and Materials Science M. Springborg, Ed., John Wiley & Sons, New York (1997).

M. Ernzerhof, J. P Perdew, K. burke, D. J. W. Geldart, A. Holas, N. H. March, R. van Leeuwen, O. V. Gritsenko, E. J. Baerends, E. V. Ludeña, R. López-Boada, *Density Functional Theory I* Springer, Berlin (1996).

Chemical Applications of Density-Functional Theory B. B. Laird, R. B. Ross, T. Ziegler, Eds., ACS, Washington (1996).

Recent Advances in Density Functional Theory D. P. Chong, Ed., World Scientific, Singapore (1995).

Modern Density Functional Theory J. M. Seminario, P. Politzer, Eds., Elsevier, Amsterdam (1995).

Density Functional Theory E. K. U. Gross, R. M. Dreizler, Eds., Plenum, New York (1995).

Density Functional Theory of Molecules, Clusters, and Solids D. E. Ellis, Ed., Kluwer, Dordrecht (1995).

Density Functional Methods in Chemistry J. K. Labanowski, J. W. Andzelm, Eds., Springer-Verlag, New York (1991).

R. G. Parr, W. Yang *Density-Functional Theory of Atoms and Molecules* Oxford, Oxford (1989).

Density Functional Methods in Physics R. M. Dreizler, J. du Providencia, Eds., Plenum, New York (1985).

Local Density Approximation in Quantum Chemistry and Solid State Physics J. P. Dahl, J. Avery, Eds., Plenum, New York (1984).

Review articles are

W. Koch, R. H. Hertwig, *Encycl. Comput. Chem.* **1**, 689 (1998).

O. N. Ventura, M. Kieninger, K. Irving, *Adv. Quantum Chem.* **28**, 294 (1997).

L. J. Bartolotti, K. Flurchick, *Rev. Comput. Chem.* **7**, 187 (1996).

A. St-Amant, *Rev. Comput. Chem.* **7**, 187 (1996).

R. Neumann, R. H. Nobes, N. C. Handy, *Molecular Physics* **87**, 1 (1996).

W. Kohn, A. D. Becke, R. G. Parr, *J. Phys. Chem.* **100**, 12974 (1996).

R. G. Parr, W. Yang, *Ann. Rev. Phys. Chem.* **46**, 701 (1995).

A. D. Becke, *Modern Electronic Structure Theory Part 2* D. R. Yarkony, Ed., 1022, World Scientific, Singapore (1995).

S. R. Gadre, R. K. Patha, *Adv. Quantum Chem.* **22**, 212 (1991).

T. Ziegler, *Chem. Rev.* **91**, 651 (1991).

Adv. Quantum Chem. Volume **21** (1990).

R. O. Jones, *Adv. Chem. Phys.* **67**, 413 (1987).

R. G. Parr, *Ann. Rev. Phys. Chem.* **34**, 631 (1983).

D. A. Case, *Ann. Rev. Phys. Chem.* **33**, 151 (1982).

K. H. Johnson, *Ann. Rev. Phys. Chem.* **26**, 39 (1975).

Sources comparing the accuracy of results are

C.-H. Hu, D. P. Chong, *Encycl. Comput. Chem.* **1**, 664 (1998).

L. A. Curtiss, P. C. Redfern, K. Raghavachari, J. A. Pople, *J. Chem. Phys.* **109**, 42 (1998).

E. R. Davidson, *Int. J. Quantum Chem.* **69**, 241 (1998).

A. C. Scheiner, J. Baker, J. W. Andzelm, *J. Comput. Chem.* **18**, 775 (1997).

W. J. Hehre, *Practical Strategies for Electronic Structure Calculations* Wavefunction, Irvine (1995).

G. G. Hoffman, L. R. Pratt, *Mol. Phys.* **82**, 245 (1994).

B. G. Johnson, P. M. W. Gill, J. A. Pople, *J. Chem. Phys.* **98**, 5612 (1993).

G. L. Laming, N. C. Handy, R. D. Amos, *Mol. Phys.* **80**, 1121 (1993).

A. M. Lee, N. C. Handy, *J. Chem. Soc. Faraday Trans.* **89**, 3999 (1993).

6 Molecular Mechanics

The most severe limitation of *ab initio* methods is the limited size of the molecule that can be modeled on even the largest computers. Semiempirical calculations can be used for large organic molecules, but are also too computation-intensive for most biomolecular systems. If a molecule is so big that a semiempirical treatment cannot be used effectively, it is still possible to model its behavior avoiding quantum mechanics totally by using molecular mechanics.

6.1 BASIC THEORY

The molecular mechanics energy expression consists of a simple algebraic equation for the energy of a compound. It does not use a wave function or total electron density. The constants in this equation are obtained either from spectroscopic data or *ab initio* calculations. A set of equations with their associated constants is called a force field. The fundamental assumption of the molecular mechanics method is the transferability of parameters. In other words, the energy penalty associated with a particular molecular motion, say, the stretching of a carbon–carbon single bond, will be the same from one molecule to the next. This gives a very simple calculation that can be applied to very large molecular systems. The performance of this technique is dependent on four factors:

1. The functional form of the energy expression
2. The data used to parameterize the constants
3. The technique used to optimize constants from that data
4. The ability of the user to apply the technique in a way consistent with its strengths and weaknesses

In order for the transferability of parameters to be a good description of the molecule, force fields use atom types. This means that a sp^3 carbon will be described by different parameters than a sp^2 carbon, and so on. Usually, atoms in aromatic rings are treated differently from sp^2 atoms. Some force fields even parameterize atoms for specific functional groups. For example, the carbonyl oxygen in a carboxylic acid may be described by different parameters than the carbonyl oxygen in a ketone.

The energy expression consists of the sum of simple classical equations. These equations describe various aspects of the molecule, such as bond stretching, bond bending, torsions, electrostatic interactions, van der Waals forces, and hydrogen bonding. Force fields differ in the number of terms in the energy expression, the complexity of those terms, and the way in which the constants were obtained. Since electrons are not explicitly included, electronic processes cannot be modeled.

Terms in the energy expression that describe a single aspect of the molecular shape, such as bond stretching, angle bending, ring inversion, or torsional motion, are called valence terms. All force fields have at least one valence term and most have three or more.

Terms in the energy expression that describe how one motion of the molecule affects another are called cross terms. A cross term commonly used is a stretch-bend term, which describes how equilibrium bond lengths tend to shift as bond angles are changed. Some force fields have no cross terms and may compensate for this by having sophisticated electrostatic functions. The MM4 force field is at the opposite extreme with nine different types of cross terms.

Force fields may or may not include an electrostatic term. The electrostatic term most often used is the Coulombs law term for the energy of attraction or repulsion between charged centers. These charges are usually obtained from non-orbital-based algorithms designed for use with molecular mechanics. These charges are meant to be the partial charges on the nuclei. The modeling of molecules with a net charge is described best by using atom types parameterized for describing charged centers. A dielectric constant is sometimes included to model solvation effects. These charge calculation methods are described further in Chapter 12.

Bond stretching is most often described by a harmonic oscillator equation. It is sometimes described by a Morse potential. In rare cases, bond stretching will be described by a Leonard–Jones or quartic potential. Cubic equations have been used for describing bond stretching, but suffer from becoming completely repulsive once the bond has been stretched past a certain point.

Bond bending is most often described by a harmonic equation. Bond rotation is generally described by a cosine expression (Figure 6.1). Intermolecular forces, such as van der Waals interactions and hydrogen bonding, are often described by Leonard–Jones equations. Some force fields also use a combined stretch-bend term. The choice of equation functional forms is particularly important for computing energies of molecules distorted from the equilibrium geometry, as evidenced by the difference between a harmonic potential and a Morse potential shown in Figure 6.2.

Table 6.1 gives the mathematical forms of energy terms often used in popular force fields. The constants may vary from one force field to another according to the designer's choice of unit system, zero of energy, and fitting procedure.

All the constants in these equations must be obtained from experimental data or an *ab initio* calculation. The database of compounds used to parameterize the method is crucial to its success. A molecular mechanics method may be parameterized against a specific class of molecules, such as proteins or nucleotides. Such a force field would only be expected to have any relevance in

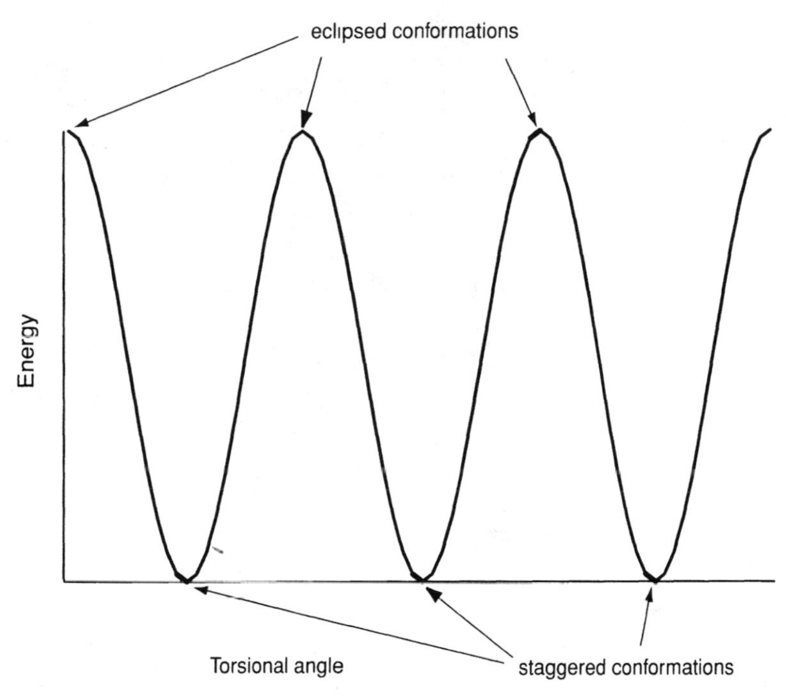

FIGURE 6.1 The energy due to conformation around a single bond represented by a cosine function.

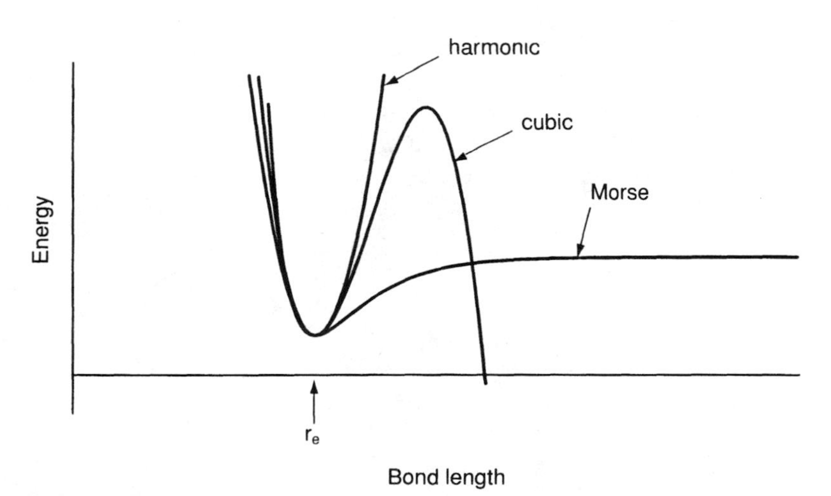

FIGURE 6.2 Harmonic, cubic, and Morse potential curves used to describe the energy due to bond stretching in molecular mechanics force fields.

TABLE 6.1 Common Force Field Terms

Name	Use	Energy Term
Harmonic	Bond stretch	$k(l - l_0)^2$
Harmonic	Angle bend	$k(\theta - \theta_0)^2$
Cosine	Torsion	$k[1 + \cos(n\theta)]$
Leonard–Jones 6–12	van der Waals	$4k\left(\frac{A}{r}\right)^{12} - \left(\frac{B}{r}\right)^{6}$
Leonard–Jones 10–12	van der Waals	$4k\left(\frac{A}{r}\right)^{12} - \left(\frac{B}{r}\right)^{10}$
Coulomb	Electrostatic	$\frac{q_1 q_2}{4\pi\varepsilon_0 r}$
Taylor	Stretch-bend	$k(\theta - \theta_0)[(l_1 - l_{1_0})(l_2 - l_{2_0})]$
Morse	Bond stretch	$D_e[1 - e^{-\alpha(l - l_0)}]^2$

l-bond length.

θ-bond angle.

k, α, A, B-constants particular to the elements in a certain hybridization state.

n-an integer.

r-nonbond distance.

q-charge.

D_e-dissociation energy.

describing other proteins or nucleotides. Other force fields are parameterized to give a reasonable description of a wide variety of organic compounds. A few force fields have even been parameterized for all the elements.

Some force fields simplify the complexity of the calculations by omitting most of the hydrogen atoms. The parameters describing each backbone atom are then modified to describe the behavior of the atoms with the attached hydrogens. Thus, the calculation uses a CH_2 group rather than a sp^3 carbon bonded to two hydrogens. These are called united atom force fields or intrinsic hydrogen methods. This calculation is most often employed to describe very large biomolecules. It is not generally applied if the computer hardware available is capable of using the more accurate explicit hydrogen force fields. Some force fields have atom types for use with both implicit and explicit hydrogens.

The way in which the force field parameters are obtained from this original data is also important. Bond stretching and bending are relatively stiff motions. Thus, they can often be described very well by using equilibrium values obtained from X-ray diffraction results and force constants from vibrational spectroscopy. On the other hand, torsional behavior is sensitive to both the torsional behavior of the isolated bond and the nonbonded interactions between distant sections of the molecule and surrounding molecules. The choice of fitting procedure becomes important because it determines how much of the energy is from each contributing process. A force field can also be parameterized to best predict vibrational motion or intermolecular forces.

The energies computed by molecular mechanics are usually conformational energies. This means that the energy computed is meant to be an energy that will reliably predict the difference in energy from one conformation to the next. The effect of strained bond lengths or angles is also included in this energy. This is not the same as the total energies obtained from *ab initio* programs or the heat of formation from semiempirical programs. The actual value of the conformational energy does not necessarily have any physical meaning and is not comparable between different force fields. Molecular mechanics methods can be modified to compute heats of formation by including a database or computation scheme to yield bond energies that might be added to the conformational energy and account for the zero of energy.

Molecular mechanics methods are not generally applicable to structures very far from equilibrium, such as transition structures. Calculations that use algebraic expressions to describe the reaction path and transition structure are usually semiclassical algorithms. These calculations use an energy expression fitted to an *ab initio* potential energy surface for that exact reaction, rather than using the same parameters for every molecule. Semiclassical calculations are discussed further in Chapter 19.

6.2 EXISTING FORCE FIELDS

Most researchers do not parameterize force fields because many good force fields have already been developed. On rare occasions, a researcher will add an additional atom as described in Chapter 29. The following are some commonly used molecular mechanics force fields. Many of these have been implemented in more than one software package. There tend to be minor differences in the implementation leading to small differences in results from one software package to another.

6.2.1 AMBER

Assisted model building with energy refinement (AMBER) is the name of both a force field and a molecular mechanics program. It was parameterized specifically for proteins and nucleic acids. AMBER uses only five bonding and nonbonding terms along with a sophisticated electrostatic treatment. No cross terms are included. Results are very good for proteins and nucleic acids, but can be somewhat erratic for other systems.

6.2.2 CHARMM

Chemistry at Harvard macromolecular mechanics (CHARMM) is the name of both a force field and a program incorporating that force field. The academic version of this program is designated CHARMM and the commercial version is called CHARMm. It was originally devised for proteins and nucleic acids. It has now been applied to a range of biomolecules, molecular dynamics, solvation, crystal packing, vibrational analysis, and QM/MM studies. CHARMM uses five valence terms, one of which is an electrostatic term.

6.2.3 CFF

The consistent force field (CFF) was developed to yield consistent accuracy of results for conformations, vibrational spectra, strain energy, and vibrational enthalpy of proteins. There are several variations on this, such as the Ure–Bradley version (UBCFF), a valence version (CVFF), and Lynghy CFF. The quantum mechanically parameterized force field (QMFF) was parameterized from *ab initio* results. CFF93 is a rescaling of QMFF to reproduce experimental results. These force fields use five to six valence terms, one of which is an electrostatic term, and four to six cross terms.

6.2.4 CHEAT

Carbohydrate hydroxyls represented by external atoms (CHEAT) is a force field designed specifically for modeling carbohydrates.

6.2.5 DREIDING

DREIDING is an all-purpose organic or bio-organic molecule force field. It has been most widely used for large biomolecular systems. It uses five valence terms, one of which is an electrostatic term. The use of DREIDING has been dwindling with the introduction of improved methods.

6.2.6 ECEPP

Empirical conformational energy program for peptides (ECEPP) is the name of both a computer program and the force field implemented in that program. This is one of the earlier peptide force fields that has seen less use with the introduction of improved methods. It uses three valence terms that are fixed, a van der Waals term, and an electrostatic term.

6.2.7 EFF

Empirical force field (EFF) is a force field designed just for modeling hydrocarbons. It uses three valence terms, no electrostatic term and five cross terms.

6.2.8 GROMOS

Gronigen molecular simulation (GROMOS) is the name of both a force field and the program incorporating that force field. The GROMOS force field is popular for predicting the dynamical motion of molecules and bulk liquids. It is also used for modeling biomolecules. It uses five valence terms, one of which is an electrostatic term.

6.2.9 MM1, MM2, MM3, MM4

MM1, MM2, MM3, and MM4 are general-purpose organic force fields. There have been many variants of the original methods, particularly MM2. MM1 is seldom used since the newer versions show measurable improvements. The MM3 method is probably one of the most accurate ways of modeling hydrocarbons. At the time of this book's publication, the MM4 method was still too new to allow any broad generalization about the results. However, the initial published results are encouraging. These are some of the most widely used force fields due to the accuracy of representation of organic molecules. MMX and MM+ are variations on MM2. These force fields use five to six valence terms, one of which is an electrostatic term and one to nine cross terms.

6.2.10 MMFF

The Merck molecular force field (MMFF) is one of the more recently published force fields in the literature. It is a general-purpose method, particularly popular for organic molecules. MMFF94 was originally intended for molecular dynamics simulations, but has also seen much use for geometry optimization. It uses five valence terms, one of which is an electrostatic term, and one cross term.

6.2.11 MOMEC

MOMEC is a force field for describing transition metal coordination compounds. It was originally parameterized to use four valence terms, but not an electrostatic term. The metal–ligand interactions consist of a bond-stretch term only. The coordination sphere is maintained by nonbond interactions between ligands. MOMEC generally works reasonably well for octahedrally coordinated compounds.

6.2.12 OPLS

Optimized potentials for liquid simulation (OPLS) was designed for modeling bulk liquids. It has also seen significant use in modeling the molecular dynamics of biomolecules. OPLS uses five valence terms, one of which is an electrostatic term, but no cross terms.

6.2.13 Tripos

Tripos is a force field created at Tripos Inc. for inclusion in the Alchemy and SYBYL programs. It is sometimes called the SYBYL force field. Tripos is designed for modeling organic and bio-organic molecules. It is also often used for CoMFA analysis, a 3D QSAR technique. Tripos uses five valence terms, one of which is an electrostatic term.

6.2.14 UFF

UFF stands for universal force field. Although there have been a number of universal force fields, meaning that they include all elements, there has only been one actually given this name. This is the most promising full periodic table force field available at this time. UFF is most widely used for systems containing inorganic elements. It was designed to use four valence terms, but not an electrostatic term.

UFF was originally designed to be used without an electrostatic term. The literature accompanying one piece of software recommends using charges obtained with the Q-equilibrate method. Independent studies have found the accuracy of results to be significantly better without charges.

6.2.15 YETI

YETI is a force field designed for the accurate representation of nonbonded interactions. It is most often used for modeling interactions between biomolecules and small substrate molecules. It is not designed for molecular geometry optimization so researchers often optimize the molecular geometry with some other force field, such as AMBER, then use YETI to model the docking process. Recent additions to YETI are support for metals and solvent effects.

6.3 PRACTICAL CONSIDERATIONS

Force fields for describing inorganic elements have not yet seen as much development as organic molecule force fields. A number of organic methods have been extended to full periodic table applicability, but the results have been less than spectacular. The best available force field for inorganics is probably the UFF force field. Many inorganic studies in the past have not used preexisting force fields, such as UFF. What has often been done is parameterizing a new atom to describe the behavior of the inorganic element in an organic force field. It is usually parameterized to describe that element in a specific compound or class of compounds. It is too soon to predict whether these specifically parameterized techniques will be replaced by full periodic force fields. Inorganic compound modeling is discussed more thoroughly in Chapters 37 and 41.

Molecular mechanics calculations are deceptively simple to perform. Many software packages now make molecular mechanics as easy as specifying a molecular structure and saying "go," at which point the calculation will run and very soon give a result. The difficulty is in knowing which results to trust.

The most reliable results are energy differences between conformers. Another popular usage is to examine intermolecular binding. The difference in binding energy between two sites or two orientations is usually fairly reliable. The absolute binding energy (separating the molecules to infinite distance) is not so reliably predicted. Computing properties other than energy and geometry is discussed in Chapter 13.

6.4 RECOMMENDATIONS

The advantage of molecular mechanics is that it allows the modeling of enormous molecules, such as proteins and segments of DNA. This is why it is the primary tool of computational biochemists. It also models intermolecular forces well.

The disadvantage of molecular mechanics is that there are many chemical properties that are not even defined within the method, such as electronic excited states. Since chemical bonding terms are explicitly included in the force field, it is not possible without some sort of mathematical manipulation to examine reactions in which bonds are formed or broken. In order to work with extremely large and complicated systems, molecular mechanics software packages often have powerful and easy-to-use graphic interfaces. Because of this, mechanics is sometimes used because it is an easy, but not necessarily a good, way to describe a system.

Due to their sensitivity to parameterization, the best technique for choosing a force field is to look for similar studies in the literature and validate test results against experimental results. The references listed in the bibliography at the end of this chapter and in Chapter 16 give some excellent starting points for finding relevant accuracy comparisons. A generalization of the results for studies comparing force field accuracies is as follows:

1. The MM2, MM3, and Merck (MMFF) force fields perform best for a wide range of organic molecules.
2. The AMBER and CHARMM force fields are best suited for protein and nucleic acid studies.
3. Most existing molecular mechanics studies of inorganic molecules required careful customization of force field parameters.
4. UFF is the most reliable force field to be used without modification for inorganic systems.
5. Molecular dynamics studies are best done with a force field designed for that purpose.
6. The rings in sugars pose a particular problem to general-purpose force fields and should be modeled using a force field designed specifically for carbohydrates.

BIBLIOGRAPHY

Introductory descriptions are in

F. Jensen, *Introduction to Computational Chemistry* John Wiley & Sons, New York (1999).

J. P. Doucet, *Computer Aided Molecular Design* Academic Press, San Diego (1996).

A. R. Leach, *Molecular Modeling Principles and Applications* Longman, Essex (1996).

G. H. Grant, W. G. Richards, *Computational Chemistry* Oxford, Oxford (1995).

T. Clark, *A Handbook of Computational Chemistry* John Wiley & Sons, New York (1985).

Books about molecular mechanics are

M. F. Schlecht, *Molecular Modeling on the PC* Wiley-VCH, New York (1998).

A. K. Rappé, C. J. Casewit, *Molecular Mechanics across Chemistry* University Science Books, Sausalito (1997).

U. Burkert, N. L. Allinger, *Molecular Mechanics* American Chemical Society, Washington (1982).

Review articles are

S. Profeta, Jr., *Kirk-Othmer Encyclopedia of Chemical Technology Supplement* J. I. Kroschwitz Ed. 315, John Wiley & Sons, New York (1998).

A. G. Császár, *Encycl. Comput. Chem.* **1**, 13 (1998).

B. Reindl, *Encycl. Comput. Chem.* **1**, 196 (1998).

R. J. Woods, *Encycl. Comput. Chem.* **1**, 220 (1998).

A. D. MacKerell, Jr., B. Brooks, C. L. Brooks, III, L. Nilsson, B. Roux, Y. Won, M. Karplus, *Encycl. Comput. Chem.* **1**, 271 (1998).

N. L. Allinger, *Encycl. Comput. Chem.* **2**, 1013 (1998).

J. R. Maple, *Encycl. Comput. Chem.* **2**, 1015 (1998).

J. R. Maple, *Encycl. Comput. Chem.* **2**, 1025 (1998).

T. A. Halgren, *Encycl. Comput. Chem.* **2**, 1033 (1998).

W. F. van Gunsteren, X. Daura, A. E. Mark, *Encycl. Comput. Chem.* **2**, 1211 (1998).

J. Tai, N. Nevins, *Encycl. Comput. Chem.* **3**, 1671 (1998).

P. Cieplak, *Encycl. Comput. Chem.* **3**, 1922 (1998).

W. L. Jorgensen, *Encycl. Comput. Chem.* **3**, 1986 (1998).

A. D. MacKerell, Jr., *Encycl. Comput. Chem.* **3**, 2191 (1998).

M. Zimmer, *Chem. Rev.* **95**, 2629 (1995).

P. Comba, *Coord. Chem. Rev.* **123**, 1 (1993).

B. P. Hay, *Coord. Chem. Rev.* **126**, 177 (1993).

C. E. Dykstra, *Chem. Rev.* **93**, 2339 (1993).

N. L. Allinger, *Accurate Molecular Structures* 19 A. Domenicano, I. Hargittal, Eds., Oxford, Oxford (1992).

P. Kollman, *Ann. Rev. Phys. Chem.* **38**, 303 (1987).

G. C. Maitland, E. B. Smith, *Chem. Soc. Rev.* **2**, 181 (1973).

T. Shimanouchi, I. Nakagawa, *Ann. Rev. Phys. Chem.* **23**, 217 (1972).

Comparisons of force field accuracy are

I. Pettersson, T. Liljefors, *Rev. Comput. Chem.* **9**, 167 (1996).

K. Gundertofte, T. Liljefors, P.-O. Norby, I. Pettersson, *J. Comput. Chem.* **17** 429 (1996).

W. D. Corness, M. P. Ha, Y. Sun, P. A. Kollman, *J. Comput. Chem.* **17**, 1541 (1996).

T. A. Halgren, *J. Comput. Chem.* **17**, 520 (1996).

D. M. Ferguson, I. R. Gould, W. A. Glauser, S. Schroeder, P. A. Kolman, *J. Comput. Chem.* **13,** 525 (1992).

V. S. Allured, C. M. Kelly, C. R. Landis, *J. Am. Chem. Soc.* **113**, 1 (1991).

M. Saunders, *J. Comput. Chem.* **12**, 645 (1991).

K. Gundertofte, J. Palm, I. Pettersson, A. Stamvik, *J. Comput. Chem.* **12**, 200 (1991).

M. Clark, R. D. Cramer, III, N. Van Opdenbosch, *J. Comput. Chem.* **10**, 982 (1989).

I. K. Roterman, K. D. Gibson, H. A. Scheraga, *J. Biomol. Struct. Dynam.* **7**, 391 (1989).

I. K. Roterman, M. H. Lambert, K. D. Gibson, H. A. Scheraga, *J. Biomol. Struct. Dynam.* **7**, 421 (1989).

C. Altona, D. H. Faber, *Top. Curr. Chem.* **45**, 1 (1974).

E. M. Engler, J. D. Andose, P. v. R. Schleyer, *J. Am. Chem. Soc.* **95**, 8005 (1973).

Listings of references to all published force field parameters are

M. Jalaie, K. B. Lipkowitz, *Rev. Comput. Chem.* **14**, 441 (2000).

E. Osawa, K. B. Lipkowitz, *Rev. Comput. Chem.* **6**, 355 (1995).

7 Molecular Dynamics and Monte Carlo Simulations

In Chapter 2, a brief discussion of statistical mechanics was presented. Statistical mechanics provides, in theory, a means for determining physical properties that are associated with not one molecule at one geometry, but rather, a macroscopic sample of the bulk liquid, solid, and so on. This is the net result of the properties of many molecules in many conformations, energy states, and the like. In practice, the difficult part of this process is not the statistical mechanics, but obtaining all the information about possible energy levels, conformations, and so on. Molecular dynamics (MD) and Monte Carlo (MC) simulations are two methods for obtaining this information

7.1 MOLECULAR DYNAMICS

Molecular dynamics is a simulation of the time-dependent behavior of a molecular system, such as vibrational motion or Brownian motion. It requires a way to compute the energy of the system, most often using a molecular mechanics calculation. This energy expression is used to compute the forces on the atoms for any given geometry. The steps in a molecular dynamics simulation of an equilibrium system are as follows:

1. Choose initial positions for the atoms. For a molecule, this is whatever geometry is available, not necessarily an optimized geometry. For liquid simulations, the molecules are often started out on a lattice. For solvent–solute systems, the solute is often placed in the center of a collection of solvent molecules, with positions obtained from a simulation of the neat solvent.
2. Choose an initial set of atom velocities. These are usually chosen to obey a Boltzmann distribution for some temperature, then normalized so that the net momentum for the entire system is zero (it is not a flowing system).
3. Compute the momentum of each atom from its velocity and mass.
4. Compute the forces on each atom from the energy expression. This is usually a molecular mechanics force field designed to be used in dynamical simulations.
5. Compute new positions for the atoms a short time later, called the time step. This is a numerical integration of Newton's equations of motion using the information obtained in the previous steps.

6. Compute new velocities and accelerations for the atoms.
7. Repeat steps 3 through 6.
8. Repeat this iteration long enough for the system to reach equilibrium. In this case, equilibrium is not the lowest energy configuration; it is a configuration that is reasonable for the system with the given amount of energy.
9. Once the system has reached equilibrium, begin saving the atomic coordinates every few iterations. This information is typically saved every 5 to 25 iterations. This list of coordinates over time is called a trajectory.
10. Continue iterating and saving data until enough data have been collected to give results with the desired accuracy.
11. Analyze the trajectories to obtain information about the system. This might be determined by computing radial distribution functions, diffusion coefficients, vibrational motions, or any other property computable from this information.

In order for this to work, the force field must be designed to describe intermolecular forces and vibrations away from equilibrium. If the purpose of the simulation is to search conformation space, a force field designed for geometry optimization is often used. For simulating bulk systems, it is more common to use a force field that has been designed for this purpose, such as the GROMOS or OPLS force fields.

There are several algorithms available for performing the numerical integration of the equations of motion. The Verlet algorithm is widely used because it requires a minimum amount of computer memory and CPU time. It uses the positions and accelerations of the atoms at the current time step and positions from the previous step to compute the positions for the next time step. The velocity Verlet algorithm uses positions, velocities, and accelerations at the current time step. This gives a more accurate integration than the Verlet algorithm. The Verlet and velocity Verlet algorithms often have a step in which the velocities are rescaled in order to correct for minor errors in the integration, thus simulating a constant-temperature system. Beeman's algorithm uses positions, velocities, and accelerations from the previous time step. It gives better energy conservation at the expense of computer memory and CPU time. A Gear predictor-corrector algorithm predicts the next set of positions and accelerations, then compares the accelerations to the predicted ones to compute a correction for the step. Each step can thus be refined iteratively. Predictor-corrector algorithms give an accurate integration but are seldom used due to their large computational needs.

The choice of a time step is also important. A time step that is too large will cause atoms to move too far along a given trajectory, thus poorly simulating the motion. A time step that is too small will make it necessary to run more iterations, thus taking longer to run the simulation. One general rule of thumb is that the time step should be one order of magnitude less than the timescale of the shortest motion (vibrational period or time between collisions). This gives a time step on the order of tens of femtoseconds for simulating a liquid of rigid molecules and tenths of a femtosecond for simulating vibrating molecules.

It is important to verify that the simulation describes the chemical system correctly. Any given property of the system should show a normal (Gaussian) distribution around the average value. If a normal distribution is not obtained, then a systematic error in the calculation is indicated. Comparing computed values to the experimental results will indicate the reasonableness of the force field, number of solvent molecules, and other aspects of the model system.

The algorithm described above is for a system with a constant volume, number of particles, and temperature. It is also possible to set up a calculation in which the velocities are rescaled slightly at each step to simulate a changing temperature. For solvent–solute systems, this can lead to the problem of having a "hot solvent, cold solute" situation because energy transfer takes many collisions, thus a relatively long time. Raising the temperature very slowly fixes this problem but leads to extremely long simulation times. A slightly artificial, but more efficient solution to this problem is to scale solvent and solute velocities separately. Constant pressure calculations can be obtained by automatically varying the box size to maintain the pressure.

7.2 MONTE CARLO SIMULATIONS

There are many types of calculations that are referred to as Monte Carlo calculations. All Monte Carlo methods are built around some sort of a random sampling, which is simulated with a random-number-generating algorithm. In this context, a Monte Carlo simulation is one in which the location, orientation, and perhaps geometry of a molecule or collection of molecules are chosen according to a statistical distribution. For example, many possible conformations of a molecule could be examined by choosing the conformation angles randomly. If enough iterations are done and the results are weighted by a Boltzmann distribution, this gives a statistically valid result. The steps in a Monte Carlo simulation are as follows:

1. Choose an initial set of atom positions. The same techniques used for molecular dynamics simulations are applicable.
2. Compute the energy for the system.
3. Randomly choose a trial move for the system. This could be moving all atoms, but it more often involves moving one atom or molecule for efficiency reasons.
4. Compute the energy of the system in the new configuration.
5. Decide whether to accept the new configuration. There is an acceptance criteria based on the old and new energies, which will ensure that the re-

sults reproduce a Boltzmann distribution. Either keep the new configuration or restore the atoms to their previous positions.

6. Iterate steps 3 through 5 until the system has equilibrated.
7. Continue iterating and collecting data to compute the desired property. The expectation value of any property is its average value (sum divided by the number of iterations summed). This is correct as long as the acceptance criteria in step 5 ensured that the probability of a configuration being accepted is equal to the probability of it being included in a Boltzmann distribution. If one atom is moved at a time, summing configurations into the average every few iterations will prevent the average from overrepresenting some configurations.

There are a few variations on this procedure called importance sampling or biased sampling. These are designed to reduce the number of iterations required to obtain the given accuracy of results. They involve changes in the details of how steps 3 and 5 are performed. For more information, see the book by Allen and Tildesly cited in the end-of-chapter references.

The size of the move in step 3 of the above procedure will affect the efficiency of the simulation. In this case, an inefficient calculation is one that requires more iterations to obtain a given accuracy result. If the size is too small, it will take many iterations for the atom locations to change. If the move size is too large, few moves will be accepted. The efficiency is related to the acceptance ratio. This is the number of times the move was accepted (step 5 above) divided by the total number of iterations. The most efficient calculation is generally obtained with an acceptance ratio between 0.5 and 0.7.

Monte Carlo simulations require less computer time to execute each iteration than a molecular dynamics simulation on the same system. However, Monte Carlo simulations are more limited in that they cannot yield time-dependent information, such as diffusion coefficients or viscosity. As with molecular dynamics, constant NVT simulations are most common, but constant NPT simulations are possible using a coordinate scaling step. Calculations that are not constant N can be constructed by including probabilities for particle creation and annihilation. These calculations present technical difficulties due to having very low probabilities for creation and annihilation, thus requiring very large collections of molecules and long simulation times.

7.3 SIMULATION OF MOLECULES

In order to analyze the vibrations of a single molecule, many molecular dynamics steps must be performed. The data are then Fourier-transformed into the frequency domain to yield a vibrational spectrum. A given peak can be selected and transformed back to the time domain. This results in computing the vibrational motion at that frequency. Such a technique is one of the most reliable ways to obtain the very-low-frequency or anharmonic motions of large molecules.

Another reason for simulating molecules with these techniques is as a conformation search technique. Many Monte Carlo iterations can be run, with the lowest-energy conformation being saved because it will likely be near a very-low-energy minimum. In some cases, a few optimization steps are performed after each Monte Carlo step. Likewise, a molecular dynamics simulation with a sufficiently high temperature will allow the molecule to pass over energy barriers with a statistical preference for lower-energy conformations. Molecular dynamics is often used for conformation searching by employing an algorithm called simulated annealing, in which the temperature is slowly decreased over the course of the simulation. These techniques are discussed further in Chapter 21.

7.4 SIMULATION OF LIQUIDS

The application of molecular dynamics to liquids or solvent–solute systems allows the computation of properties such as diffusion coefficients or radial distribution functions for use in statistical mechanical treatments. A liquid is simulated by having a number of molecules (perhaps 1000) within a specific volume. This volume might be cube, a parallelepiped, or a hexagonal cylinder. Even with 1000 molecules, a significant fraction would be against the wall of the box. In order to avoid such severe edge effects, periodic boundary conditions are used to make it appear as though the fluid is infinite. Actually, the molecules at the edge of the next box are a copy of the molecules at the opposite edge of the box. These simulations are discussed in more detail in Chapter 39.

7.5 PRACTICAL CONSIDERATIONS

Running molecular dynamics and Monte Carlo calculations is often more difficult than running single-molecule calculations. The input must specify not only the molecular structure, but also the temperature, pressure, density, boundary conditions, time steps, annealing schedule, and more. The actual calculations can be easily as computationally intensive as *ab initio* calculations due to the large amount of information being simulated and the large number of iterations needed to obtain a good statistical description of the system.

There is a big difference in the software packages available for performing these computations. The most complex software packages require an input specifying many details of the computation and may require the use of multiple input files and executable programs. The advantage of this scheme is that a knowledgeable researcher can run very sophisticated simulations. The most user-friendly software packages require little more work than a molecular mechanics calculation. The price for this ease of use is that the program uses many defaults, which may not be the most appropriate for the needs of a given research project.

Recently, molecular dynamics and Monte Carlo calculations with quantum mechanical energy computation methods have begun to appear in the literature. These are probably some of the most computationally intensive simulations being done in the world at this time.

Multiphase and nonequilibrium simulations are extremely difficult. These usually entail both a large amount of computing resources and a lot of technical expertise on the part of the researcher. Readers of this book are urged to refer such projects to specialists in this area.

BIBLIOGRAPHY

An introductory discussion is in

G. H. Grant, W. G. Richards, *Computational Chemistry* Oxford, Oxford (1995).

Books with more detailed treatments are

P. B. Balbuena, J. M. Seminario, *Molecular Dynamics* Elsevier, Amsterdam (1999).

G. D. Billing, K. V. Mikkelsen, *Advanced Molecular Dynamics and Chemical Kinetics* John Wiley & Sons, New York (1997).

D. C. Rapaport, *The Art of Molecular Simulation* Cambridge, Cambridge (1997).

J. M. M. Haile, *Molecular Dynamics Simulation: Elementary Methods* John Wiley & Sons, New York (1997).

D. Frenkel, B. Smit, *Understanding Molecular Simulation* Academic Press, San Diego (1996).

A. R. Leach, *Molecular Modeling Principles and Applications* Longman, Essex (1996).

M. P. Allen, D. J. Tildesley, *Computer Simulation of Liquids* Oxford, Oxford (1987).

J. A. McCammon, S. C. Harvey, *Dynamics of Proteins and Nucleic Acids* Cambridge University Press, Cambridge (1987).

M. H. Kalos, P. A. Whitlock, *Monte Carlo Methods, Volume I: Basics* John Wiley & Sons, New York (1986).

D. A. McQuarrie, *Statistical Mechanics* Harper Collins, New York (1976).

Molecular dynamics review articles are

D. L. Beveridge, *Encycl. Comput. Chem.* **3**, 1620 (1998).

P. Auffinger, E. Westhof, *Encycl. Comput. Chem.* **3**, 1628 (1998).

H. J. C. Berendsen, D. P. Tielemen, *Encycl. Comput. Chem.* **3**, 1639 (1998).

L. Pedersen, T. Darden, *Encycl. Comput. Chem.* **3**, 1650 (1998).

S. Profeta, Jr., *Kirk-Othmer Encyclopedia of Chemical Technology Supplement* J. I. Kroschwitz 315, John Wiley & Sons, New York (1998).

R. A. Marcus, *Adv. Chem. Phys.* **101**, 391 (1997).

R. M. Whitnell, K. B. Wilson, *Rev. Comput. Chem.* **4**, 67 (1993).

R. T. Skodje, *Ann. Rev. Phys. Chem.* **44**, 145 (1993).

J. N. Murrell, S. D. Bosanac, *Chem. Soc. Rev.* **21**, 17 (1992).

W. F. van Gunsteren, H. J. C. Berendsen, *Angew. Chem. Int. Ed. Eng.* **29**, 992 (1990).

W. W. Evans, G. J. Evans, *Adv. Chem. Phys.* **63**, 377 (1985).

D. Fincham, D. M. Heyes, *Adv. Chem. Phys* **63**, 493 (1985).

J. N. L. Conner, *Chem. Soc. Rev.* **5**, 125 (1976).

Simulating liquids is reviewed in

B. M. Ladanyi, M. S. Skaf, *Ann. Rev. Phys. Chem.* **44**, 335 (1993).

I. Ohmine, T. Tanaka, *Chem. Rev.* **93**, 2545 (1993).

J.-J. Barrat, M. L. Klein, *Ann. Rev. Phys. Chem.* **42**, 23 (1991).

Using Monte Carlo to simulate liquids is reviewed in

B. Smit, *Encycl. Comput. Chem.* **3**, 1743 (1998).

W. L. Jorgensen, *Encycl. Comput. Chem.* **3**, 1754 (1998).

Monte Carlo methods are reviewed in

J. J. de Pablo, F. A. Escobedo, *Encycl. Comput Chem.* **3**, 1763 (1998).

Molecular dynamics with quantum mechanical methods is reviewed in

M. E. Tuckerman, P. J. Ungar, T. von Rosenvivge, M. L. Klein, *J. Phys Chem.* **100**, 12878 (1996).

E. Clementi, G. Corongiu, D. Bahattacharya, B. Feuston, D. Frye, A. Preiskorn, A. Rizzo, W. Xue, *Chem. Rev.* **91**, 679 (1991).

Other reviews are

B. J. Berne, *Encycl. Comput. Chem.* **3**, 1614 (1998).

H. C. Kang, W. H. Weinberg, *Chem. Rev.* **95**, 667 (1995).

R. Parthasarathy, K. J. Rao, C. N. R. Rao, *Chem. Soc. Rev.* **12**, 361 (1983).

D. Frenkel, J. P. McTague, *Ann. Rev Phys. Chem* **31**, 419 (1980).

W. W. Wood, J. J. Erpenbeck, *Ann. Rev. Phys. Chem.* **27**, 319 (1976).

W. G. Hoover, W. T. Ashurst, *Theoretical Chemistry Advances and Perspectives* **1**, 2, Academic Press, New York (1975).

P. A. Egelstaff, *Ann. Rev. Phys. Chem.* **24**, 159 (1973).

J. A. Barker, D. Henderson, *Ann. Rev. Phys. Chem.* **23**, 439 (1972).

The original Monte Carlo paper is

N. Metropolis, A. W. Rosenbluth, M. N. Rosenbluth, A. H. Teller, E. Teller, *J. Chem. Phys.* **21**, 1087 (1953).

8 Predicting Molecular Geometry

Computing the geometry of a molecule is one of the most basic functions of a computational chemistry program. However, it is not trivial process. The user of the program will be able to get their work done more quickly if they have some understanding of the various algorithms within the software. The user must first describe the geometry of the molecule. Then the program computes the energies and gradients of the energy to find the molecular geometry corresponding to the lowest energy. This chapter discusses the merits of various algorithms to be used at each of these steps.

8.1 SPECIFYING MOLECULAR GEOMETRY

One way of defining the geometry of a molecule is by using a list of bond distances, angles, and conformational angles, called a Z-matrix. A Z-matrix is a convenient way to specify the geometry of a molecule by hand. This is because it corresponds to the way that most chemists think about molecular structure: in terms of bonds, angles, and so on as shown in Figure 8.1. Constructing a Z-matrix is addressed in detail in the next chapter.

Another way to define the geometry of a molecule is as a set of Cartesian coordinates for each atom as shown in Figure 8.2. Graphic interface programs often generate Cartesian coordinates since this is the most convenient way to write those programs.

A somewhat different way to define a molecule is as a simplified molecular input line entry specification (SMILES) structure. It is a way of writing a single text string that defines the atoms and connectivity. It does not define the exact bond lengths, and so forth. Valid SMILES structures for ethane are CC, C2, and H3C-CH3. SMILES is used because it is a very convenient way to describe molecular geometry when large databases of compounds must be maintained. There is also a very minimal version for organic molecules called SSMILES.

8.2 BUILDING THE GEOMETRY

In most programs, it is still possible to input a geometry manually in an ASCII input file. If the geometry is already in a file but of the wrong format, there are several utilities for converting molecular structure files. The most popular of these is the Babel program, which is described in Appendix A.

```
C
C 1 CClength
H 1 CHlength 2 CCHangle
H 1 CHlength 2 CCHangle 3 120 0
H 1 CHlength 2 CCHangle 4 120.0
H 2 CHlength 1 CCHangle 3  60.0
H 2 CHlength 1 CCHangle 4  60.0
H 2 CHlength 1 CCHangle 5  60.0

CClength   1.5
CHlength   1.0
CCHangle 109.5
```

FIGURE 8.1 Z-matrix for ethane. The first column is the element, the second column the atom to which the length refers, the third column the length, the fourth column the atom to which the angle refers, the fifth column the angle, the sixth column the atom to which the conformation angle refers, and the seventh column the conformation angle.

```
C  0.000000   0.000000   0.750000
C  0.000000   0 000000  -0 750000
H  0.000000   0.942641   1.083807
H -0.816351  -0.471321   1.083807
H  0.816351  -0.471321   1.083807
H  0 816351   0.471321  -1.083807
H -0.816351   0.471321  -1.083807
H  0.000000  -0 942641  -1.083807
```

FIGURE 8.2 Cartesian coordinate representation of ethane. The first column is the element, the other columns are the *x*, *y*, and *z* Cartesian coordinates.

It is becoming more common to uses programs that have a graphical builder in which the user can essentially draw the molecule. There are several ways in which such programs work. Some programs allow the molecule to be built as a two-dimensional stick structure and then convert it into a three-dimensional structure. Some programs have the user draw the three-dimensional backbone and then automatically add the hydrogens. This works well for organic molecules. Some programs build up the molecule in three dimensions starting from a list of elements and hybridizations, which can be most convenient for inorganic molecules. Many programs include a library of commonly used functional groups, which is convenient if it has the functional groups needed for a particular project. A number of programs have specialized building modes for certain classes of molecules, such as proteins, nucleotides, or carbohydrates. Appendix A discusses specific software packages.

8.3 COORDINATE SPACE FOR OPTIMIZATION

The way in which geometry was specified is not necessarily the coordinate system that will be used by the algorithm which optimizes the geometry. For

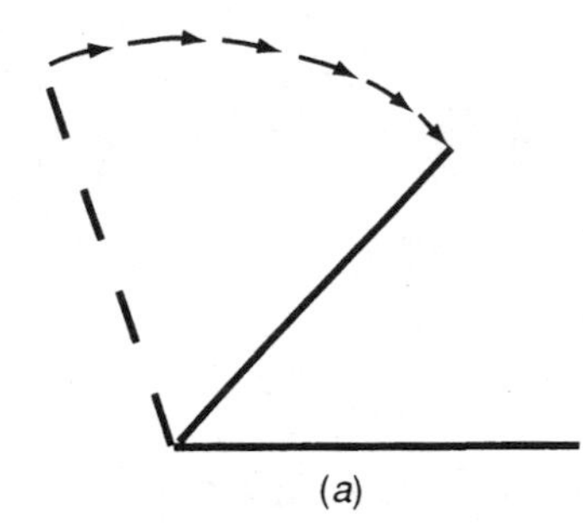

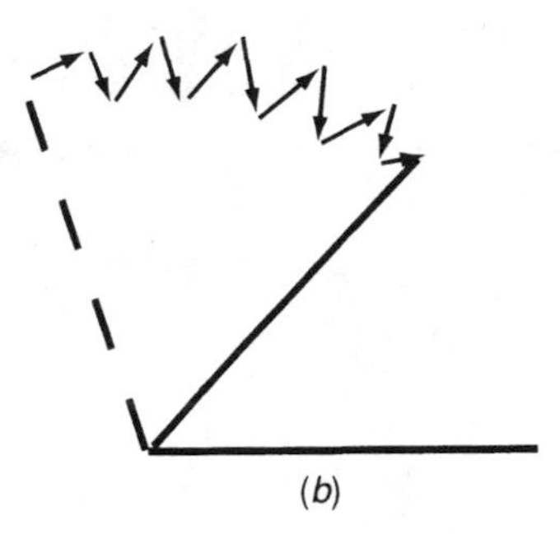

FIGURE 8.3 Example of paths taken when an angle changes in a geometry optimization. (*a*) Path taken by an optimization using a Z-matrix or redundant internal coordinates. (*b*) Path taken by an optimization using Cartesian coordinates.

example, it is very simple for a program to convert a Z-matrix into Cartesian coordinates and then use that space for the geometry optimization.

Many *ab initio* and semiempirical programs optimize the geometry of the molecule by changing the parameters in the Z-matrix. In general, this can be a very good way to change the geometry because these parameters correspond to molecular motions similar to those seen in the vibrational modes. However, if the geometry is specified in such a way that changing one of the parameters slightly could result in a large distortion to some portion of the molecule, then the geometry optimization is less efficient. Thus, a poorly constructed Z-matrix can result in a very inefficient geometry optimization. The construction of Z-matrices is addressed in Chapter 9.

Many computational chemistry programs will do the geometry optimization in Cartesian coordinates. This is often the only way to optimize geometry in molecular mechanics programs and an optional method in orbital-based programs. A Cartesian coordinate optimization may be more efficient than a poorly constructed Z-matrix. This is often seen in ring systems, where a badly constructed Z-matrix will perform very poorly. Cartesian coordinates can be less efficient than a well constructed Z-matrix as shown in Figure 8.3. Cartesian coordinates are often preferable when simulating more than one molecule since they allow complete freedom of motion between separate molecules.

In order to have the advantages of a well-constructed Z-matrix, regardless of how the geometry was defined, a system called redundant internal coordinates was created. When redundant internal coordinates are used, the input geometry is first converted to a set of Cartesian coordinates. The algorithm then checks the distances between every pair of atoms to determine which are within a reasonable bonding distance. The program then generates a list of atom distances and angles for nearby atoms. This way, the algorithm does the job of constructing a sort of Z-matrix that has more coordinates than are necessary to completely specify the geometry. This is usually the most efficient way to optimize geometry. The exception is when the automated algorithm did not include a critical coordinate. This can happen with particularly long bonds, such as when the bond is formed or broken in a transition state calculation or intermolecular interactions. In this case, the calculation will run very poorly unless the user has manually defined the extra coordinate. Geometry optimizations that run poorly either take a large number of iterations or fail to find an optimized geometry.

8.4 OPTIMIZATION ALGORITHM

There are many different algorithms for finding the set of coordinates corresponding to the minimum energy. These are called optimization algorithms because they can be used equally well for finding the minimum or maximum of a function.

If only the energy is known, then the simplest algorithm is one called the simplex algorithm. This is just a systematic way of trying larger and smaller variables for the coordinates and keeping the changes that result in a lower energy. Simplex optimizations are used very rarely because they require the most CPU time of any of the algorithms discussed here. A much better algorithm to be used when only energy is known is the Fletcher–Powell (FP) algorithm. This algorithm builds up an internal list of gradients by keeping track of the energy changes from one step to the next. The Fletcher–Powell algorithm is usually the method of choice when energy gradients cannot be computed.

If the energy and the gradients of energy can be computed, there are a number of different algorithms available. Some of the most efficient algorithms are the quasi-Newton algorithms, which assume a quadratic potential surface. One of the most efficient quasi-Newton algorithms is the Berny algorithm, which internally builds up a second derivative Hessian matrix. Steepest decent and scaled steepest decent algorithms can be used if this is not a reasonable assumption. Another good algorithm is the geometric direct inversion of the iterative subspace (GDIIS) algorithm. Molecular mechanics programs often use the conjugate gradient method, which finds the minimum by following each coordinate in turn, rather than taking small steps in each direction. The Polak–Ribiere algorithm is a specific adaptation of the conjugate gradient for molecular mechanics problems. The details of these procedures are discussed in the sources listed in the bibliography of this chapter.

Algorithms using both the gradients and second derivatives (Hessian matrix) often require fewer optimization steps but more CPU time due to the time necessary to compute the Hessian matrix. In some cases, the Hessian is computed numerically from differences of gradients. These methods are sometimes used when the other algorithms fail to optimize the geometry. Some of the most often used are eigenvector following (EF), Davidson–Fletcher–Powell (DFP), and Newton–Raphson.

8.5 LEVEL OF THEORY

The entire discussion thus far has focused on the efficient specification and computation of molecular geometries. Regardless of whether or not this process is efficient, the final geometry obtained will be what is predicted by the level of theory being used to compute the energy. The accuracy of various levels of theory is discussed in the sections of this book addressing the individual levels of theory and in Chapter 16. In general, there tends to be a trade-off between methods that are faster and more approximate and methods that are very accurate, but also very computationally intensive. In addition, there are methods that are both fast and accurate, but only applicable to limited classes of molecules.

In order to obtain the best accuracy results as quickly as possible, it is often advantageous to do two geometry optimizations. The first geometry optimization should be done with a faster level of theory, such as molecular mechanics or a semiempirical method. Once a geometry close to the correct geometry has been obtained with this lower level of theory, it is used as the starting geometry for a second optimization at the final, more accurate level of theory.

8.6 RECOMMENDATIONS

There is no one best way to specify geometry. Usually, a Z-matrix is best for specifying symmetry constraints if properly constructed. Cartesian coordinate input is becoming more prevalent due to its ease of generation by graphical user interface programs.

For the coordinate system used for optimization, redundant internal coordinates are usually best, followed by a well constructed Z-matrix, then Cartesian coordinates, then a poorly constructed Z-matrix. For simulating multiple molecules, Cartesian coordinates are often best. Most programs that generate a Z-matrix automatically from Cartesian coordinates make a poorly constructed Z-matrix.

The choice of a geometry optimization algorithm has a very large influence on the amount of computer time necessary to optimize the geometry. The gradient-based methods are most efficient, with quasi-Newton methods usually a bit better than GDIIS. The exception is for molecular mechanics calculations where the conjugate gradient algorithm can be implemented very efficiently. The Fletcher–Powell algorithm usually works best when gradients are not available.

BIBLIOGRAPHY

Geometry specification is usually addressed in the software manual as well as the following books

A. R. Leach, *Molecular Modeling Principles and Applications* Longman, Essex (1996).

J. B. Foresman, Æ. Frisch, *Exploring Chemistry with Electronic Structure Methods* Gaussian, Pittsburgh (1996).

T. Clark, *A Handbook of Computational Chemistry* John Wiley & Sons, New York (1985).

Books addressing optimization strategies and algorithms are

W. J. Hehre, *Practical Strategies for Electronic Structure Calculations* Wavefunction, Irvine (1995).

P. Willett, *Three-Dimensional Chemical Structure Handling* Research Studies Press, Baldock (1991).

I. N. Levine, *Quantum Chemistry* Prentice Hall, Englewood Cliffs (1991).

W. H. Press, B. P. Flannery, S. A. Teukolsky, W. T. Vetterling, *Numerical Recipies, The Art of Scientific Computing* Cambridge University Press, Cambridge (1989).

Adv. Chem. Phys. vol. 12 (1967).

Articles addressing the merits of various optimization algorithms are

H. B. Schlegel, *Encycl. Comput. Chem.* **2**, 1136 (1998).

T. Schlick, *Encycl. Comput. Chem.* **2**, 1143 (1998).

H. B. Schlegel, *Modern Electronic Structure Theory Part 1* D. R. Yarkony, Ed., 459, World Scientific, Singapore (1995).

C. Peng, P. Y. Ayala, H. B. Schlegel, M. J. Frisch, *J. Comput. Chem.* **17**, 49 (1996).

J. Baker, *J. Comput Chem.* **14**, 1085 (1993).

T. Schlick, *Rev. Comput. Chem.* **3**, 1 (1992).

C. Lemarechal, *Modelling of Molecular Structures and Properties* J.-L. Rivail, Ed., 63, Elsevier, Amsterdam (1990).

P. L. Cummins, J. E. Gready, *J. Comput. Chem.* **10**, 939 (1989).

J. D. Head, M. C. Zerner, *Adv. Quantum Chem.* **20**, 241 (1989).

Information on SMILES can be found at

D. Weininger, *J. Chem. Inf. Comput Sci.* **28**, 31 (1988).

http://www.daylight.com/dayhtml/smiles/

9 Constructing a Z-Matrix

The previous chapter discussed the merits of specifying the molecular geometry using a Z-matrix versus Cartesian coordinates. This chapter describes the construction of a Z-matrix. The use of a Z-matrix geometry specification is slowly declining with the increasing availability of graphic user input programs and the increasing availability of redundant internal coordinate algorithms. However, Z-matrix geometry specification is a skill still necessary for using some software programs and it remains the best way of incorporating symmetry constraints. Furthermore, a well-constructed Z-matrix can often help a program run more efficiently, thus allowing more work to be done in a given amount of time. The examples in this chapter show the construction of a Z-matrix in the format used by the Gaussian program. Other programs may require slightly different formats.

9.1 Z-MATRIX FOR A DIATOMIC MOLECULE

Here is a Z-matrix for a carbon monoxide molecule:

```
line 1   C
line 2   O 1 R
line 3
line 4   R 0.955
```

Line 1: "C" specifies that the first atom is a carbon atom.

Line 2: "O 1 R" specifies that an oxygen atom occurs at a distance R from the first atom (the carbon).

Line 3: There must be a blank line between the list of atoms and the list of variables.

Line 4: R is defined (in Angstroms or Å).

9.2 Z-MATRIX FOR A POLYATOMIC MOLECULE

Here is a Z-matrix for a formaldehyde molecule:

```
line 1   C
line 2   O 1 OC
line 3   H 1 HC 2 A
```

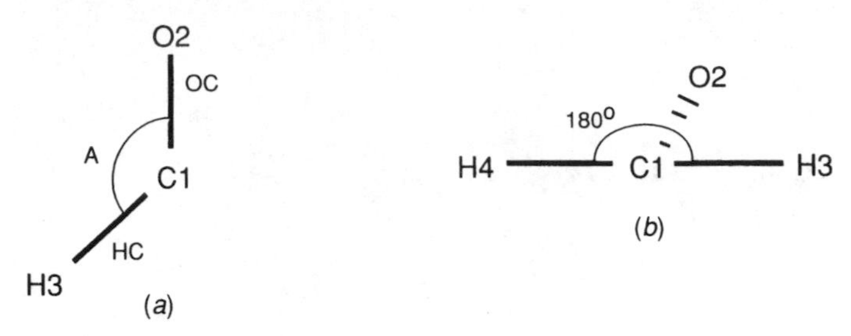

FIGURE 9.1 Illustration of the formaldehyde Z-matrix example. (*a*) First three atoms and associated variables. (*b*) Dihedral angle.

```
line 4    H 1 HC 2 A 3 180.0
line 5
line 6    OC 1.2
line 7    A 120.0
line 8
line 9    HC 1.08
```

Line 1: "C" indicates that the first atom is a carbon.

Line 2: "O 1 OC" indicates that the second atom is an oxygen with a distance of OC to the first atom.

Line 3: "H 1 HC 2 A" indicates that the third atom is a hydrogen with a distance of HC to the first atom and an angle between the third, first, and second atoms of A (in degrees).

Line 4: "H 1 HC 2 A 3 180.0" indicates that the fourth atom is a hydrogen with a distance of HC to the first atom and an angle between the third, first, and second atoms of A. The dihedral angle between the first, second, third, and fourth atoms is 180° (see Figure 9.1).

Line 5: There must be a blank line between the list of atoms and the list of variables.

Line 8: The second blank line sets aside variables that are not to be optimized in the geometry optimization.

If an optimization were being done, the parameters OC and A would be optimized, but HC would be held fixed and the molecule would be kept planar. Note that parameters can be used more than once in the Z-matrix. This makes the geometry optimization run more quickly because fewer parameters are being optimized. Additional atoms are added by appending lines like line 4 consisting of distance, angle, and dihedral angle specifications.

9.3 LINEAR MOLECULES

A linear molecule, such as CO_2, presents an additional difficulty. If an angle of 180° is specified, then the dihedral angle referenced to that angle will be math-

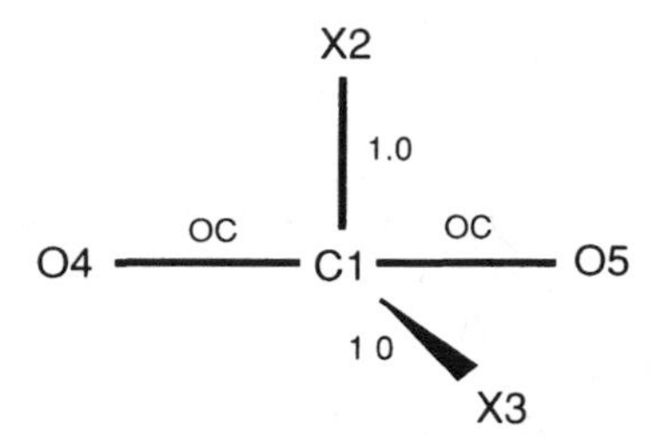

FIGURE 9.2 Illustration of the geometry formed from the CO_2 Z-matrix example.

ematically undefined. This is avoided by using a dummy atom, denoted by an element of X type. A dummy atom is not an atom at all. It is a way of defining a point in space from which geometry can be specified. A dummy atom does not have any associated nucleus or basis functions. Here is a carbon dioxide input with two dummy atoms, which is also shown in Figure 9.2:

```
C
X 1 1.0
X 1 1.0 2 90.0
O 1 OC  2 90.0 3  90.0
O 1 OC  2 90.0 3 -90.0

OC 1.2
```

Note that the distance to the dummy atoms is held fixed at 1.0 Å. This value was chosen arbitrarily. The calculation would likely fail if told to optimize this distance because there is no energy associated with it.

Enforcing the molecular symmetry will also help orbital-based calculations run more quickly. This is because some of the integrals are equivalent by symmetry and thus need be computed only once and used several times.

9.4 RING SYSTEMS

It is possible to specify a ring system by specifying the atoms sequentially. Each atom can be referenced to the previous atom. In this case, a small change in angle between, say, the 3rd and 4th atoms specified would result in a significant change in the distance between the first and last atoms specified. This makes the calculation run inefficiently if it is successful at all.

Molecules with rings should always be given a dummy atom in the center of the ring. The atoms in the ring should then be referenced to the central dummy atom rather than each other. Here is a Z-matrix for a benzene molecule enforcing D_{6h} symmetry:

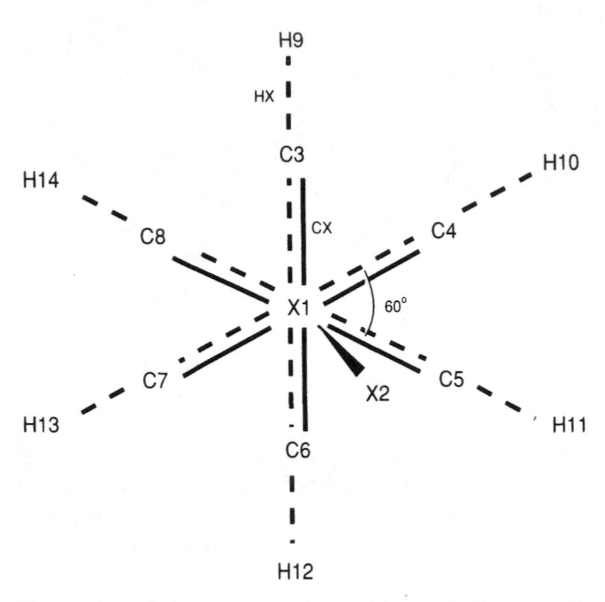

FIGURE 9.3 Illustration of the geometry formed from the benzene Z-matrix example.

```
X
X 1 1.0
C 1 CX  2 90.0
C 1 CX  2 90.0 3 60.0
C 1 CX  2 90.0 4 60.0
C 1 CX  2 90.0 5 60.0
C 1 CX  2 90.0 6 60.0
C 1 CX  2 90.0 7 60.0
H 1 HX  2 90.0 3  0.0
H 1 HX  2 90.0 4  0.0
H 1 HX  2 90.0 5  0.0
H 1 HX  2 90.0 6  0.0
H 1 HX  2 90.0 7  0.0
H 1 HX  2 90.0 8  0.0

CX 1.3
HX 2.3
```

It is often convenient to use two dummy atoms: one in the center of the ring and one perpendicular to the ring as shown here and in Figure 9.3. Even if the actual optimization is being done in redundant internal coordinates, the presence of a dummy atom in the center of the ring can give the redundant internals a better point from which to reference bond lengths and angles. Note that only two parameters need be optimized when the symmetry is used correctly.

BIBLIOGRAPHY

Geometry specification is usually addressed in the software manual as well as the following books

A. R. Leach, *Molecular Modeling Principles and Applications* Longman, Essex (1996).

J. B. Foresman, Æ. Frisch, *Exploring Chemistry with Electronic Structure Methods* Gaussian, Pittsburgh (1996).

T. Clark, *A Handbook of Computational Chemistry* John Wiley & Sons, New York (1985).

10 Using Existing Basis Sets

A basis set is a set of functions used to describe the shape of the orbitals in an atom. Molecular orbitals and entire wave functions are created by taking linear combinations of basis functions and angular functions. Most semiempirical methods use a predefined basis set. When *ab initio* or density functional theory calculations are done, a basis set must be specified. Although it is possible to create a basis set from scratch, most calculations are done using existing basis sets. The type of calculation performed and basis set chosen are the two biggest factors in determining the accuracy of results. This chapter discusses these standard basis sets and how to choose an appropriate one.

10.1 CONTRACTION SCHEMES

The orbitals mentioned in Chapter 3 almost always have the functional form given in Eq. (10.1):

$$\varphi = Y_{lm} \sum_{i} C_i \sum_{j} C_{ij} e^{-\zeta_{ij} r^2} \tag{10.1}$$

The Y_{lm} function gives the orbital the correct symmetry (s, p, d, etc.). $\exp(-r^2)$ is called a Gaussian primitive function. The contraction coefficients C_{ij} and exponents ζ_{ij} are read from a database of standard functions and do not change over the course of the calculation. This predefined set of coefficients and exponents is called a basis set. An enormous amount of work is involved in optimizing a basis set to obtain a good description of an individual atom. By using such a predefined basis set, the program must only optimize the molecular orbital coefficients C_i. As seen above, each C_i may weigh a sum of typically one to nine primitive Gaussian functions, called a contraction. Basis sets of contracted functions are called *segmented basis sets*.

Before computational chemists started employing segmented basis sets, calculations were done without using contractions. These uncontracted basis functions are called *generally contracted basis functions*. The danger with segmented basis sets is that having too few contractions will result in a function with too little flexibility to properly describe the change in electron density from an individual atom to the atom in a molecule. Compared to a segmented basis set with a reasonable number of contractions, generally contracted basis set calculations require more computer resources to run in exchange for an extremely slight improvement in the accuracy of results. Also having the orbitals closest to the nucleus uncontracted often leads to SCF convergence problems. Thus, it is rare to see generally contracted calculations in the current literature.

A second issue is the practice of using the same set of exponents for several sets of functions, such as the 2*s* and 2*p*. These are also referred to as *general contraction* or more often *split valence* basis sets and are still in widespread use. The acronyms denoting these basis sets sometimes include the letters "SP" to indicate the use of the same exponents for *s* and *p* orbitals. The disadvantage of this is that the basis set may suffer in the accuracy of its description of the wave function needed for high-accuracy calculations. The advantage of this scheme is that integral evaluation can be completed more quickly. This is partly responsible for the popularity of the Pople basis sets described below.

An issue affecting calculation runtime is how the integrals are evaluated. There are several common methods: *conventional*, *direct*, *in core*, and *semidirect*. A conventional calculation is one in which all the integrals are evaluated at the beginning of the calculation and stored in a file on the computer hard drive. This file is then accessed as the integrals are needed on each iteration of the self-consistent field calculation. Over time, the speed of computer processors has increased more than the size and access speed of hard drives. In order to obtain the best overall performance, many calculations are now done with a direct algorithm in which the integrals are evaluated as they are needed and not stored at all. Direct calculations are not hindered by slow disk access or limited disk space. However, direct calculations often take more CPU time than conventional calculations because the program must do extra work to evaluate the same integral every time it is needed. Some programs use a semidirect algorithm that stores some of the integrals on disk to decrease disk use without increasing CPU time as much as is the case with direct calculations. An "in core" algorithm is one that computes all the needed integrals and then keeps them in RAM memory rather than in a disk file. In-core calculations are always the fastest calculations because RAM memory can be accessed much faster than disk files and there is no extra work. However, the higher price and subsequently smaller size of RAM compared to hard drive space mean that in core calculations can be done only for much smaller molecular systems than can be computed using the other algorithms.

The choice of basis set also has a large effect on the amount of CPU time required to perform a calculation. In general, the amount of CPU time for Hartree–Fock calculations scales as N^4. This means that making the calculation twice as large will make the calculation take 16 times (2^4) as long to run. Making the calculation twice as large can occur by switching to a molecule with twice as many electrons or by switching to a basis set with twice as many functions. Disk use for conventional calculations scales as N^4 and the amount of RAM use scales as N^2 for most algorithms. Some of the largest CI calculations scale as N^8 or worse. Computer resource use is covered in more detail in Chapter 15.

The orbitals in Eq. (10.1) are referred to as Gaussian type orbitals, or GTO,

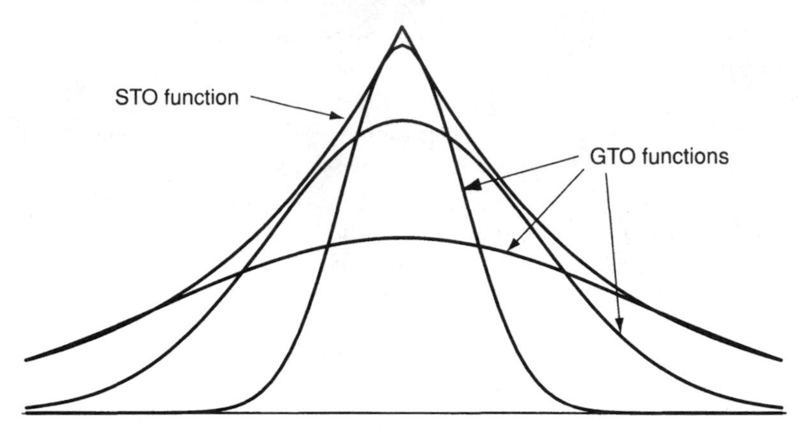

FIGURE 10.1 Approximating a Slater-type orbital with several Gaussian-type orbitals.

since they incorporate Gaussian functions, $\exp(-\zeta r^2)$. The exact solution to the Schrödinger equation for the hydrogen atom is a Slater type orbital, or STO, of the form $\exp(-\zeta r)$. GTO basis sets require more primitives to describe the wave function than are needed for STO calculations, as shown in Figure 10.1. However, the integrals over GTO primitives can be computed analytically, which is so much faster than the numeric integrals over STO functions that any given accuracy can be obtained most quickly using GTO functions. As such, STO basis sets are sometimes used for high-accuracy work, but most calculations are now done with GTO basis sets.

A complication arises for functions of d or higher symmetry. There are five real d orbitals, which transform as xy, xz, yz, x^2-y^2, and z^2, that are called pure d functions. The orbital commonly referred to as z^2 is actually $2z^2-x^2-y^2$. An alternative scheme for the sake of fast integral evaluation is to use the six Cartesian orbitals, which are xy, xz, yz, x^2, y^2, and z^2. These six orbitals are equivalent to the five pure d functions plus one additional spherically symmetric function $(x^2+y^2+z^2)$. Calculations using the six d functions often yield a very slightly lower energy due to this additional function. Some *ab initio* programs give options to control which method is used, such as 5*d*, 6*d*, pure-*d*, or Cartesian. Pure-*d* is equivalent to 5*d* and Cartesian is equivalent to 6*d*. Similarly, 7*f* and 10*f* are equivalent to pure-*f* and Cartesian *f* functions, respectively.

Choosing a standard GTO basis set means that the wave function is being described by a finite number of functions. This introduces an approximation into the calculation since an infinite number of GTO functions would be needed to describe the wave function exactly. Differences in results due to the quality of one basis set versus another are referred to as *basis set effects*. In order to avoid the problem of basis set effects, some high-accuracy work is done with numeric basis sets. These basis sets describe the electron distribution without using functions with a predefined shape. A typical example of such a basis set might be a cubic spline set in which a large number of third-order polynomials are used. Each polynomial would describe the wave function for just a small range of distances from the nucleus. The coefficients of these polynomials are then chosen so that the wave function and its derivatives will be continuous as well as describing the shape of the wave function.

10.2 NOTATION

Most calculations today are done by choosing an existing segmented GTO basis set. These basis sets are identified by one of a number of notation schemes. These abbreviations are often used as the designator for the basis set in the input to *ab initio* computational chemistry programs. The following is a look at the notation for identifying some commonly available contracted GTO basis sets.

The smallest basis sets are called minimal basis sets. The most popular minimal basis set is the **STO−3G** set. This notation indicates that the basis set approximates the shape of a STO orbital by using a single contraction of three GTO orbitals. One such contraction would then be used for each orbital, which is the definition of a minimal basis. Minimal basis sets are used for very large molecules, qualitative results, and in certain cases quantitative results. There are **STO−nG** basis sets for $n = 2–6$. Another popular minimal basis set is the MINI set described below.

Another family of basis sets, commonly referred to as the Pople basis sets, are indicated by the notation **6−31G**. This notation means that each core orbital is described by a single contraction of six GTO primitives and each valence shell orbital is described by two contractions, one with three primitives and the other with one primitive. These basis sets are very popular, particularly for organic molecules. Other Pople basis sets in this set are 3−21G, 4−31G, 4−22G, 6−21G, 6−311G, and 7−41G.

The Pople basis set notation can be modified by adding one or two asterisks, such as 6−31G* or 6−31G**. A single asterisk means that a set of *d* primitives has been added to atoms other than hydrogen. Two asterisks mean that a set of *p* primitives has been added to hydrogen as well. These are called polarization functions because they give the wave function more flexibility to change shape. Adding polarization functions usually decreases the variational total energy by about the same amount as adding another contraction. However, this energy change is almost completely systematic, so it changes the relative energies very little. Polarization functions are used because they often result in more accurate computed geometries and vibrational frequencies.

The **3−21G*** basis is an exception to the notation above. In this particular case, the *d* functions are added only to 2nd row atoms, Al through Ar. In order to indicate this difference, this basis is sometimes given the notation **3−21G(*)**.

One or two plus signs can also be added, such as **6−31+G*** or **6−31++G***. A single plus sign indicates that diffuse functions have been added to atoms other than hydrogen. The second plus sign indicates that diffuse functions are being used for all atoms. These diffuse functions are primitives with small exponents, thus describing the shape of the wave function far from the nucleus. Diffuse functions are used for anions, which have larger electron density distributions. They are also used for describing interactions at long distances, such as van der Waals interactions. The effect of adding diffuse functions is usually to change the relative energies of the various geometries associated with these systems. Basis sets with diffuse functions are also called augmented basis sets. Very diffuse orbitals are called Rydberg orbitals since they are used to describe Rydberg states of molecules.

As the Pople basis sets have further expanded to include several sets of polarization functions, *f* functions and so on, there has been a need for a new notation. In recent years, the types of functions being added have been indicated in parentheses. An example of this notation is **6–31G(dp,p)** which means that extra sets of *p* and *d* functions have been added to nonhydrogens and an extra set of *p* functions have been added to hydrogens. Thus, this example is synonymous with **6–31+G****.

Many basis sets are just identified by the author's surname and the number of primitive functions. Some examples of this are the Huzinaga, Dunning, and Duijneveldt basis sets. For example, **D95** and **D95V** are basis sets created by Dunning with nine *s* primitives and five *p* primitives. The **V** implies one particular contraction scheme for the valence orbitals. Another example would be a basis set listed as "Duijneveldt 13s8p".

In order to describe the number of primitives and contractions more directly, the notation **(6s,5p)→(1s,3p)** or **(6s,5p)/(1s,3p)** is sometimes used. This example indicates that six *s* primitives and five *p* primitives are contracted into one *s* contraction and three *p* contractions. Thus, this might be a description of the 6–311G basis set. However, this notation is not precise enough to tell whether the three *p* contractions consist of three, one, and one primitives or two, two, and one primitives. The notation **(6,311)** or **(6,221)** is used to distinguish these cases. Some authors use round parentheses ( ) to denote the number of primitives and square brackets [ ] to denote the number of contractions.

An older, but still used, notation specifies how many contractions are present. For example, the acronym **TZV** stands for triple-zeta valence, meaning that there are three valence contractions, such as in a 6–311G basis. The acronyms **SZ** and **DZ** stand for single zeta and double zeta, respectively. A **P** in this notation indicates the use of polarization functions. Since this notation has been used for describing a number of basis sets, the name of the set creator is usually included in the basis set name (i.e., Ahlrichs VDZ). If the author's name is not included, either the Dunning–Hay set is implied or the set that came with the software package being used is implied.

An extension of this last notation is **aug–cc–pVDZ**. The "aug" denotes that this is an augmented basis (diffuse functions are included). The "cc" denotes that this is a correlation-consistent basis, meaning that the functions were optimized for best performance with correlated calculations. The "p" denotes that polarization functions are included on all atoms. The "VDZ" stands for valence double zeta, meaning that the valence orbitals are described by two contractions. There is a family of correlation consistent basis sets created by Dunning and coworkers. These sets have become popular for high-accuracy correlated calculations. They have shown that large basis sets with high-angular-momentum polarization functions have a greater effect on the accuracy of correlated calculations than HF calculations. Because this family of basis sets was developed in a systematic way, many properties converge asymptotically as larger basis sets are chosen. This fact has been exploited by fitting the results of the same calculation with several basis sets to an exponential decay to predict the infinite basis set limit for HF calculations. A similar procedure has been used to predict the full CI limit from multireference valence CI calculations. This extrapolation has only been tested for total energies and a few other properties. The technique should be applicable to any property showing asymptotic convergence, but there is not yet a sufficient volume of literature to predict how accurate this extrapolation will be.

The Gaussian theories Gaussian-1 and Gaussian-2, abbreviated as **G1** and **G2**, are not basis sets, but they are similar to the basis set extrapolation mentioned in the previous paragraph. These model chemistries arose from the observation that certain levels of theory with certain basis sets tended always to give results with systematic errors for the equilibrium geometries of main group compounds. The procedure for obtaining these results consists of running a series of calculations with different basis sets and levels of theory and then plugging the energies into an equation that is meant to correct for systematic errors so energies are closer to the exact energy than with any of the individual methods. The results from this procedure have been good for equilibrium geometries of main group compounds. Results for other calculations such as transition structures or nonbonded interactions have been less encouraging. Gaussian theory is discussed in more detail in Chapter 4.

The complete basis set (**CBS**) scheme is a series of basis sets designed to extrapolate energies to the infinite basis set limit. The earlier methods used Pople basis sets or modifications of them. CBS calculations are actually a set of calculations with different numbers of basis functions and levels of theory. The results from these calculations are used to give an extrapolation to the complete basis set, fully correlated limit. The extrapolation equations were derived using perturbation theory. The extrapolation to complete correlation uses a summation of wave function coefficients and overlaps times an empirically determined scaling factor. Some of the CBS methods correct for spin contamination in open-shell calculations using the amount of spin contamination times an empirically determined constant. They have applied this technique to energies, but do not address molecular properties other than those directly related to energies, such as the ionization potential. The smaller CBS methods give accuracy comparable to the G1 method with one-tenth of the CPU time. The CBS–APNO method yields results significantly more accurate than those with the G2 method.

10.3 TREATING CORE ELECTRONS

Unlike semiempirical methods that are formulated to completely neglect the core electrons, *ab initio* methods must represent all the electrons in some manner. However, for heavy atoms it is desirable to reduce the amount of computation necessary. This is done by replacing the core electrons and their basis functions in the wave function by a potential term in the Hamiltonian. These are called core potentials, effective core potentials (ECP), or relativistic effective core potentials (RECP). Core potentials must be used along with a valence basis set that was created to accompany them. As well as reducing the computation time, core potentials can include the effects of the relativistic mass defect and spin coupling terms that are significant near the nuclei of heavy atoms. This is often the method of choice for heavy atoms, Rb and up.

The energy obtained from a calculation using ECP basis sets is termed valence energy. Also, the virial theorem no longer applies to the calculation. Some molecular properties may no longer be computed accurately if they are dependent on the electron density near the nucleus.

There are several issues to consider when using ECP basis sets. The core potential may represent all but the outermost electrons. In other ECP sets, the outermost electrons and the last filled shell will be in the valence orbital space. Having more electrons in the core will speed the calculation, but results are more accurate if the $n-1$ shell is outside of the core potential. Some ECP sets are designated as shape-consistent sets, which means that the shape of the atomic orbitals in the valence region matches that for all electron basis sets. ECP sets are usually named with an acronym that stands for the authors' names or the location where it was developed. Some common core potential basis sets are listed below. The number of primitives given are those describing the valence region.

- *CREN* Available for SC(4*s*) through Hs(0*s*6*p*6*d*), this is a shape-consistent basis set developed by Ermler and coworkers that has a large core region and small valence. This is also called the CEP–4G basis set. The CEP–31G and CEP–121G sets are related split valence sets.
- *SBKJC VDZ* Available for Li(4*s*4*p*) through Hg(7*s*7*p*5*d*), this is a relativistic basis set created by Stevens and coworkers to replace all but the outermost electrons. The double-zeta valence contraction is designed to have an accuracy comparable to that of the 3–21G all-electron basis set.
- *Hay–Wadt MB* Available for K(5*s*5*p*) through Au(5*s*6*p*5*d*), this basis set contains the valence region with the outermost electrons and the previous shell of electrons. Elements beyond Kr are relativistic core potentials. This basis set uses a minimal valence contraction scheme. These sets are also given names starting with "LA" for Los Alamos, where they were developed.
- *Hay–Wadt VDZ* Available for K(5*s*5*p*) through Au(5*s*6*p*5*d*), this basis

set is similar to Hay–Wadt MB, but it has a double-zeta valence contraction. This set is popular for transition metal modeling.

- *LANL2DZ* Available for H(4*s*) through Pu(7*s*6*p*2*d*2*f*), this is a collection of double-zeta basis sets, which are all-electron sets prior to Na.
- *CRENBL* Available for H(4*s*) through Hs(0*s*3*p*6*d*5*f*), this is a collection of shape-consistent sets, which use a large valence region and small core region.
- *Dolg* Also called Stuttgart sets, this is a collection of ECP sets currently under development by Dolg and coworkers. These sets are popular for heavy main group elements.

10.4 COMMON BASIS SETS

This section gives a listing of some basis sets and some notes on when each is used. The number of primitives is listed as a simplistic measure of basis set accuracy (bigger is always slower and usually more accurate). The contraction scheme is also important since it determines the basis set flexibility. Even two basis sets with the same number of primitives and the same contraction scheme are not completely equivalent since the numerical values of the exponents and contraction coefficients determine how well the basis describes the wave function.

There are several types of basis functions listed below. Over the past several decades, most basis sets have been optimized to describe individual atoms at the HF level of theory. These basis sets work very well, although not optimally, for other types of calculations. The atomic natural orbital, ANO, basis sets use primitive exponents from older HF basis sets with coefficients obtained from the natural orbitals of correlated atom calculations to give a basis that is a bit better for correlated calculations. The correlation-consistent basis sets have been completely optimized for use with correlated calculations. Compared to ANO basis sets, correlation consistent sets give a comparable accuracy with significantly fewer primitives and thus require less CPU time.

There have been a few basis sets optimized for use with DFT calculations, but these give little if any increase in efficiency over using HF optimized basis sets for these calculations. In general, DFT calculations do well with moderate-size HF basis sets and show a significant decrease in accuracy when a minimal basis set is used. Other than this, DFT calculations show only a slight improvement in results when large basis sets are used. This seems to be due to the approximate nature of the density functional limiting accuracy more than the lack of a complete basis set.

Several basis schemes are used for very-high-accuracy calculations. The highest-accuracy HF calculations use numerical basis sets, usually a cubic spline method. For high-accuracy correlated calculations with an optimal amount of computing effort, correlation-consistent basis sets have mostly replaced ANO basis sets. Complete basis set, or CBS, calculations go a step beyond this in estimating the infinite basis set limit. STO basis sets (Slater orbitals, not STO–nG) are now most often used for the extremely high-accuracy calculations done with quantum Monte Carlo methods, which use a correlation function in addition to an STO basis to describe the wave function.

Below is a listing of commonly used basis sets. The few most widely used are listed at the end of this section.

- *STO–nG* ($n = 2$–6) n primitives per shell per occupied angular momentum (s, p, d). STO–3G is heavily used for large systems and qualitative results. The STO–3G functions have been made for H with three primitives ($3s$) through Xe($15s12p6d$). STO–2G is seldom used due to the poor quality of its results. The larger STO–nG sets are seldom used because they have too little flexibility.
- *MINI–i* ($i = 1$–4) These four sets have different numbers of primitives per contraction, mostly three or four. These are minimal basis sets with one contraction per orbital. Available for Li through Rn.
- *MIDI–i* Same primitives as the MINI basis sets with two contractions to describe the valence orbitals for greater flexibility.
- *MAXI–i and MIDI!* Are higher-accuracy basis sets derived from the MIDI basis set.
- *3–21G* Same number of primitives as STO–3G, but more flexibility in the valence orbitals. Available for H through Cs. Popular for qualitative and sometimes quantitative results for organic molecules.
- *6–31G* Available for H($4s$) through Ar($16s10p$). Very popular for quantitative results for organic molecules.
- *6–311G* Available for H($5s$) through Kr($14s12p5d$). Very popular for quantitative results for organic molecules.
- *DET* Created by Koga, Tatewaki, and Thakkar, available for He($4s$) through Xe($13s12p8d$).
- *D95 and D95V* Available for H($4s$) and B through F($9s5p$). Used for quantitative results.
- *Dunning–Hay SV* Available for H($4s$) through Ne($9s5p$). SVP adds one polarization function. If this notation is used without an author's name, this is the set that is usually implied.
- *Dunning–Hay DZ* Available for H($4s$) through Cl($11s7p$). DZP adds one polarization function. If this notation is used without an author's name, this is the set that is usually implied.
- *Dunning–Hay TZ* Available for H($5s$) through Ne($10s6p$). If this notation is used without an author's name, this is the set that is usually implied.
- *Duijneveldt* A range of sets for H through Ne. H sets range from ($2s$) to ($10s$) and Ne sets range from ($4s2p$) to ($14s9p$). The larger sets are used for accurate work on organic systems.

- *Huzinaga* A range of sets for Li through Ne. Li sets range from (6*s*) to (11*s*) and Ne sets range from (6*s*3*p*) to (11*s*7*p*). The larger sets are used for accurate work on organic systems.
- *Sadlej pVTZ* Available for H(6*s*4*p*) through Ca(15*s*13*p*4*d*), Br, Rb through Sr, I(19*s*15*p*12*d*4*f*). Optimized to reproduce experimental polarizabilities.
- *Chipman DZP+diffuse* Available for H(6*s*1*p*) through F(10*s*6*p*2*d*). Optimized to reproduce high-accuracy spin density results.
- *Roos & Siegbahn* Available for Na through Ar(10*s*6*p*).
- *GAMESS VTZ* Available for H(5*s*) through Ar(12*s*9*p*). PVTZ adds one polarization function. This is a combination of the Dunning and McClean/Chandler sets.
- *Koga, Saito, Hoffmeyer, Thakkar* Available for Na through Ar(12*s*8*p*) and (12*s*9*p*).
- *McLean/Chandler VTZ* Available for Na through Ar(12*s*8*p*) and (12*s*9*p*) with several contraction schemes.
- *Veillard* Available for Na through Ar(12*s*9*p*).
- *Roos, Veillard, Vinot* Available for Sc through Cu(12*s*6*p*4*d*).
- *STD−SET(1)* Available for Sc through Zn(12*s*6*p*3*d*). Seldom used due to a poor description of the core.
- *DZC−SET(1)* Available for Sc through Zn(12*s*6*p*4*d*). Seldom used due to a poor description of the core.
- *Hay* Available for Sc through Cu(12*s*6*p*4*d*) and (14*s*9*p*5*d*). The larger set is popular for transition metal calculations.
- *Ahlrichs VDZ, pVDZ, VTZ* Available for Li(4*s*) to (11*s*) through Kr(14*s*10*p*5*d*) to (17*s*13*p*6*d*). These have been used for many high-accuracy calculations.
- *Binning/Curtiss SV, VDZ, SVP, VTZP, VTZ* Available for Ga through Kr(14*s*11*p*5*d*).
- *Huzinaga* Available for K(14*s*9*p*) through Cd(17*s*11*p*8*d*). Balch, Bauschlicker, and Nein have published additional functions to augment the Y through Ag functions in this sets.
- *Basch* Available for Sc through Cu(15*s*8*p*5*d*) with several contraction schemes. The transition metal set yields slightly higher energy than Wachters' set.
- *Wachters* Available for K through Zn(14*s*9*p*5*d*). Often used for transition metals.
- *Stromberg* Available for In through Xe(15*s*11*p*6*d*).
- *WTBS* Well-tempered basis set for high-accuracy results. Available for He(17*s*) through Rn(28*s*24*p*18*d*12*f*).
- *Partridge uncontracted sets 1–3* Available for Li(14*s*) to (18*s*) through Sr(24*s*16*p*10*d*). The larger sets only go up to V or Zn.

- *Castro & Jorge universal* Available for H(20s) through Lr(32s25p20d15f). For actually reaching the infinite basis set limit to about seven digits of accuracy.
- *Almlöf, Taylor ANO* Available for H(8s), (8s6p), and (8s6p4d); N and O(13s8p6d) and (13s8p6d4f); Ne(9s5p) and (13s8p); S(20s16p10d).
- *Roos augmented double- and triple-zeta ANO* Available for H(8s4p) to (8s4p3d) through Zn(21s15p10d6f) to (21s15p10d6f4g).
- *NASA Ames ANO* Available for H(8s6p4d3f) through P(18s13p6d4f2g). Ti, Fe, and Ni functions are available. Collection of functions from various authors.
- *Bauschlicker ANO* Available for Sc through Cu (20s15p10d6f4g).
- *cc−pVnZ* ($n = D, T, Q, 5, 6$) Correlation-consistent basis sets that always include polarization functions. Atoms H through Ar are available. The 6Z set goes up to Ne only. The various sets describe H with from (2s1p) to (5s4p3d2f1g) primitives. The Ar atoms is described by from (4s3p1d) to (7s6p4d3f2g1h) primitives. One to four diffuse functions are denoted by prepending the notation with "aug-" or "n-aug-", where $n = d, t, q$.
- *cc−pCVnZ* ($n = D, T, Q, 5$) Correlation-consistent basis set designed to describe the correlation of the core electrons as well as the valence electrons. Available for H through Ne. These basis sets were created from the cc−pVnZ sets by adding from 2 to 14 additional primitives starting at the inner shells. Augmented in the same manner as the cc−pVnZ sets.
- *CBS−n* ($n = 4$, Lq, Q, APNO) Available for H through Ne. For estimating the infinite basis set limit. This implies a series of calculations with different basis sets, some of which are large sets.
- *DZVP, DZVP2, TZVP* DFT-optimized functions. Available for H(5s) through Xe(18s14p9d) plus polarization functions.
- *Dgauss A1* DFT Coulomb and exchange fitting. Available for H(4s) through Xe(10s5p5d).
- *Dgauss A2* DFT Coulomb and exchange fitting. Available for H(4s1p1d) through Zn(10s5p5d).
- *DeMon Coulomb fitting* Available for H(4s1p) through Xe(10s5p5d) for DFT calculations.
- *ADF AE SZ, DZ, & TZ* STO sets for DFT. Available for H(1s) to (3s) through Lr(7s5p4d2f) to (13s10p9d5f).
- *ADF I-V* Fixed-core STO sets for DFT. Available for H(1s) to (3s1p1d) through Kr(12s10p4d) to (14s12p5d1f).
- *DN* Numerical, cubic spline, DFT set.
- *Froese–Fischer* HF numerical sets for He through Rn.
- *Bange, Barrientos, Bunge, Cogordan STO* Available for He(4s) through Xe(13s12p8d).

- *Koga, Watanabe, Kanayama, Yasuda, Thakkar STO* Available for He(4*s*) through Xe(13*s*12*p*8*d*).
- *Koga, Tatewaki, Thakkar STO* Available for He(5*s*) through Xe(11*s*9*p*5*d*).
- *Clementi STO* Available for He(5*s*) through Kr(10*s*9*p*5*d*).
- *Clementi & Roetti STO* Available for He(5*s*) through Xe(11*s*9*p*5*d*). Often used when STO functions are desired, such as accurate descriptions of the wave function very near the nucleus.

There have been many more basis sets developed, but the list above enumerates the most widely used ones. Some of these sets were the work of many different authors and later improved. This sometimes results in different programs using the same name for two slightly different sets. It is also possible to combine basis sets or modify them, which can result in either poor or excellent results, depending on how expertly it is done. How to customize basis sets is discussed in Chapter 28.

Some of the basis sets discussed here are used more often than others. The STO–3G set is the most widely used minimal basis set. The Pople sets, particularly, 3–21G, 6–31G, and 6–311G, with the extra functions described previously are widely used for quantitative results, particularly for organic molecules. The correlation consistent sets have been most widely used in recent years for high-accuracy calculations. The CBS and G2 methods are becoming popular for very-high-accuracy results. The Wachters and Hay sets are popular for transition metals. The core potential sets, particularly Hay–Wadt, LANL2DZ, Dolg, and SBKJC, are used for heavy elements, Rb and heavier.

Experience has shown that is better to obtain basis sets in electronic form than paper form since even slight errors in transposition will affect the calculation results. Some basis sets are included with most computer programs that require them. There is also a form page on the Web that allows a user to choose a basis and specify a format consistent with the input of several popular computational chemistry programs at **http://www.emsl.pnl.gov:2080/forms/basisform.htm**. The basis set is then sent to the user in the form of an e-mail message.

10.5 STUDIES COMPARING RESULTS

For many projects, a basis set cannot be chosen based purely on the general rules of thumb listed above. There are a number of places to obtain a much more quantitative comparison of basis sets. The paper in which a basis set is published often contains the results of test calculations that give an indication of the accuracy of results. Several books, listed in the references below, contain extensive tabulations of results for various methods and basis sets. Every year, a bibliography of all computational chemistry papers published in the previous year is printed in the *Journal of Molecular Structure* and indexed by molecular formula. Using this bibliography, it will be fairly easy to find all previous computational work on a given compound.

BIBLIOGRAPHY

References to the individual basis sets have not been included here. Most can be readily found in the review articles.

Some books discussing basis sets

F. Jensen, *Introduction to Computational Chemistry* John Wiley & Sons, New York (1999).

P. W. Atkins, R. S. Friedman, *Molecular Quantum Mechanics* Oxford, Oxford (1997).

J. Simons, J. Nichols, *Quantum Mechanics in Chemistry* Oxford, Oxford (1997).

A. R. Leach, *Molecular Modelling Principles and Applications* Longman, Essex (1996).

I. N. Levine, *Quantum Chemistry* Prentice Hall, Englewood Cliffs (1991).

L. Szasz, *Pseudopotential Theory of Atoms and Molecules* John Wiley & Sons, New ork (1985).

R. Poirer, R. Kari, I. G. Csizmadia, *Handbook of Gaussian Basis Sets: A Compendium for Ab Initio Molecular Orbital Calculations* Elsevier Science Publishing, New York (1985).

Gaussian basis sets for molecular calculations S. Huzinaga, Ed., Elsevier, Amsterdam (1984).

Review articles covering basis set construction and performance are

T. H. Dunning, Jr., K. A. Peterson, D. E. Woon, *Encycl. Comput. Chem.* **1**, 88 (1998).

S. Wilson D. Moncrieff, *Adv. Quantum Chem.* **28**, 47 (1997).

C. W. Bauschlicker, Jr., *Theor. Chim. Acta* **92**, 183 (1995).

J. R. Thomas, B.-J. DeLeeuw, G. Vacek, T. D. Crawford, Y. Yamaguchi, H. F. Schaefer III, *J. Chem. Phys.* **91**, 403 (1993).

T. L. Sordo, C. Barrientos, J. A. Sordo, *Computational Chemistry: Structure, Interactions and Reactivity* S. Fraga, Ed., 23, Elsevier, Amsterdam (1992).

J. Almlöf, P. R. Taylor, *Adv. Quantum Chem.* **22**, 301 (1991).

D. Feller, E. R. Davidson, *Rev. Comput. Chem.* 1, 1 (1990).

E. R. Davidson, D. Feller, *Chem. Rev.* **86**, 681 (1986).

S. Huzinaga, *Comp. Phys. Reports* **2**, 279 (1985).

S. Wilson, *Methods in Computational Molecular Physics* 71 G. H. Diercksen, S. Wilson, Eds., Dordrecht, Reidel (1983).

T. H. Dunning, P. J. Hay, *Modern Theoretical Chemistry vol. 4* H. F. Schaefer III Ed., 1, Plenum, New York (1977).

K. A. R. Mitchell, *Chem. Rev.* **69**, 157 (1969).

J. D. Weeks, A. Hazi, S. A. Rice, *Adv. Chem. Phys.* **16**, 283 (1969).

L. A. Leland, A. M. Caro, *Rev. Mod. Phys.* **32**, 275 (1960).

J. K. Labanowski, http://ccl.osc.edu/pub/documents/basis-sets/

Reviews of ECP basis sets are

M. Klobukowski, S. Huzinaga, Y. Sakai, *Computational Chemistry Reviews of Current Trends Volume 3* 49, J. Leszczynski, Ed., World Scientific, Singapore (1999).

M. Dolg, H. Stoll, *Handbook on the Physics and Chemistry of Rare Earths* K. A. Gschneidner, Jr., L. Eyring, Eds., **22**, 607, Elsevier, Amsterdam (1994).

W. C. Ermler, R. B. Ross, P. A. Christiansen, *Adv. Quantum Chem.* **19**, 139 (1988).

K. Balasubramanian, K. S. Pitzer, *Adv. Chem. Phys.* **67**, 287 (1987).

P. A. Christiansen, W. C. Ermler, K. S. Pitzer, *Annu. Rev. Phys. Chem.* **36**, 407 (1985).

M. Krauss, W. J. Stevens, *Ann. Rev. Phys. Chem.* **35**, 357 (1984).

Numerical basis sets are discussed in

E. A. McCullough, Jr., *Encycl. Comput. Chem.* **3**, 1941 (1998).

Comparisons of results are in

W. J. Hehre, *Practical Strategies for Electronic Structure Calculations* Wavefunction, Irvine (1995).

J. W. Ochterski, G. A. Petersson, K. B. Wiberg, *J. Am. Chem. Soc.* **117**, 11299 (1995).

W. J. Hehre, L. Radom, P. v. R. Schleyer, J. A. Pople, *Ab Initio Molecular Orbital Theory* Wiley-Interscience, New York (1986).

Web based forms for obtaining basis sets

http://www.emsl.pnl.gov:2080/forms/basisform.html

http://www.theochem.uni-stuttgart.de/pseudopotentiale/

11 Molecular Vibrations

The vibrational states of a molecule are observed experimentally via infrared and Raman spectroscopy. These techniques can help to determine molecular structure and environment. In order to gain such useful information, it is necessary to determine what vibrational motion corresponds to each peak in the spectrum. This assignment can be quite difficult due to the large number of closely spaced peaks possible even in fairly simple molecules. In order to aid in this assignment, many workers use computer simulations to calculate the vibrational frequencies of molecules. This chapter presents a brief description of the various computational techniques available.

Different motions of a molecule will have different frequencies. As a general rule of thumb, bond stretches are the highest energy vibrations. Bond bends are somewhat lower energy vibrations and torsional motions are even lower. The lowest frequencies are usually torsions between substantial pieces of large molecules and breathing modes in very large molecules.

11.1 HARMONIC OSCILLATOR APPROXIMATION

The simplest description of a vibration is a harmonic oscillator, which describes springs exactly and pendulums with small amplitudes fairly well. A harmonic oscillator is defined by the potential energy proportional to the square of the distance displaced from an equilibrium position. In a classical treatment of a vibrating object, the motion is fastest at the equilibrium position and comes to a complete stop for an instant at the turning points, where all the energy is potential energy. The probability of finding the object is highest at the turning point and lowest at the equilibrium point.

A quantum mechanical description of a harmonic oscillator uses the same potential energy function, but gives radically different results. In a quantum description, there are no turning points. There is some probability of finding the object at any displacement, but it becomes very small (decreasing exponentially) at large distances. The energy is quantized, with a quantum number describing each possible energy state and only certain possible energies. Very small objects, such as atoms, behave according to the quantum description with low quantum numbers.

The vibration of molecules is best described using a quantum mechanical approach. A harmonic oscillator does not exactly describe molecular vibra-

TABLE 11.1 Vibrational Frequency Correction Factors

Correction	Method
0.86–0.9	HF
0.9085	HF/3–12G
0.8953	HF/6–31G*
0.893	ROHF/6–31G*
0.8970	HF/6–31+G*
0.8992	HF/6–31G**
0.9051	HF/6–311G**
0.9	HF/aug–cc–pVTZ
0.9	MNDO
0.95	AM1
0.993	SAM1
0.92	MP2
0.9434, 0.95	MP2/6–31G*
0.9427, 0.9370	MP2/6–31G**
0.9496	MP2/6–311G**
0.96	DFT
0.9945	BLYP/6–31G*
0.9914	BP86/6–31G*
0.9558	B3P86/6–31G*
0.9573	B3PW91/6–31G*
0.9614, 1.0	B3LYP/6–31G*
0.976	B3LYP/cc–pVDZ
None	CC with large basis sets

tions. Bond stretching is better described by a Morse potential and conformational changes have sine-wave-type behavior. However, the harmonic oscillator description is very useful as an approximate treatment for low vibrational quantum numbers. A harmonic oscillator approximation is most widely used for computing molecular vibrational frequencies because more accurate methods require very large amounts of CPU time.

Frequencies computed with the Hartree–Fock approximation and a quantum harmonic oscillator approximation tend to be 10% too high due to the harmonic oscillator approximation and lack of electron correlation. The exception is the very low frequencies, below about 200 cm^{-1}, which are often quite far from the experimental values. Many studies are done using *ab initio* methods and multiplying the resulting frequencies by about 0.9 to obtain a good estimate of the experimental results. A list of correction factors is given in Table 11.1. Some researchers take this idea one step further by using different correction factors for stretching modes, angle bends, and so on.

Vibrational frequencies from semiempirical calculations tend to be qualitative in that they reproduce the general trend mentioned in the introduction here. However, the actual values are erratic. Some values will be close, whereas others are often too high. SAM1 is generally the most accurate semiempirical method for predicting frequencies. PM3 is generally more accurate than AM1 with the exception of S–H and P–H bonds, for which AM1 is superior.

Some density functional theory methods occasionally yield frequencies with a bit of erratic behavior, but with a smaller deviation from the experimental results than semiempirical methods give. Overall systematic error with the better DFT functionals is less than with HF.

A molecular mechanics force field can be designed to describe the geometry of the molecule only or specifically created to describe the motions of the atoms. Calculation of the vibrational frequencies using a harmonic oscillator approximation can yield usable results if the force field was designed to reproduce the vibrational frequencies. Note that many of the force fields in use today were not designed to reproduce vibrational frequencies in this manner. When using this method, there is not necessarily a systematic error between the results and the experiments. This is because the parameters may have been created by determining what harmonic parameters would reproduce the experimental results, thus building in the correction. As a general rule of thumb, mechanics methods give qualitatively reasonable frequencies if the compound being examined is similar to those used to create the parameters. Molecular mechanics does not do so well if the structure is significantly different from the compounds in the parameterization set.

Some computer programs will output a set of frequencies containing six values near zero for the three degrees of translation and three degrees of rotation of the molecule. Any number within a range of about -20 to $20\ \text{cm}^{-1}$ is essentially zero within the numerical accuracy of typical software packages. This range is larger if second derivatives are computed numerically. Other programs will use a more sophisticated technique to avoid computing these extra values, thus reducing the computation time.

Before frequencies can be computed, the program must compute the geometry of the molecule because the normal vibrational modes are centered at the equilibrium geometry. Harmonic frequencies have no relevance to the vibrational modes of the molecule, unless computed at the exact same level of theory that was used to optimize the geometry.

Orbital-based methods can be used to compute transition structures. When a negative frequency is computed, it indicates that the geometry of the molecule corresponds to a maximum of potential energy with respect to the positions of the nuclei. The transition state of a reaction is characterized by having one negative frequency. Structures with two negative frequencies are called second-order saddle points. These structures have little relevance to chemistry since it is extremely unlikely that the molecule will be found with that structure.

11.2 ANHARMONIC FREQUENCIES

For very-high-accuracy *ab initio* calculations, the harmonic oscillator approximation may be the largest source of error. The harmonic oscillator frequencies are obtained directly from the Hessian matrix, which contains the second derivative of energy with respect to the movement of the nuclei. One of the most direct ways to compute anharmonic corrections to the vibrational frequencies is to compute the higher-order derivatives (3rd, 4th, etc.). The frequency for a potential represented by a polynomial of this order can then be computed. This requires considerably more computer resources than harmonic oscillator calculations; thus, it is much more seldom done.

Both harmonic oscillator and higher-order derivative calculations represent the potential energy surface near the optimized geometry only. It is possible to compute vibrational frequencies taking the entire potential energy surface into account. For a diatomic molecule, this requires computing the entire bond dissociation curve, which requires far more computer time than computing higher-order derivatives. Likewise, computing anharmonic frequencies for any molecule requires computing at least a sampling of all possible nuclear positions. Due to the enormous amount of time necessary to compute all these energies, this sort of calculation is very seldom done. The advantage of this technique is that it is applicable to very anharmonic vibrational modes and also high-energy modes.

A technique built around molecular mechanics is a dynamics simulation. The vibrational motion seen in molecular dynamics is a superposition of all the normal modes of vibration so frequencies cannot be determined directly from this simulation. However, the spectrum can be determined by applying a Fourier transform to these motions. The motion corresponding to a peak in this spectrum is determined by taking just that peak and doing the inverse Fourier transform to see the motion. This technique can be used to calculate anharmonic modes, very low frequencies, and frequencies corresponding to conformational transitions. However, a fairly large amount of computer time may be necessary to obtain enough data from the dynamics simulation to get a good spectrum.

11.3 PEAK INTENSITIES

Another related issue is the computation of the intensities of the peaks in the spectrum. Peak intensities depend on the probability that a particular wavelength photon will be absorbed or Raman-scattered. These probabilities can be computed from the wave function by computing the transition dipole moments. This gives relative peak intensities since the calculation does not include the density of the substance. Some types of transitions turn out to have a zero probability due to the molecules' symmetry or the spin of the electrons. This is where spectroscopic selection rules come from. *Ab initio* methods are the preferred way of computing intensities. Although intensities can be computed using semiempirical methods, they tend to give rather poor accuracy results for many chemical systems.

There have been a few studies comparing *ab initio* intensities. In nearly all cases, the addition of polarization functions to the basis set leads to a significant improvement in results. HF-computed intensities are usually scaled by a constant factor prior to making comparisons. In general, results from scaled HF, DFT, and MP2 all give similar accuracies. Out of this group, some of the hybrid DFT functionals seem to perform best. Higher-level correlated calculations, such as CISD and CCSD, give an improvement in results. Intensities also improve with higher orders of perturbation theory.

11.4 ZERO-POINT ENERGIES AND THERMODYNAMICS CORRECTIONS

The total energy computed by a geometry optimization is the minimum on the potential energy curve. However, a molecule can never actually have this energy because it must always have some vibrational motion. Many programs compute the zero-point energy correction due to being in the lowest-energy vibrational mode along with the vibrational frequencies. For accurate work, the zero-point energy correction will be added to the total energy for the optimized geometry. This corrected value can then be used for computing the relative energies of various conformers, isomers, and the like and should be slightly closer to the experimental results.

Molecular enthalpies and entropies can be broken down into the contributions from translational, vibrational, and rotational motions as well as the electronic energies. These values are often printed out along with the results of vibrational frequency calculations. Once the vibrational frequencies are known, a relatively trivial amount of computer time is needed to compute these. The values that are printed out are usually based on ideal gas assumptions.

11.5 RECOMMENDATIONS

It is possible to use computational techniques to gain insight into the vibrational motion of molecules. There are a number of computational methods available that have varying degrees of accuracy. These methods can be powerful tools if the user is aware of their strengths and weaknesses. The user is advised to use *ab initio* or DFT calculations with an appropriate scale factor if at all possible. Anharmonic corrections should be considered only if very-high-accuracy results are necessary. Semiempirical and molecular mechanics methods should be tried cautiously when the molecular system prevents using the other methods mentioned.

BIBLIOGRAPHY

For introductory discussions see the texts

J. Simons, J. Nichols, *Quantum Mechanics in Chemistry* Oxford, Oxford (1997).

P. W. Atkins, R. S. Friedman, *Molecular Quantum Mechanics* Oxford, Oxford (1997).

I. N. Levine, *Quantum Chemistry* Prentice Hall, Englewood Cliffs (1991).

D. A. McQuarrie, *Quantum Chemistry* University Science Books, Mill Valley (1983).

M. Karplus, R. N. Porter, *Atoms & Molecules: An Introduction for Students of Physical Chemistry* W. A. Benjamin, Menlo Park (1970).

Books on molecular vibrations are

C. E. Dykstra, *Quantum Chemistry & Molecular Spectroscopy* Prentice Hall, Englewood Cliffs (1992).

I. B. Bersuker, V. Z. Polinger, *Vibronic Interactions in Molecules and Crystals* Springer-Verlag, Berlin (1989).

G. Fischer, *Vibronic Coupling* Academic Press, London (1984).

I. B. Bersuker, *The Jahn-Teller Effect and Vibronic Interactions in Mondern Chemistry* Plenum, New York (1984).

E. B. Wilson Jr., J. C. Decius, P. C. Cross, *Molecular Vibrations: The Theory of Infrared and Raman Vibrational Spectra* Dover, New York (1980).

Reviews are

S. Krimm, K. Palmo, *Encycl. Comput. Chem.* **2**, 1360 (1998).

W. Cornell, S. Louise-May, *Encycl. Comput. Chem.* **3**, 1904 (1998).

J. J. P. Stewart, *Encycl. Comput. Chem.* **4**, 2579 (1998).

T. Carrington, Jr., *Encycl. Comput. Chem.* **5**, 3157 (1998).

H. Köppel, W. Domcke, *Encycl. Comput. Chem.* **5**, 3166 (1998).

J. Cremer, J. A. Larsson, E. Kraka, *Theoretical Organic Chemistry* C. Párkányi, ed., 259, Elsevier, Amsterdam (1998).

J. Tennyson, S. Miller, *Chem. Soc Rev.* **21**, 91 (1992).

H. Ågren, A. Cesar, C.-M. Liegener, *Adv. Quantum Chem.* **23**, 4 (1992).

C. W. Bauschlicker, Jr., S. R. Langhoff, *Chem. Rev.* **91**, 701 (1991).

C. B. Harris, D. E. Smith, D. J. Russell, *Chem. Rev.* **90**, 481 (1990).

K. Balasubramanian, *Chem. Rev.* **90**, 93 (1990).

Z. Bai, J. C. Light, *Ann. Rev. Phys. Chem.* **40**, 469 (1989).

R. D. Amos, *Adv. Chem. Phys.* **67**, 99 (1987).

B. A. Hess, Jr., L. J. Schaad, P. Čársky, R. Zahradník, *Chem. Rev.* **86**, 709 (1986).

W. J. Hehre, L. Radom, P. v.R. Schleyer, J. A. Pople, *Ab Initio Molecular Orbital Theory* John Wiley & Sons, New York (1986).

H. Köppel, W. Domcke, L. S. Ceederbaum, *Adv. Chem. Phys.* **57**, 59 (1984).

I. B. Bersuker, V. Z. Polinger, *Adv. Quantum Chem.* **15**, 85 (1982).

D. W. Oxtoby, *Ann. Rev. Phys. Chem.* **32**, 77 (1981).

R. T. Bailey, F. R. Cruickshank, *Gas Kinetics and Energy Transfer vol. 3* 127, P. G. Ashmore, R. J. Donovan (Eds.) Chemical Society, London (1978).

D. Rapp, T. Kassal, *Chem. Rev.* **69**, 61 (1969).

Adv. Chem. Phys. vol. **85**, part 3. (1994).

Papers comparing the accuracy of computed intensities are

M. D. Halls, H. B. Schlegel, *J. Chem. Phys.* **109**, 10587 (1998).

J. R. Thomas, B. J. DeLeeuw, G. Vacek, T. D. Crawford, Y. Yamaguchi, H. F. Schaefer III, *J. Chem. Phys.* **99**, 403 (1993).

J. F. Stanton, W. N. Lipscomb, D. H. Magers, R. J. Bartlett, *J. Chem. Phys.* **90**, 3241 (1989).

A. Holder, R. Dennington, *Theochem* **401**, 207 (1997).

Correction factors are published in

http://srdata.nist.gov/cccbdb/

J. Baker, A. A. Jarzecki, P. Pulay, *J. Phys. Chem. A* **102**, 1412 (1998).

J. R. Wright, *Jaguar User's Guide* Schrödinger, Portland OR (1998).

A. P. Scott, L. Radom, *J. Phys. Chem.* **100**, 16502 (1996).

J. M. L. Martin, J. El-Yazal, J.-P. Francois, *J Phys. Chem.* **100**, 15358 (1996).

D. Bakowies, W. Thiel, *Chem. Phys* **151**, 309 (1991).

Harmonic and anharmonic vibrations are discussed at

http://suzy.unl.edu/bruno/CHE484/irvib.html

12 Population Analysis

Chemists are able to do research much more efficiently if they have a model for understanding chemistry. Population analysis is a mathematical way of partitioning a wave function or electron density into charges on the nuclei, bond orders, and other related information. These are probably the most widely used results that are not experimentally observable.

Atomic charges cannot be observed experimentally because they do not correspond to any unique physical property. In reality, atoms have a positive nucleus surrounded by negative electrons, not partial charges on each atom. However, condensing electron density and nuclear charges down to partial charges on the nucleus results in an understanding of the electron density distribution. These are not integer formal charges, but rather fractions of an electron corresponding to the percentage of time an electron is near each nucleus. Although this is an artificial assignment, it is very effective for predicting sites susceptible to nucleophilic or electrophilic attack and other aspects of molecular interaction. These partial charges correspond well to the chemist's view of ionic or covalent bonds, polarity, and so on. Only the most ionic compounds, such as alkali metal halides, will have nearly whole number charges. Organometallics typically have charges on the order of ± 0.5. Organic compounds often have charges around ± 0.2 or less.

12.1 MULLIKEN POPULATION ANALYSIS

One of the original and still most widely used population analysis schemes is the Mulliken population analysis. The fundamental assumption used by the Mulliken scheme for partitioning the wave function is that the overlap between two orbitals is shared equally. This does not completely reflect the electronegativity of the individual elements. However, it does give one a means for partitioning a wave function and has been found to be very effective for small basis sets. For large basis sets, results can be very unreasonable. This is due to diffuse functions describing adjacent atoms more than they describe the atom on which they are centered. In some cases, Mulliken analysis can assign an electron population to an orbital that is negative or more than two electrons. It also tends to underestimate the charge separation in ionic bonded systems.

In spite of its deficiencies, the Mulliken population scheme is very popular. One reason is that it is very easy to implement so it is available in many software packages. Probably the most important reason for its popularity is the fact that the method is easy to understand. This is a great advantage because population analysis is often used for the purpose of understanding chemistry rather than quantitatively predicting experimental results.

There is some ambiguity about Mulliken population analysis in the literature. This is because various software packages print different portions of the analysis and may name them slightly differently. The description here follows some of the more common conventions.

A molecular orbital is a linear combination of basis functions. Normalization requires that the integral of a molecular orbital squared is equal to 1. The square of a molecular orbital gives many terms, some of which are the square of a basis function and others are products of basis functions, which yield the overlap when integrated. Thus, the orbital integral is actually a sum of integrals over one or two center basis functions.

In Mulliken analysis, the integrals from a given orbital are not added. Instead, the contribution of a basis function in all orbitals is summed to give the net population of that basis function. Likewise, the overlaps for a given pair of basis functions are summed for all orbitals in order to determine the overlap population for that pair of basis functions. The overlap populations can be zero by symmetry or negative, indicating antibonding interactions. Large positive overlaps between basis functions on different atoms are one indication of a chemical bond.

Gross populations are determined by starting with the net populations for a basis function, then adding half of every overlap population to which that basis function contributes. The Gross populations for all orbitals centered on a given atom can be summed in order to obtain the gross atomic population for that atom. The gross atomic population can be subtracted from the nuclear charge in order to obtain a net charge. Further analysis of overlap populations can yield bond orders.

12.2 LÖWDIN POPULATION ANALYSIS

The Löwdin population analysis scheme was created to circumvent some of the unreasonable orbital populations predicted by the Mulliken scheme, which it does. It is different in that the atomic orbitals are first transformed into an orthogonal set, and the molecular orbital coefficients are transformed to give the representation of the wave function in this new basis. This is less often used since it requires more computational work to complete the orthogonalization and has been incorporated into fewer software packages. The results are still basis-set-dependent.

12.3 NATURAL BOND-ORDER ANALYSIS

Natural bond order analysis (NBO) is the name of a whole set of analysis techniques. One of these is the natural population analysis (NPA) for obtaining occupancies (how many electrons are assigned to each atom) and charges. Some researchers use the acronyms NBO and NPA interchangeably.

Rather than using the molecular orbitals directly, NBO uses the natural orbitals. Natural orbitals are the eigenfunctions of the first-order reduced density matrix. These are then localized and orthogonalized. The localization procedure allows orbitals to be defined as those centered on atoms and those encompassing pairs of atoms. These can be integrated to obtain charges on the atoms. Analysis of the basis function weights and nodal properties allows these transformed orbitals to be classified as bonding, antibonding, core, and Rydberg orbitals. Further decomposition into three-body orbitals will yield a characterization of three center bonds. There is also a procedure that searches for the π bonding patterns typical of a resonant system. This is not a rigorous assignment as there may be some electron occupancy of antibonding orbitals, which a simple Lewis model would predict to be unoccupied.

This results in a population analysis scheme that is less basis set dependent than the Mulliken scheme. However, basis set effects are still readily apparent. This is also a popular technique because it is available in many software packages and researchers find it convenient to use a method that classifies the type of orbital.

12.4 ATOMS IN MOLECULES

A much less basis set dependent method is to analyze the total electron density. This is called the atoms in molecules (AIM) method. It is designed to examine the small effects due to bonding in the primarily featureless electron density. This is done by examining the gradient and Laplacian of electron density. AIM analysis incorporates a number of graphic analysis techniques as well as population analysis. The population analysis will be discussed here and the graphic techniques in the next chapter.

The first step in this process is to examine the total electron density to find the critical point in the middle of each bond. This is the point of minimum electron density along the line connecting the atoms. It reflects atomic sizes by being closer to the smaller atom. From the critical point, the gradient vector path (path of fastest electron density decrease) can be followed in all directions, which is nearly perpendicular to the line connecting atoms at the critical point. The gradient vector path defines surfaces in three-dimensional space, which will separate that space into regions around each nucleus. The number of electrons in this region can be integrated in order to find an electron population and thus an atomic charge. The bond order can be predicted, based on the magnitude of the electron density at the bond critical point.

The AIM scheme is popular due to its reliability with large basis sets for which some other schemes fail. Unfortunately, the numerical surface finding and integration involved in this scheme are not completely robust. For example, non-nuclear attractor compounds like Li_2 and Na clusters have maxima in the middle of the bonds, which the AIM method does not assign to either atom. Thus, the analysis sometimes fails to give a result. Also, the amount of charge separation in polar bonds is greater than what is generally accepted as reasonable.

12.5 ELECTROSTATIC CHARGES

If one is to choose a chemically relevant set of partial charges on the nuclei, it would probably be those that most reflect the way that the electron density distribution interacts with other molecules. Electrostatic charges, also called ESP charges, are computed from the electrostatic potential. The electrostatic potential is evaluated at a series of points, usually on the van der Waals surface around the molecule. A curve-fitting procedure is then used to determine the set of partial charges on the nuclei that would most closely result in generating that electrostatic potential. This gives a very good description of charge interactions with other species. Because of this, electrostatic charges are often used as point charges for more approximate calculations, such as molecular mechanics calculations.

Several electrostatic charge calculation methods have been devised. These vary primarily in how the electrostatic potential points are chosen. Some software packages include the ability to further constrain the charge calculation procedure to only compute charges that reproduce the dipole moment. Some of the common algorithms are Merz–Singh–Kollman (MK), Chelp, and ChelpG. Perhaps the most popular electrostatic charge computational scheme is the ChelpG method.

ESP charges are not without problems, particularly when they are to be used for molecular mechanics calculations. The charges predicted by ESP methods will vary as the conformation of the molecule changes. This results in atoms that should be equivalent within a molecular mechanics methodology having different charges, such as the three hydrogens in a methyl group. This can be corrected by averaging atoms that should be equivalent. This average can be determined either for one conformer or it can be an average over multiple conformations.

12.6 CHARGES FROM STRUCTURE ONLY

Molecular mechanics methods often include a Coulombic interaction term. However, a molecular mechanics model does not have a wave function or electron density from which to compute charges. Sometimes, these charges are obtained from the types of calculations above, particularly electrostatic charges. When this is done, molecular mechanics is usually used to optimize the molecular geometry without charges included and then an orbital-based calculation is done without geometry optimization to obtain the charges, which can be used in subsequent molecular mechanics calculations. Sometimes, the molecule is too big for any type of orbital-based calculation of charges. There are several methods for determining charges without any type of orbital-based calculation.

The Del Re charge calculation method uses several parameters for each element and for describing interactions between various elements. A simple set of equations incorporating these parameters and the distance between atoms is used to compute charges. The method was only designed to describe molecules with σ bonds, but describes π-bonded molecules fairly reasonably. The disadvantage of the Del Re scheme is that it cannot be used if parameters are not available for the elements. It is parameterized for describing organic compounds. Since it is a parameterized method, it only works well for systems similar to those used for the parameterization: typical organic molecules.

The Pullman method is a combination of the Del Re method for computing the σ component of the charge and a semiempirical Hückel calculation for the π portion. It has been fairly successful in describing dipole moments and atomic charges for nucleic acids and proteins.

The Gasteiger charge calculation method is based on a simple relationship between charge and electronegativity. It still has parameters for each element, but not parameters for interactions between elements. The original method has been extended to describe aromatic compounds by optimizing first σ and then π charges. This is used for organic molecules only.

The Q-equilibrate method is applicable to the widest range of chemical systems. It is based on atomic electronegativities only. An iterative procedure is used to adjust the charges until all charges are consistent with the electronegativities of the atoms. This is perhaps the most often used of these methods.

12.7 RECOMMENDATIONS

There are cases where each of these methods excels. However, the literature does indicate a preference for certain methods that obtain the most consistent results. Below are some of the suggestions based on a review of the literature:

- For molecular mechanics, the charge calculation method used in parameterizing the force field should be used if possible. Otherwise, use Q-equilibrate or electrostatic charges.
- For examining the interactions between molecules, use electrostatic charges.
- For gaining a detailed understanding of orbital interactions, use the Mulliken analysis with a minimal basis set.
- For large basis sets, use AIM, NBO, or electrostatic charges.
- Mulliken analysis is most often used with semiempirical wave functions.

Table 12.1 gives the partial charges for the atoms in acetic acid computed with a number of different methods and basis sets. All calculations use the molecular

TABLE 12.1 Charges For Acetic Acid

Wave function	Method	C1	O2	C3	O4	H5	H6
AM1	Mulliken	0.35	−0.38	−0.38	−0.36	0.27	0.17
PM3	Mulliken	0.39	−0.40	−0.36	−0.31	0.24	0.15
HF/STO−3G	Mulliken	0.33	−0.27	−0.21	−0.30	0.21	0.08
HF/6−31G*	Mulliken	0.74	−0.56	−0.57	−0.70	0.47	0.21
HF/6−311++G**	Mulliken	0.33	−0.39	−0.52	−0.25	0.31	0.16
B3LYP/STO−3G	Mulliken	0.24	−0.24	−0.25	−0.25	0.22	0.10
B3LYP/6−31G*	Mulliken	0.56	−0.45	−0.51	−0.56	0.41	0.19
B3LYP/6−311++G**	Mulliken	0.15	−0.31	−0.48	−0.17	0.28	0.17
HF/STO−3G	NBO	0.42	−0.31	−0.20	−0.34	0.23	0.07
HF/6−31G*	NBO	0.99	−0.70	−0.75	−0.80	0.51	0.25
HF/6−311++G**	NBO	0.96	−0.69	−0.62	−0.76	0.49	0.21
B3LYP/STO−3G	NBO	0.31	−0.26	−0.24	−0.27	0.23	0.08
B3LYP/6−31G*	NBO	0.82	−0.60	−0.78	−0.72	0.50	0.26
B3LYP/6−311++G**	NBO	0.80	−0.60	−0.68	−0.70	0.48	0.23
AM1	CHELPG	0.19	−0.20	−1.01	−0.31	0.26	0.34
PM3	CHELPG	−0.61	0.52	−4.44	0.11	0.39	1.35
HF/STO−3G	CHELPG	0.80	−0.46	−0.50	−0.56	0.33	0.12
HF/6−31G*	CHELPG	0.86	−0.62	−0.34	−0.67	0.44	0.09
HF/6−311++G**	CHELPG	0.92	−0.66	−0.35	−0.69	0.45	0.10
B3LYP/STO−3G	CHELPG	0.65	−0.40	−0.52	−0.47	0.33	0.13
B3LYP/6−31G*	CHELPG	0.71	−0.53	−0.32	−0.58	0.41	0.09
B3LYP/6−311++G**	CHELPG	0.82	−0.59	−0.34	−0.63	0.42	0.10
HF/STO−3G	AIM	1.47	−1.05	0.17	−0.98	0.47	−0.02
HF/6−31G*	AIM	1.84	−1.38	0.07	1.28	0.62	0.04
HF/6−311++G**	AIM	1.75	−1.32	0.13	−1.27	0.64	0.02
B3LYP/STO−3G	AIM	1.38	−1.00	0.08	−0.91	0.46	−0.003
B3LYP/6−31G*	AIM	1.61	−1.22	−0.02	−1.12	0.58	0.05
B3LYP/6−311++G**	AIM	1.50	−1.14	0.01	−1.08	0.58	0.04

geometry predicted by a B3LYP/6−31G* calculation. This is a molecule with some charge separation, but not an extreme case as ionic molecules would be. The assignment of atom numbers is shown in Figure 12.1. Charges on hydrogens 7 and 8 are nearly identical to those on hydrogen 6.

FIGURE 12.1 Acetic acid atom assignments for Table 12.1.

For large molecules, computation time becomes a consideration. Orbital-based techniques, such as Mulliken, Löwdin, and NBO, take a negligible amount of CPU time relative to the time required to obtain the wave function. Techniques based on the charge distribution, such as AIM and ESP, require a significant amount of CPU time. The GAPT method, which was not mentioned above, requires a second derivative evaluation, which can be prohibitively expensive.

BIBLIOGRAPHY

Textbooks discussing population analysis are

F. Jensen, *Introduction to Computational Chemistry* John Wiley & Sons, New York (1999).

R. A. Albright, *Orbital Interactions in Chemistry* John Wiley & Sons, New York (1998).

A. R. Leach, *Molecular Modelling Principles and Applications* Longman, Essex (1996).

J. B. Foresman, Æ. Frisch, *Exploring Chemistry with Electronic Structure Methods* Gaussian, Pittsburgh (1996).

W. J. Hehre, *Practical Strategies for Electronic Structure Calculations* Wavefunction, Irvine (1995).

G. H. Grant, W. G. Richards, *Computational Chemistry* Oxford, Oxford (1995).

I. N. Levine, *Quantum Chemistry* Prentice Hall, Englewood Cliffs (1991).

R. F. W. Bader, *Atoms in Molecules, A Quantum Theory* Oxford, Oxford (1990).

W. J. Hehre, L. Radom, P. v.R. Schleyer, J. A. Pople, *Ab Initio Molecular Orbital Theory* John Wiley & Sons, New York (1986).

T. Clark, *A Handbook of Computational Chemistry* John Wiley & Sons, New York (1985).

S. Fliszár, *Charge Distributions and Chemical Effects* Springer-Verlag, New York (1983).

Review articles are

R. F. W. Bader, *Encycl. Comput. Chem.* **1**, 64 (1998).

W. D. Cornell, C. Chipot, *Encycl Comput. Chem.* **1**, 258 (1998).

J. Cioslowski, *Encycl. Comput. Chem.* **2**, 892 (1998).

F. Weinhold, *Encycl. Comput. Chem* **3**, 1793 (1998).

K. Jug, *Encycl. Comput. Chem.* **3**, 2150 (1998).

M. S. Gordon, J. H. Jenson, *Encycl. Comput. Chem.* **5**, 3198 (1998).

S. M. Barach, *Rev. Comput. Chem.* **5**, 171 (1994).

K. B. Wiberg, P. R. Rablen, *J. Comput. Chem.* **14**, 1504 (1993).

A. B. Sannigrahi, *Adv. Quantum Chem.* **23**, 302 (1992).

R. F. W. Bader, *Chem. Rev* **91**, 893 (1991).

D. E. Williams, *Rev. Comput. Chem.* **2**, 219 (1991).

R. F. W. Bader, T. T. Nguyen-Dang, *Adv. Quantum Chem.* **14**, 63 (1981).

M. Randi, Z. B. Maksi, *Chem. Rev.* **72**, 43 (1972).

Some of the original papers are

U. C. Singh, P. A. Kollman, *J. Comput. Chem.* **5**, 129 (1984).

A. E. Reed, F. Weinhold, *J. Chem. Phys* **78**, 4066 (1983).

J. P. Foster, F. Weinhold, *J. Am. Chem. Soc.* **102**, 7211 (1980).

R. S. Mulliken, *J. Chem. Phys.* **23**, 1833 (1955).

R. S. Mulliken, *J. Chem. Phys.* **23**, 1841 (1955).

R. S. Mulliken, *J. Chem. Phys.* **23**, 2338 (1955).

R. S. Mulliken, *J Chem. Phys.* **23**, 2343 (1955).

13 Other Chemical Properties

This chapter covers a number of concepts or properties that did not relate to material discussed in earlier chapters. Some of these techniques are seldom needed. Others just do not merit a chapter of their own because they are easy to apply.

The first section of this chapter discusses various ways that chemical properties are computed. Then a number of specific properties are addressed. The final section is on visualization, which is not so much a property as a way of gaining additional insight into the electronic structure and motion of molecules.

13.1 METHODS FOR COMPUTING PROPERTIES

The reliability and accuracy of property results vary greatly. There is no generalization that says any given method will compute every property best. However, there are some generalizations to be made. One of these generalizations is that a given type of algorithm will tend to have certain strengths and weaknesses in spite of the type of property being computed. Below are the most common techniques.

13.1.1 From the Energy

Some of the most important information about chemistry is the energy or relative energetics associated with various species or processes. A few of these are mentioned specifically in this chapter. The accuracy of computed energies is mentioned many other places in this book. Energy is an integral part of most computational techniques. However, some energies are easier to compute than others. For example, the difference in energy between two conformers is one of the easiest energies to compute, whereas reaction barriers are much more difficult to compute accurately.

13.1.2 From Molecular Geometry

Some properties, such as the molecular size, can be computed directly from the molecular geometry. This is particularly important, because these properties are accessible from molecular mechanics calculations. Many descriptors for quantitative structure activity or property relationship calculations can be computed from the geometry only.

13.1.3 From the Wave Function or Electron Density

Many molecular properties can be related directly to the wave function or total electron density. Some examples are dipole moments, polarizability, the electrostatic potential, and charges on atoms.

13.1.4 Group Additivity

Group additivity methods have been developed specifically because they can be applied to a simple pen-and-paper calculation. Many of these methods have been incorporated in software packages also. These methods all involve adding up weights from a table of various functional groups in order to obtain an estimate of some property of a molecule. These also have the advantage of quantifying some intuitive understanding of molecular behavior.

Group additivity methods must be derived as a consistent set. It is not correct to combine fragments from different group additivity techniques, even for the same property. This additivity approximation essentially ignores effects due to the location of one functional group relative to another. Some of these methods have a series of corrections for various classes of compounds to correct for this. Other methods use some sort of topological description.

13.1.5 QSAR or QSPR

Quantitative structure property relationships (QSPR) and, when applied to biological activity, quantitative structure activity relationships (QSAR) are methods for determining properties due to very sophisticated mechanisms purely by a curve fit of that property to aspects of the molecular structure. This allows a property to be predicted independent of having a complete knowledge of its origin. For example, drug activity can be predicted without knowing the nature of the binding site for that drug. QSPR is covered in more detail in Chapter 30.

13.1.6 Database Searching

There are now extensive databases of molecular structures and properties. There are some research efforts, such as drug design, in which it is desirable to find all molecules that are very similar to a molecule which has the desired property. Thus, there are now techniques for searching large databases of structures to find compounds with the highest molecular similarity. This results in finding a collection of known structures that are most similar to a specific compound.

Molecular similarity is also useful in predicting molecular properties. Programs that predict properties from a database usually first search for compounds in the database that are similar to the unknown compound. The property of the unknown is probably close in value to the property for the known compounds. The software will then have some scheme for correcting for the effect of differences in structure between the known and unknown compounds, or it may use the exact structure if it is found in the database. This can often be viewed as a group additivity method, which is reparameterized from the most similar compounds in the database. This has proven to be a very reliable method for predicting NMR spectra and works well for partition coefficients, boiling points, and other properties for organic compounds. This method may give poor results if the unknown is unlike any of the structures in the database.

13.1.7 Artificial Intelligence

Algorithms originally designed by artificial intelligence (AI) researchers have been used for predicting molecular properties. These programs are not necessarily "intelligent" in the way that a person is, but they incorporate some of the characteristics of intelligence, such as learning from new data or developing some type of understanding. These techniques are in their infancy now and may become much more developed in the future.

Some AI-based programs are qualitative. These are usually rule-based decision systems. For example, one program will ask the user about the characteristics of the polymer to be designed. After obtaining enough information, the program will suggest that the polymer should be a condensation or thermosetting or block copolymer, and so on. This determination is based on qualitative descriptions rather than numerical computations.

One variation of rule-based systems are fuzzy logic systems. These programs use statistical decision-making processes in which they can account for the fact that a specific piece of data has a certain chance of indicating a particular result. All these probabilities are combined in order predict a final answer.

Some systems can give quantitative results from known pieces of data complete with proper units. For example, these systems can take all the starting information and then determine a set of equations from the available list that can yield the desired result. The program could subsequently convert units or algebraically solve the equations if necessary.

Neural networks are programs that simulate a brain's structure: With their many simple units (functions) that can communicate with each other and do very simple jobs, they work similarly to the way that neurons in the brain do. The neural network is trained by giving it data on systems for which the results are known. Then the network can be given unknown data to make a prediction. This is a sort of curve-fitting technique and is good for the interpolation of existing data, but generally poor for the extrapolation of results. Neural networks can predict nonlinear data well. Neural networks can be overtrained, thus fitting to anomalies in the training set at the expense of poorer performance when predicting properties of unknowns.

One class of AI-based computational chemistry programs are *de novo* programs. These programs generally try to efficiently automate tedious tasks by using some rational criteria to guide a trial-and-error process. For example, finding a molecule that will bind well in a particular binding site requires testing many molecules in many orientations within that site. A *de novo* program will examine the binding site to determine that it is only reasonable to try molecular orientations in which a nucleophilic group is oriented in a certain way, and so forth.

13.1.8 Statistical Processes

It is important to realize that many important processes, such as retention times in a given chromatographic column, are not just a simple aspect of a molecule. These are actually statistical averages of all possible interactions of that molecule and another. These sorts of processes can only be modeled on a molecular level by obtaining many results and then using a statistical distribution of those results. In some cases, group additivities or QSPR methods may be substituted.

13.2 MULTIPOLE MOMENTS

The unique multipole moment of a molecule gives a description of the separation of charge of the molecule. Which multipole is unique depends on both the charge and the geometry of the molecule. For a charged ion, the charge, its monopole, is the only unique multipole. The higher-order multipoles, such as the dipole moment, the quadrupole moment, and the like, can still be computed but will be dependent on the origin used for that computation.

Many molecules, such as carbon monoxide, have unique dipole moments. Molecules with a center of inversion, such as carbon dioxide, will have a dipole moment that is zero by symmetry and a unique quadrupole moment. Molecules of T_d symmetry, such as methane, have a zero dipole and quadrupole moment and a unique octupole moment. Likewise, molecules of octahedral symmetry will have a unique hexadecapole moment.

Multipole moments are most accurately computed from *ab initio* calculations. HF calculations with minimal basis sets often give good results. Correlated calculations can yield high-accuracy results. Some semiempirical methods also give reasonable results. For very large molecules, the multipoles can be computed from atomic charges used by molecular mechanics calculations. Computed multipoles can be very sensitive to the geometry at which they are computed, particularly if the value is fairly small in magnitude. It is generally advisable to use multipole moments that were computed with the same level of theory used to optimize the molecular geometry.

13.3 FERMI CONTACT DENSITY

The Fermi contact density is defined as the electron density at the nucleus of an atom. This is important due to its relationship to analysis methods dependent on electron density at the nucleus, such as EPR and NMR spectroscopy. Fermi contact densities are computed with *ab initio* methods.

13.4 ELECTRONIC SPATIAL EXTENT AND MOLECULAR VOLUME

The electronic spatial extent is a single number that attempts to describe the size of a molecule. This number is computed as the expectation value of electron density times the distance from the center of mass of a molecule. Because the information is condensed down to a single number, it does not distinguish between long chains and more globular molecules.

Molecular volumes are usually computed by a nonquantum mechanical method, which integrates the area inside a van der Waals or Connolly surface of some sort. Alternatively, molecular volume can be determined by choosing an isosurface of the electron density and determining the volume inside of that surface. Thus, one could find the isosurface that contains a certain percentage of the electron density. These properties are important due to their relationship to certain applications, such as determining whether a molecule will fit in the active site of an enzyme, predicting liquid densities, and determining the cavity size for solvation calculations.

The solvent-excluded volume is a molecular volume calculation that finds the volume of space which a given solvent cannot reach. This is done by determining the surface created by running a spherical probe over a hard sphere model of molecule. The size of the probe sphere is based on the size of the solvent molecule.

A convex hull is a molecular surface that is determined by running a planar probe over a molecule. This gives the smallest convex region containing the molecule. It also serves as the maximum volume a molecule can be expected to reach.

13.5 ELECTRON AFFINITY AND IONIZATION POTENTIAL

The electron affinity (EA) and ionization potential (IP) can be computed as the difference between the total energies for the ground state of a molecule and for the ground state of the appropriate ion. The difference between two calculations such as this is often much more accurate than either of the calculations since systematic errors will cancel. Differences of energies from correlated quantum mechanical techniques give very accurate results, often more accurate than might be obtained by experimental methods.

The electron affinity and ionization potential can be either for vertical excitations or adiabatic excitations. For adiabatic potentials, the geometry of both ions is optimized. For vertical transitions, both energies are computed for the same geometry, optimized for the starting state.

Another technique for obtaining an ionization potential is to use the negative of the HOMO energy from a Hartree–Fock calculation. This is called Koopman's theorem; it estimates vertical transitions. This does not apply to methods other than HF but gives a good prediction of the ionization potential for many classes of compounds.

13.6 HYPERFINE COUPLING

Traditional wisdom has been that correlated *ab initio* calculations with large basis sets are necessary to accurately predict hyperfine coupling constants. More recently, some researchers have begun using the B3LYP functional with a moderate-size basis set (6–31G* or larger). UHF semiempirical calculations were used at one time, but have now been mostly replaced by more accurate methods. The most rigorous calculations include vibronic coupling in order to determine the average of the results for the expected vibrational level occupation at some temperature.

13.7 DIELECTRIC CONSTANT

The dielectric constant is a property of a bulk material, not an individual molecule. It arises from the polarity of molecules (static dipole moment), and the polarizability and orientation of molecules in the bulk medium. Often, it is the relative permitivity ε_s that is computed rather than the dielectric constant κ, which is the constant of proportionality between the vacuum permitivity ε_0 and the relative permitivity.

$$\varepsilon_s = \kappa\varepsilon_0 \tag{13.1}$$

For fluids, this is computed by a statistical sampling technique, such as Monte Carlo or molecular dynamics calculations. There are a number of concerns that must be addressed in setting up these calculations, such as

- The choice of boundary conditions
- Whether an adequate sampling of phase space is obtained
- Whether the system size is large enough to represent the bulk material
- Whether the errors in calculation have been estimated correctly

Another way to obtain a relative permitivity is using some simple equations that relate relative permitivity to the molecular dipole moment. These are derived from statistical mechanics. Two of the more well-known equations are the Clausius–Mossotti equation and the Kirkwood equation. These and others are discussed in the review articles referenced at the end of this chapter. The computation of dielectric constants is also discussed in the books by Leach, and Allen and Tildesley.

13.8 OPTICAL ACTIVITY

Molecular chirality is most often observed experimentally through its optical activity, which is the effect on polarized light. The spectroscopic techniques for measuring optical activity are optical rotary dispersion (ORD), circular dichroism (CD), and vibrational circular dichroism (VCD).

The measurements are predicted computationally with orbital-based techniques that can compute transition dipole moments (and thus intensities) for transitions between electronic states. VCD is particularly difficult to predict due to the fact that the Born–Oppenheimer approximation is not valid for this property. Thus, there is a choice between using the wave functions computed with the Born–Oppenheimer approximation giving limited accuracy, or very computationally intensive exact computations. Further technical difficulties are encountered due to the gauge dependence of many techniques (dependence on the coordinate system origin).

The most reliable results are obtained using *ab initio* methods with moderate- to large-sized polarized basis sets. The use of gauge-independent atomic orbitals (GIAO) removes gauge dependency problems.

For transition metal complexes, techniques derived from a crystal-field theory or ligand-field theory description of the molecules have been created. These tend to be more often qualitative than quantitative.

Recent progress in this field has been made in predicting individual atoms' contribution to optical activity. This is done using a wave-functioning, partitioning technique roughly analogous to Mulliken population analysis.

13.9 BIOLOGICAL ACTIVITY

There is great commercial interest in predicting the activity of a compound in a biological system. This includes both desired properties, such as drug activity, and undesired properties, such as toxicity. Such a prediction poses some very difficult problems due to the complexity of biological systems. No method in existence is capable of automatically computing all the interactions between a given molecule and every molecule found in a single cell, let alone an entire organism. Such an attempt is completely beyond the capabilities of any computer hardware available today by many orders of magnitude.

Molecular simulation techniques can be used to predict how a compound will interact with a particular active site of a biological molecule. This is still not trivial because the molecular orientation must be considered along with whether the active site shifts geometry as it approaches.

One very popular technique is to use QSAR. It is, in essence, a curve-fitting technique for creating an equation that predicts biological activity from the properties of the individual molecule only. Once this equation has been created using many compounds of known activity, it can be used to predict the activity of new compounds. QSAR is discussed further in Chapter 30.

Another technique is to use pattern recognition routines. Whereas QSAR relates activity to properties such as the dipole moment, pattern recognition examines only the molecular structure. It thus attempts to find correlations between the functional groups and combinations of functional groups and the biological activity.

Expert systems have also been devised for predicting biological activity. Predicting biological activity is discussed further in Chapter 38.

13.10 BOILING POINT AND MELTING POINT

Several methods have been successfully used to predict the normal boiling point of liquids. Group additivity methods give an approximate estimate. Some group additivity methods gain accuracy at the expense of being applicable to a narrow range of chemical systems. Techniques that use a database to parameterize a group additivity method are significantly more accurate.

QSPR methods have yielded the most accurate results. Most often, they use large expansions of parameters obtainable from semiempirical calculations along with other less computationally intensive properties. This is often the method of choice for small molecules.

Molecular dynamics and Monte Carlo simulations can be used, but these methods involve very complex calculations. They are generally only done when more information than just the boiling point is desired and they are not calculations for a novice.

Melting points are much more difficult to predict accurately. This is because of their dependence on crystal structure. Seemingly similar compounds can have significantly different melting points due to one geometry being able to pack into a crystal with stronger intermolecular interactions. Some group additivity methods have been designed to give a rough estimate of the melting point.

13.11 SURFACE TENSION

Surface tension is usually predicted using group additivity methods for neat liquids. It is much more difficult to predict the surface tension of a mixture, especially when surfactants are involved. Very large molecular dynamics or Monte Carlo simulations can also be used. Often, it is easier to measure surface tension in the laboratory than to compute it.

13.12 VAPOR PRESSURE

Different compounds can display a very large difference in vapor pressure, depending on what type of intermolecular forces is present. Because of this, different prediction schemes are used, depending on whether the molecule is nonpolar, polar, or hydrogen-bonding. These methods are usually derived from thermodynamics with an empirical correction factor incorporated. The correction factors usually depend on the type of compound, that is, alcohol, keytone, and, and so forth. These methods are applicable to a wide range of temperatures so long as they are not too close to the temperature at which a phase change occurs. Constants for Henry's law are computed from vapor pressure, log *P*, and group additivity methods.

13.13 SOLUBILITY

A significant amount of research has focused on deriving methods for predicting log *P*, where *P* is the octanol–water partition coefficient. Other solubility and adsorption properties are generally computed from the log *P* value. There are some group additivity methods for predicting log *P*, some of which have extremely complex rules. QSPR techniques are reliably applicable to the widest range of compounds. Neural network based methods are very accurate so long as the unknown can be considered an interpolation between compounds in the training set. Database techniques are very accurate for organic compounds. The solvation methods discussed in chapter 24 can also be used.

13.14 DIFFUSIVITY

The rate of chemical diffusion in a nonflowing medium can be predicted. This is usually done with an equation, derived from the diffusion equation, that incorporates an empirical correction parameter. These correction factors are often based on molar volume. Molecular dynamics simulations can also be used.

Diffusion in flowing fluids can be orders of magnitude faster than in nonflowing fluids. This is generally estimated from continuum fluid dynamics simulations.

13.15 VISUALIZATION

Data visualization is the process of displaying information in any sort of pictorial or graphic representation. A number of computer programs are available to apply a colorization scheme to data or to work with three-dimensional representations. In recent years, this functionality has been incorporated in many

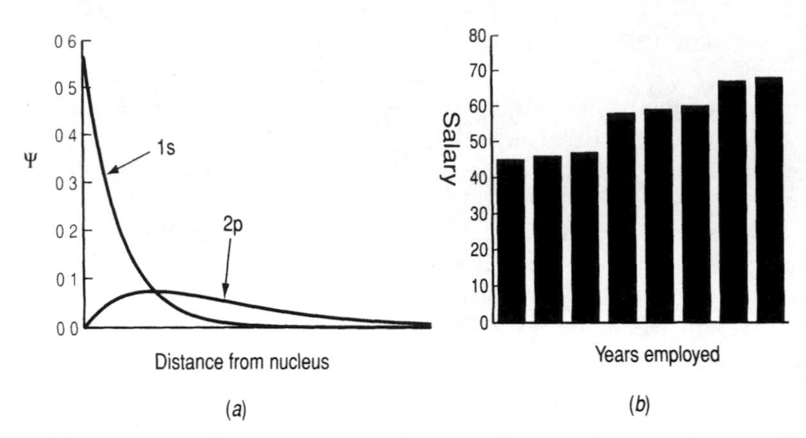

FIGURE 13.1 Graphs that have a one-dimensional data space. (*a*) Radial portion of the wave function for the hydrogen atom in the 1*s* ground state and 2*p* excited state. (*b*) Hypothetical salary chart.

of the graphic user interface programs used to set up and run chemical calculations. The term "visualization" usually refers to the graphic display of numerical results of experimental data, or computational chemistry results, not an artist's representations of the molecule.

13.15.1 Coordinate Space

A typical plot of x vs. $f(x)$ is considered to have one coordinate dimension, the x, and one data dimension, $f(x)$. These data sets are plotted as line graphs, bar graphs, and so forth. These types of plots are readily made with most spreadsheet programs as well as dedicated graphing programs. Figure 13.1 shows two graphs that are considered to have a one-dimensional data space.

There are also plots that have two coordinate dimensions and one data dimension. Examples of this would be a topographical map or the electron density in one plane. These data sets can be displayed as colorizations (Figure 13.2) or contour plots (Figure 13.3). Colorizations assign a color to each point in the plane according to the value at that point. Contour plots connect all the points having a particular value. Contour plots are perhaps more quantitative in their ability to show the shape of regions with various values. Colorizations are more complete in that no spots are left out. Another technique is to use the third dimension to plot the data values. This is called a mesh plot (Figure 13.4).

Many functions, such as electron density, spin density, or the electrostatic potential of a molecule, have three coordinate dimensions and one data dimension. These functions are often plotted as the surface associated with a particular data value, called an isosurface plot (Figure 13.5). This is the three-dimensional analog of a contour plot.

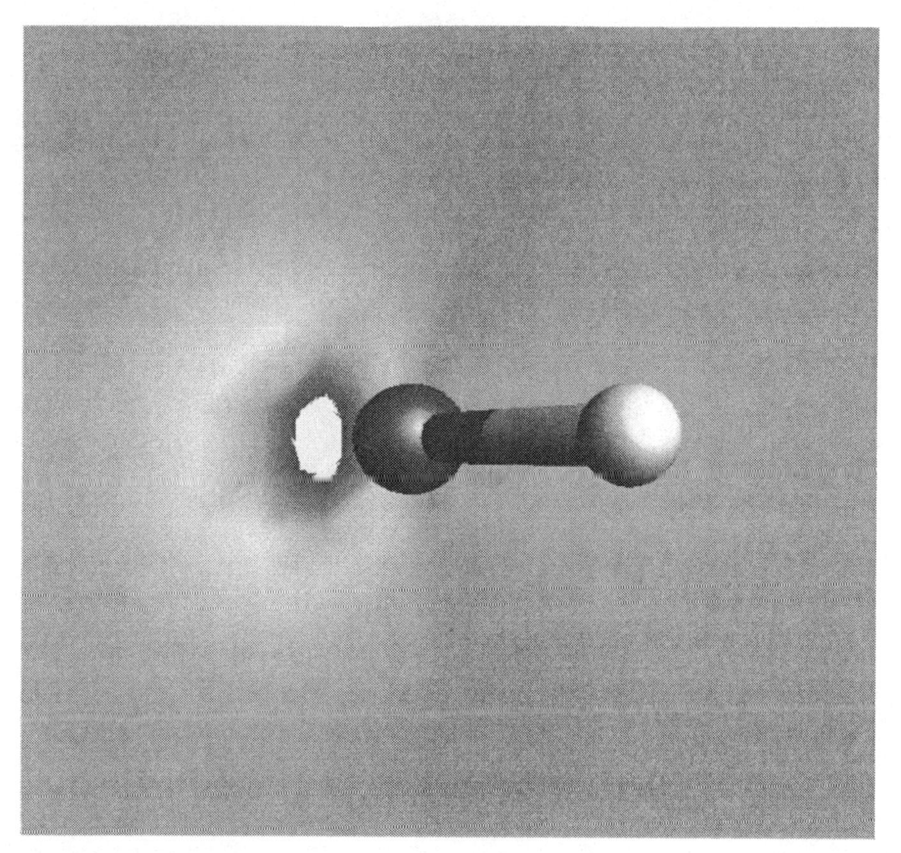

FIGURE 13.2 Colorizaton of the HOMO-1 orbital of H_2O. Colorizations often use a rainbow palette of colors.

13.15.2 Data Space

There are ways to plot data with several pieces of data at each point in space. One example would be an isosurface of electron density that has been colorized to show the electrostatic potential value at each point on the surface (Figure 13.6). The shape of the surface shows one piece of information (i.e., the electron density), whereas the color indicates a different piece of data (i.e., the electrostatic potential). This example is often used to show the nucleophilic and electrophilic regions of a molecule.

Vector quantities, such as a magnetic field or the gradient of electron density, can be plotted as a series of arrows. Another technique is to create an animation showing how the path is followed by a hypothetical test particle. A third technique is to show flow lines, which are the path of steepest descent starting from one point. The flow lines from the bond critical points are used to partition regions of the molecule in the AIM population analysis scheme.

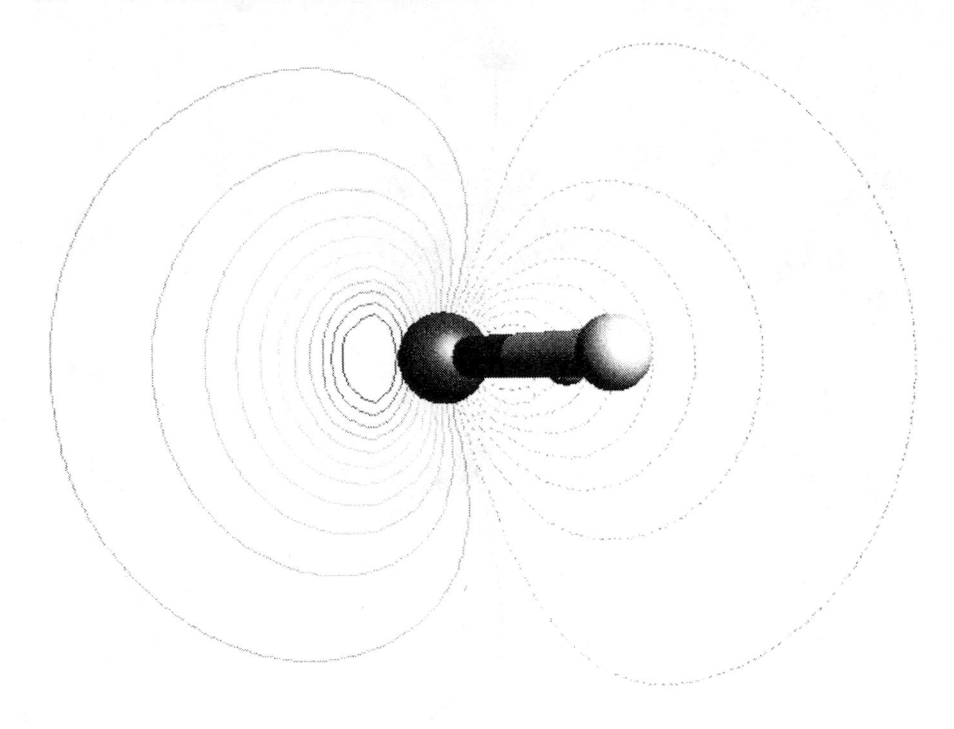

FIGURE 13.3 Contour plot of the HOMO-1 orbital of H_2O.

One technique for high dimensional data is to reduce the number of dimensions being plotted. For example, one slice of a three-dimensional data set can be plotted with a two-dimensional technique. Another example is plotting the magnitude of vectors rather than the vectors themselves.

13.15.3 Software Concerns

The quality of a final image depends on a number of things. Most visualization techniques draw a continuous surface or line by interpolating between data points in the input data. This rendering will be smoother and more accurate if a larger set of input data is used. Most three-dimensional rendering algorithms used in the chemistry field incorporate a smoothing algorithm that assumes surfaces are essentially smooth curves, rather than the disjoint set of points implied by a grid of input data. Figures 13.1 through 13.6 were produced using the default grid sizes, which are usually sufficient to show the shape while minimizing the drain on computational resources. These images were created with the programs UniChem, Spartan, and MOLDEN, all of which are discussed further in Appendix A. A few programs compute the molecular properties from

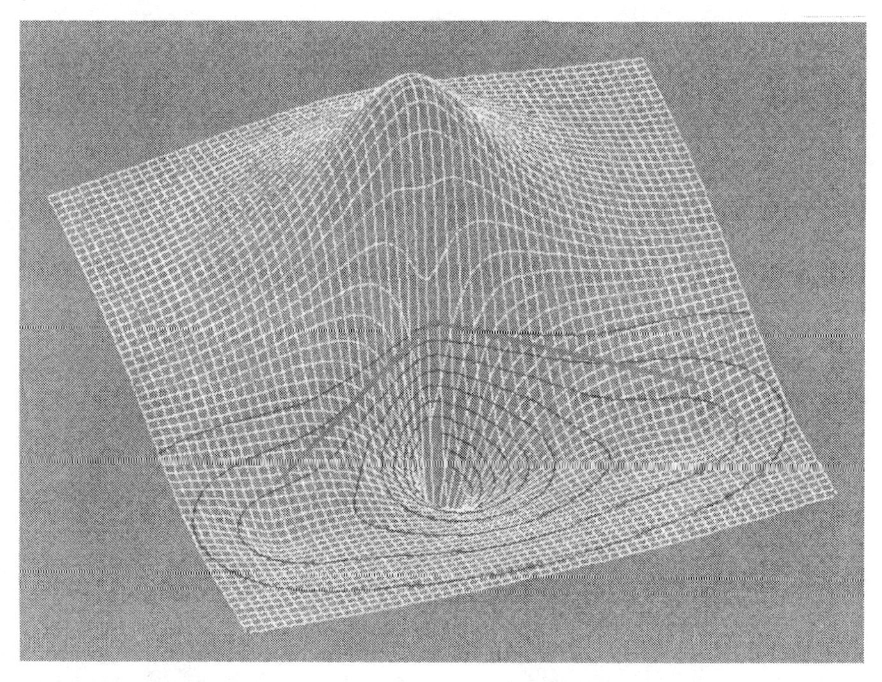

FIGURE 13.4 Mesh plot of the HOMO-1 orbital of H_2O

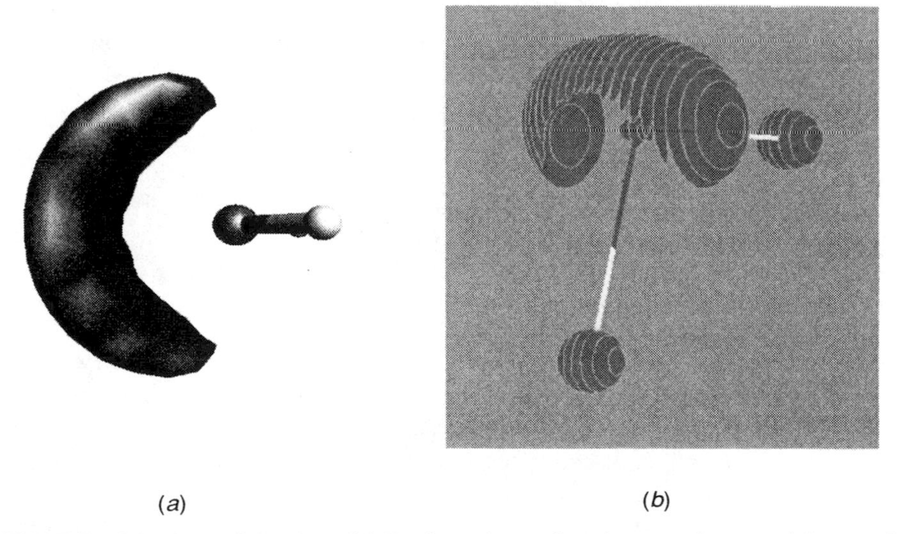

(*a*) (*b*)

FIGURE 13.5 Isosurface plots. (*a*) Region of negative electrostatic potential around the water molecule. (*b*) Region where the Laplacian of the electron density is negative. Both of these plots have been proposed as descriptors of the lone-pair electrons. This example is typical in that the shapes of these regions are similar, but the Laplacian region tends to be closer to the nucleus.

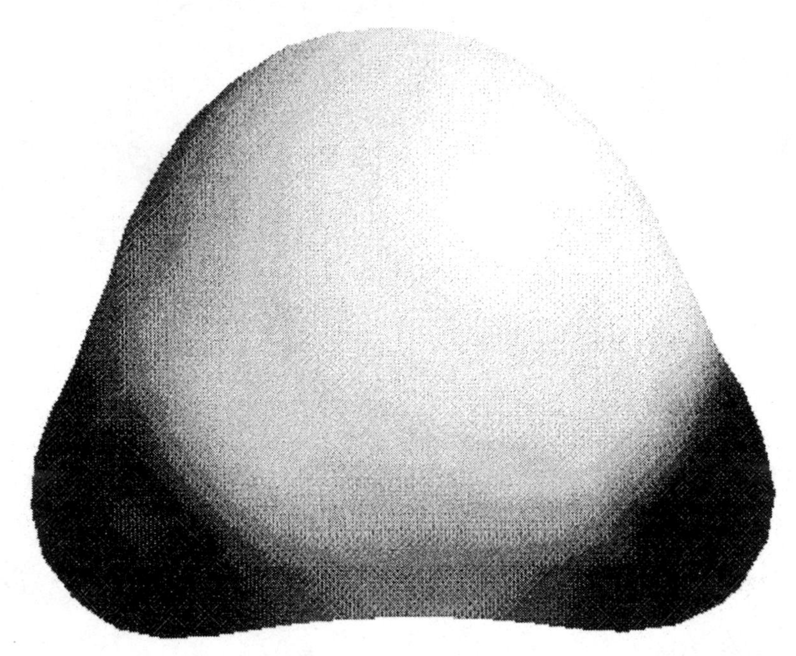

FIGURE 13.6 A plot showing two data values. The shape is an isosurface of the total electron density. The color applied to the surface is based on the magnitude of the electrostatic potential at that point in space.

the wave function and use them to create a graphical representation without the intermediate step of storing the data.

Some limitations are purely a matter of the functionality of the program. For example, some programs will not render images with a white background (generally best for publication on white paper) or include a function to save the image in a format that can be used by typical word processing software. Some programs give the user a large amount of control over the rendering settings, whereas others force the user to use a default set of options.

There is often a fundamental disparity between the graphic ability of computer monitors and that of printers. Monitors may use anywhere from 8-bit color (256 colors) to 24-bit color (16 million colors). Printers, except for dye sublimation models, use four colors, which are printed in a pattern that tricks the eye into seeing all colors. Monitors generally use about a 72-dpi (dots per inch) screen resolution, as compared to printer resolutions of 300 dpi or better.

There are also two ways to store and use image data. A raster-drawn (or bitmapped) image is one composed of many evenly spaced dots, each given a particular color. A vector-drawn image is described by lines and curves with various lengths and curvatures. Raster-drawn images are very common because they most readily created by almost any type of software. Files with gif, jpg, jpeg, and bmp extensions are raster-drawn image files. The advantage of using vector-drawn images is that they give the best possible line quality on both the computer monitor and printer. The ChemDraw program uses vector-drawn images. Postscript supports both raster and vector formats.

13.16 CONCLUSIONS

Completely *ab initio* predictions can be more accurate than any experimental result currently available. This is only true of properties that depend on the behavior of isolated molecules. Colligative properties, which are due to the interaction between molecules, can be computed more reliably with methods based on thermodynamics, statistical mechanics, structure-activity relationships, or completely empirical group additivity methods.

Empirical methods, such as group additivity, cannot be expected to be any more accurate than the uncertainty in the experimental data used to parameterize them. They can be much less accurate if the functional form is poorly chosen or if predicting properties for compounds significantly different from those in the training set.

Researchers must be particularly cautious when using one estimated property as the input for another estimation technique. This is because possible error can increase significantly when two approximate techniques are combined. Unfortunately, there are some cases in which this is the only available method for computing a property. In this case, researchers are advised to work out the error propagation to determine an estimated error in the final answer.

An example of using one predicted property to predict another is predicting the adsorption of chemicals in soil. This is usually done by first predicting an octanol water partition coefficient and then using an equation that relates this to soil adsorption. This type of property–property relationship is most reliable for monofunctional compounds. Structure–property relationships, and to a lesser extent group additivity methods, are more reliable for multifunctional compounds than this type of relationship.

BIBLIOGRAPHY

Books pertinent to this chapter are

E. J. Baum, *Chemical Property Estimation: Theory and Applications* Lewis, Boca Raton (1998).

A. R. Leach, *Molecular Modelling Principles and Applications* Longman, Essex (1996).

G. A. Jeffrey, J. F. Piniela, *The Application of Charge Density Research to Chemistry and Drug Design* Plenum, New York (1991).

W. J. Lyman, W. F. Reehl, D. H. Rosenblatt, *Handbook of Chemical Property Estimation Methods* American Chemical Society, Washington (1990).

C. E. Dykstra, *Ab Initio Calculation of the Structures and Properties of Molecules* Elsevier, Amsterdam (1988).

R. C. Reid, J. M. Prausnitz, *The Properties of Gases and Liquids* McGraw Hill, New York (1987).

M. P. Allen, D. J. Tildesley, *Computer Simulation of Liquids* Oxford, Oxford (1987).

Chemical Applications of Atomic and Molecular Electrostatic Potentials P. Politzer, D. G. Truhlar, Eds., Plenum, New York (1981).

G. J. Janz, *Thermodynamic Properties of Organic Compounds* Academic Press, New York (1967).

Reviews of using artificial intelligence to predict properties are

J. Zupan, J. Gasteiger, *Neural Networks in Chemistry and Drug Design* John Wiley & Sons, New York (1999).

D. H. Rouvray, *Fuzzy Logic in Chemistry* Academic Press, San Diego (1997).

F. J. Smith, M. Sullivan, J. Collis, S. Loughlin, *Adv. Quantum Chem.* **28**, 319 (1997).

W. Duch, *Adv. Quantum Chem.* **28**, 329 (1997).

P. L. Kilpatrick, N. S. Scott, *Adv. Quantum Chem.* **28**, 345 (1997).

H. M. Cartwright, *Applications of Artificial Intelligence in Chemistry* Oxford, Oxford (1993).

Reviews of dielectric constant prediction are

S. W. de Leeuw, J. W. Perrom, E. R. Smith, *Ann. Rev. Phys. Chem.* **37**, 245 (1986).

B. J. Alder, E. L. Pollock, *Ann. Rev. Phys. Chem.* **32**, 311 (1981).

A review of electron affinity and ionization potential prediction are

J. S. Murray, P. Politzer, *Theoretical Organic Chemistry* C. Párkányi, Ed., 189, Elsevier, Amsterdam (1998).

G. L. Gutsev, A. I. Boldyrev, *Adv. Chem. Phys.* **61**, 169 (1985).

A review of electrostatic potential analysis is

T. Brinck, *Theoretical Organic Chemistry* C. Párkányi, Ed., 51, Elsevier, Amsterdam (1998).

P. Politzer, J. S. Murray, *Rev. Comput. Chem.* **2**, 273 (1991).

E. Scrocco, J. Tomasi, *Adv. Quantum Chem.* **11**, 116 (1978).

Hardness and electronegativity are addressed in

R. G. Pearson, *Theoretical Models of Chemical Bonding Part 2* Z. B. Maksić, Ed., Springer-Verlag, Berlin (1990).

Reviews of hyperfine coupling prediction are

L. A. Eriksson, *Encycl. Comput. Chem.* **2**, 952 (1998).

S. M. Blinder, *Adv. Quantum Chem.* **2**, 47 (1965).

Reviews of multipole computation are

C. Párkányi, J.-J. Aaron, *Theoretical Organic Chemistry* C. Párkányi, Ed., 233, Elsevier, Amsterdam (1998).

I. B. Bersuker, I. Y. Ogurtsov, *Adv. Quantum Chem.* **18**, 1 (1986).

S. Green, *Adv. Chem. Phys.* **25**, 179 (1974).

Articles addressing optical activity computation are

A. Rauk, *Encycl. Comput. Chem.* **1**, 373 (1998).

R. K. Kondru, P. Wipf, D. N. Beratan, *Science* **282**, 2247 (1998).

D. Yang, A. Rouk, *Rev. Comput. Chem.* **7**, 261 (1996).

F. S. Richardson, *Chem. Rev.* **79**, 17 (1979).

A. Moscowitz, *Adv. Chem. Phys.* **4**, 67 (1962).

I. Tinoco, Jr., *Adv. Chem. Phys.* **4**, 113 (1962).

Reviews of molecular similarity are

R. Carbó, B. Calabuig, L. Vera, E. Besalú, *Adv. Quantum Chem.* **25**, 253 (1994).

Reviews of molecular size and volume are

M. L. Connolly, *Encycl. Comput. Chem.* **3**, 1698 (1998).

A. Y. Meyer, *Chem. Soc. Rev.* **15**, 449 (1986).

A review of chemical toxicity prediction is

D. F. V. Lewis, *Rev. Comput. Chem.* **3**, 173 (1992).

A review of Walsh diagram prediction is

R. J. Buenker, S. D. Peyerimhoff, *Chem. Rev.* **74**, 127 (1974).

Reviews of visualization techniques are

T. E. Ferrin, T. E. Klein, *Encycl. Comput. Chem.* **1**, 463 (1998).

J. Brickmann, T. Exner, M. Keil, R. Marhöfer, G. Moeckel, *Encycl. Comput. Chem.* **3**, 1678 (1998).

P. G. Mezey, *Encycl. Comput. Chem.* **4**, 2582 (1998).

Chemical Hardness K. D. Sen, Ed., Springer-Verlag, Berlin (1993).

P. G. Mezey, *Rev. Comput. Chem.* **1**, 265 (1990).

S. Andersson, S. T. Hyde, K. Larsson, S. Lidin, *Chem. Rev.* **88**, 221 (1988).

Electronegativity K. D. Sen, C. K. Jorgensen, Eds., Springer-Verlag, Berlin (1987).

A. Hinchliff, *Ab Initio Determination of Molecular Properties* Adam Hilger, Bristol (1987).

R. F. Hout, Jr., W. J. Pietro, W. J. Hehre, *A Pictorial Approach to Molecular Structure and Reactivity* John Wiley & Sons, New York (1984).

P. Coppens, E. D. Steven, *Adv. Quantum Chem.* **10**, 1 (1977).

A. Hinchliffe, J. C. Dobson, *Chem. Soc. Rev.* **5**, 79 (1976).

W. L. Jorgensen, *The Organic Chemists Book of Orbitals* Academic Press, New York (1973).

A review of visualizing reaction coordinates is

O. M. Becker, *J. Comput Chem.* **19**, 1255 (1998).

Other relevant reviews are

C. E. Dykstra, S.-Y. Liu, D. J. Malik, *Adv. Chem. Phys.* **75**, 37 (1989).

L. Engelbrecht, J. Hinze, *Adv. Chem. Phys.* **44**, 1 (1980).

J. R. Van Wazer, I. Absar, *Electron Densities in Molecules and Molecular Orbitals* Academic Press, New York (1975).

G. G. Hall, *Adv. Quantum Chem.* **1**, 241 (1964).

14 The Importance of Symmetry

This chapter discusses the application of symmetry to orbital-based computational chemistry problems. A number of textbooks on symmetry are listed in the bibliography at the end of this chapter.

The symmetry of a molecule is defined by determining how the nuclei can be exchanged without changing its identity, conformation, or chirality. For example, a methane molecule can be turned about the axis connecting the carbon and one of the hydrogens by 120° and it is indistinguishable from the original orientation. Alternatively, symmetry can be considered a way of determining which regions of space around the molecule are completely equivalent. This second description is important because it indicates a means for calculations to be performed more quickly.

In order to obtain this savings in the computational cost, orbitals are symmetry adapted. As various positive and negative combinations of orbitals are used, there are a number of ways to break down the total wave function. These various orbital functions will obey different sets of symmetry constraints, such as having positive or negative values across a mirror plane of the molecule. These various symmetry sets are called irreducible representations.

Molecular orbitals are not unique. The same exact wave function could be expressed an infinite number of ways with different, but equivalent orbitals. Two commonly used sets of orbitals are localized orbitals and symmetry-adapted orbitals (also called canonical orbitals). Localized orbitals are sometimes used because they look very much like a chemist's qualitative models of molecular bonds, lone-pair electrons, core electrons, and the like. Symmetry-adapted orbitals are more commonly used because they allow the calculation to be executed much more quickly for high-symmetry molecules. Localized orbitals can give the fastest calculations for very large molecules without symmetry due to many long-distance interactions becoming negligible.

Another reason that symmetry helps the calculations is that the Hamiltonian matrix will become a block diagonal matrix with one block for each irreducible representation. It is not necessary for the program to compute overlap integrals between orbitals of different irreducible representations since the overlap integrals will be zero by symmetry. Some computer programs go to the length of actually completing an SCF calculation as a number of small Hamiltonian matrices for each irreducible representation rather than one large Hamiltonian matrix. When this is done, the number of orbitals of each irreducible representation that are occupied must be defined at the beginning of the calculation

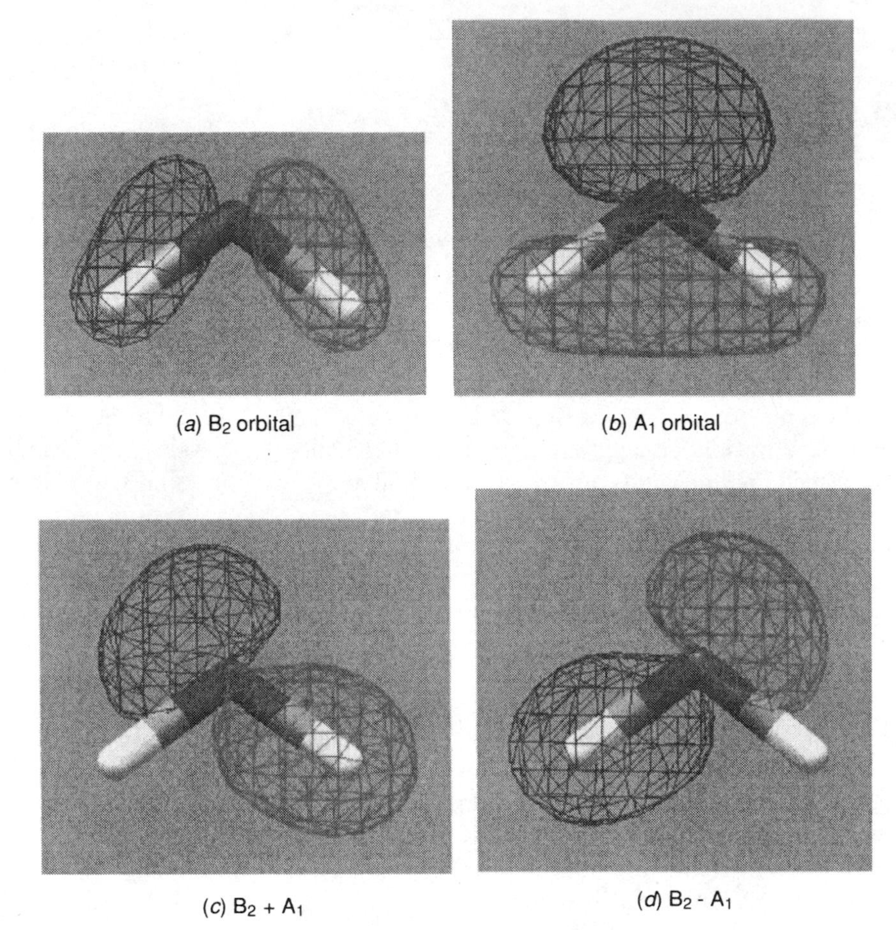

FIGURE 14.1 An illustration of symmetry-adapted vs. localized orbitals for water. (*a*, *b*) B_2 and A_1 symmetry-adapted orbitals. (*c*) Sum of these orbitals, which gives a localized orbital that is the bond between the oxygen and the hydrogen on the right. (*d*) Difference between these orbitals, which gives a localized orbital that is the bond between the oxygen and the hydrogen on the left.

either by the user or the program. Electrons will not be able to shift from one symmetry to another. Thus, the ground-state wave function cannot be determined unless the correct electron symmetry assignments were chosen originally. This can be an advantage since it is a way of obtaining excited-state wave functions, and fewer SCF convergence problems arise with this algorithm. The disadvantage is that the automated assignment algorithm may not find the correct ground state.

There is always a transformation between symmetry-adapted and localized orbitals that can be quite complex. A simple example would be for the bonding orbitals of the water molecule. As shown in Figure 14.1, localized orbitals can be constructed from positive and negative combinations of symmetry-adapted orbitals. Some computer programs have localization algorithms that are more sophisticated than this.

14.1 WAVE FUNCTION SYMMETRY

In SCF problems, there are some cases where the wave function must have a lower symmetry than the molecule. This is due to the way that the wave function is constructed from orbitals and basis functions. For example, the carbon monoxide molecule might be computed with a wave function of C_{4v} symmetry even though the molecule has a $C_{\infty v}$ symmetry. This is because the orbitals obey C_{4v} constraints.

Most programs that employ symmetry-adapted orbitals only use Abelian symmetry groups. Abelian groups are point groups in which all the symmetry operators commute. Often, the program will first determine the molecules' symmetry and then use the largest Abelian subgroup. To our knowledge, the only software package that can utilize non-Abelian symmetry groups is Jaguar.

14.2 TRANSITION STRUCTURES

Transition structures can be defined by nuclear symmetry. For example, a symmetric S_N2 reaction will have a transition structure that has a higher symmetry than that of the reactants or products. Furthermore, the transition structure is the lowest-energy structure that obeys the constraints of higher symmetry. Thus, the transition structure can be calculated by forcing the molecule to have a particular symmetry and using a geometry optimization technique.

BIBLIOGRAPHY

T. Bally, W. T. Borden, *Rev. Comput. Chem.* **13**, 1 (1999).

J. Rosen, *Symmetry in Science* Springer-Verlag, New York (1996).

G. Davidson, *Group Theory for Chemists* MacMillan, Hampshire (1991).

A. Cotton, *Chemical Applications of Group Theory* John Wiley and Sons, New York (1990).

J. R. Ferraro, J. S. Ziomek, *Introductory Group Theory* Plenum, New York (1975).

D. M. Bishop, *Group Theory and Chemistry* Clarendon, Oxford (1973).

M. Tinkham, *Group Theory and Quantum Mechanics* McGraw-Hill, New York (1964).

R. McWeeny, *Symmetry; An Introduction to Group Theory and its Applications* Pergamon, New York (1963).

15 Efficient Use of Computer Resources

Many computational chemistry techniques are extremely computer-intensive. Depending on the type of calculation desired, it could take anywhere from seconds to weeks to do a single calculation. There are many calculations, such as *ab initio* analysis of biomolecules, that cannot be done on the largest computers in existence. Likewise, calculations can take very large amounts of computer memory and hard disk space. In order to complete work in a reasonable amount of time, it is necessary to understand what factors contribute to the computer resource requirements. Ideally, the user should be able to predict in advance how much computing power will be needed.

There are often trade-offs between equivalent ways of doing the same calculation. For example, many *ab initio* programs use hard disk space to store numbers that are computed once and used several times during the course of the calculation. These are the integrals that describe the overlap between various basis functions. Instead of the above method, called conventional integral evaluation, it is possible to use direct integral evaluation in which the numbers are recomputed as needed. Direct integral evaluation algorithms use less disk space at the expense of requiring more CPU time to do the calculation. An *in-core* algorithm is one that stores all the integrals in RAM memory, thus saving on disk space at the expense of requiring a computer with a very large amount of memory. Many programs use a semidirect algorithm, which uses some disk space and a bit more CPU time to obtain the optimal balance of both.

15.1 TIME COMPLEXITY

Time complexity is a way of denoting how the use of computer resources (CPU time, memory, etc.) changes as the size of the problem changes. For example, consider a HF calculation with N orbitals. At the end of the calculation, the orbital energies must be added. Since there are N orbitals, there will be N addition operations. There are a certain number of operations, which we will call C, which have to be done regardless of the size of calculation, such as initializing variables and allocating memory. The standard matrix inversion algorithm requires N^3 operations. Computing the two-electron Coulomb and exchange integrals for a HF calculation takes N^4 operations. Thus, the total amount of CPU time required to do a HF calculation scales as $N^4 + N^3 + N + C$. However for sufficiently large N, the N^3, N, and C terms are insignificant compared to the size of the N^4 term (if $N = 10$, then N^3 is 10% of N^4, and for $N = 100$, N^3 is 1% of N^4, etc.). Computer scientists described this type of algorithm as one that has a time complexity of N^4, denoted with the notation $O(N^4)$. A similar analysis could be done for memory use, disk use, network communication volume, and so on.

These relationships can be used to estimate the amount of CPU time, disk space, or memory needed to run calculations. Let us take the example of a researcher, Jane Chemist, who would like to compute some property of a polymer. She first examines the literature to determine that an *ab initio* method with a moderately large basis set will give the desired accuracy of results. She then runs both single point and geometry optimization calculations on the monomer, which take 2 and 20 minutes, respectively. Since the calculation scales as N^4, a geometry optimization for the trimer, which has three times as many atoms, will take approximately $3^4 * 20$ minutes or about 27 hours. Jane would like to model up to a 15-unit chain, which would require $15^4 * 20$ minutes or about 2 years. Obviously, the use of *ab initio* methods for geometry optimization is not acceptable. Jane then wisely decides to stop at the 10-unit chain and use geometries optimized with molecular mechanics methods, which takes under an hour for the optimization. She then obtains the desired results with single-point *ab initio* calculations, which take $10^4 * 2$ minutes or 2 weeks for the largest molecule. This final calculation is still rather large, but it is feasible since Jane has her own work station with an uninterruptable power supply.

There are many different algorithms that can be programmed in a more or less efficient manner to obtain the exact same result. Because of this, a given method will have slightly different time complexities from one program to another. Table 15.1 gives some common time complexities. M is the number of atoms, L the length of one side of the box containing the molecules in a calculation using periodic boundary conditions, A the number of active space orbitals, and N the number of orbitals in the calculation. Thus, N can increase either by including more atoms or using a larger basis set.

Geometry optimization calculations take much longer than single-energy calculations. The reasons are two-fold: First, many calculations must be done as the geometry is changed. Second, each iteration takes longer in order to compute energy gradients. The amount of CPU time required for a geometry optimization, T_{opt}, depends on the number of degrees of freedom, denoted as D. Degrees of freedom are the geometric variables being optimized, such as bond lengths, angles, and the like. As a general rule of thumb, the amount of time for a geometry optimization can be estimated from the single-point energy CPU time, T_{single}, with the equation

$$T_{opt} \cong 5 * D^2 * T_{single} \tag{15.1}$$

For *ab initio*, semiempirical, or molecular dynamics calculations, the amount of CPU time necessary is generally the factor of greatest concern to researchers. For very large molecules, memory use is of concern for molecular mechanics

TABLE 15.1 Method Time Complexities

Method	Scaling	Comments
DFT	N	With linear scaling algorithms (very large molecules)
MM	M^2	
MD	M^2 or L^6	
Semiempiricals	N^2	For small- to medium-size molecules (limited by integrals)
HF	N^2–N^4	Depending on use of symmetry and integral accuracy cutoffs
Semiempiricals	N^3	For very large molecules (limited by matrix inversion)
HF	N^3	Pseudospectral method
DFT	N^3	
QMC	N^3	With inverse slater matrix
MP2	N^5	
CC2	N^5	
MP3, MP4(SDQ)	N^6	
CCSD	N^6	
CISD	N^6	
MP4	N^7	
CC3, CCSD(T)	N^7	
MP5	N^8	
CISDT	N^8	
CCSDT	N^8	
MP6	N^9	
MP7	N^{10}	
CISDTQ	N^{10}	
CASSCF	$A!$	A is the number of active space orbitals.
Full CI	$N!$	
QMC	$N!$	Without inverse slater matrix

calculations. Table 15.2 lists the memory requirements of a number of geometry optimization algorithms.

The time complexity only indicates how CPU time increases with larger systems. Even for a small job, more complex algorithms will take more CPU time. As a general rule of thumb, methods with a larger time complexity will also require more memory and disk storage space. However, from one software package to another, there are trade-offs between the amount of memory, CPU time, and disk space required. A few exceptions should be mentioned. HF algorithms usually have N^2 memory use, except for in-core algorithms that have N^4 memory use. QMC calculations require extremely large amounts of CPU time for even very small molecules, but require very little memory or disk space.

Because geometry optimization is so much more time-consuming than a single geometry calculation, it is common to use different levels of theory for the optimization and computing final results. For example, an *ab initio* method with a moderate-size basis set and minimal correlation may be used for opti-

TABLE 15.2 Optimization Algorithm Memory Use

Algorithm	Memory Use
Conjugate gradient	D
Fletcher–Reeves	D
Polak–Ribiere	D
Simplex	D^2
Powell	D^2
quasi-Newton	D^2
Fletcher–Powell	D^2
BFGS	D^2

mization and then a single point calculation with more correlation and a larger basis can be used for the final energy computation. This would be denoted with a notation like MP2/6–31G*//ccsd(t)/cc–pVTZ. In some cases, molecular mechanics or semiempirical calculations may be used to determine a geometry for an *ab initio* calculation. Molecular mechanics is nearly always used for conformation searching. One exception to this is that vibrational frequencies must be computed with the same level of theory used to optimize the geometry.

Note that many programs are now optimized for direct integral performance to the point that it outperforms conventional methods on current hardware configurations. An example of this is seen in Table 15.3, which was generated using the Gaussian 98 program. This program uses the available memory so the memory use is a function of the execution queue rather than the minimum

TABLE 15.3 Benzene Single Point Calculation Tests Using the cc–*p*VTZ Basis Set (run on a Cray SV1 configured with J90 CPUs computed with Gaussian 98 revision A.7)

Method	CPU (seconds)	Memory (megawords)	File Space (megabytes)
PM3	11	14.9	11
HF direct	586	14.9	42
HF nosym	3394	14.9	42
HF conv	8783	30.4	1165
HF incore	—	>1000	—
HF QC	6747	14.9	20
MP2	1921	30.4	43
MP3	6413	30.4	1465
MP4	131,363	30.4	2146
CISD	26,143	59.2	1923
CCSD	40,393	59.2	2604
G2	132,812	60.9	2462
CBS–APNO	225,079	60.9	6505
QCISD	35,527	59.2	1923
G3	73,401	60.9	1574

needed, provided that it is above a minimum size. Please note that Table 15.3 gives an example for one molecule only, not an average or expected performance.

15.2 LABOR COST

Another important consideration is the amount of labor necessary on the part of the user. One major difference between different software packages is the developer's choices between ease of use and efficiency of operation. For example, the Spartan program is extremely easy to use, but the price for this is that the algorithms are not always the most efficient available. Many chemistry users begin with software that is very simple, but when more sophisticated problems need to be solved, it is often easier to learn to use more complicated software than to purchase a supercomputer to solve a problem that could be done by a workstation with different software.

15.3 PARALLEL COMPUTERS

Mass-produced workstation-class CPUs are much cheaper than traditional supercomputer processors. Thus, a larger amount of computing power for the dollar can be purchased by buying a parallel supercomputer that might have hundreds of workstation CPUs.

Software written for single-processor computers will not automatically use multiple CPUs. At the present time, there are compilers that will attempt to parallelize computer algorithms, but these compilers are usually inefficient for sophisticated computer programs. In time, such compilers will become so good that any program will be parallelized with no additional work, just as optimizing compilers have completely eliminated the need to write applications entirely in machine node. Until that day comes, the person using computational chemistry programs will have to understand the performance of parallel software in order to efficiently do his or her work.

Ideally, a calculation that takes an hour on a single CPU would take half an hour on two CPUs. This is called linear speed-up. In practice, this is not possible because the two CPU calculations must do extra work to divide the workload between the two processors and combine results to obtain the final answer. There are a few types of algorithms that give nearly perfectly linear scaling because of the nature of the algorithm and the amount of work that the developer did to parallelize the code. Many Monte Carlo algorithms can be parallelized very efficiently. There are also a few programs for which our hypothetical hour calculation would take $1\frac{1}{2}$ hours on a two-CPU machine due to the incredibly inefficient way that the parallelization was implemented. Some of the correlated *ab initio* algorithms are very difficult to parallelize efficiently. Most parallelized programs fall somewhere between these two extremes. Different methods within a given software package are often parallelized to different degrees of efficiency.

Some software packages can be run on a networked cluster of workstations as though they were a multiple-processor machine. However, the speed of data transfer across a network is not as fast as the speed of data transfer between the CPUs of a parallel computer. Some algorithms break down the work to be done into very large chunks with a minimal amount of communication between processors. These are called large-grained algorithms, and they work as well on a cluster of workstations as on a parallel computer. Fine-grained algorithms require a significant amount of frequent communication between CPUs and run slowly on a cluster of workstations because the network speed limits the calculation more than the performance of the CPUs. There are also differences in communication speed between parallel computers made by various vendors, which can sometimes have a significant effect on how quickly calculations run.

BIBLIOGRAPHY

The manuals accompanying some software packages discuss the efficiency of algorithms used in that software. Efficiency is also discussed in

A. Frisch, M. J. Frisch, *Gaussian 98 User's Reference* Gaussian, Pittsburgh (1999).

F. Jensen, *Introduction to Computational Chemistry* John Wiley & Sons, New York (1999).

W. J. Hehre, *Practical Strategies for Electronic Structure Calculations* Wavefunction, Irvine (1995).

D. F. Feller, *MSRC Ab Initio Methods Benchmark Suite—A Measurement of Hardware and Software Performance in the Area of Electronic Structure Methods* WA Battelle Pacific Northwest Labs, Richland (1993).

E. R. Davidson, *Rev. Comput. Chem.* **1**, 373 (1990).

W. H. Press, B. P. Flannery, S. A. Teukolsky, W. T. Vetterling, *Numerical Recipes* Cambridge University Press, Cambridge (1989).

M. P. Allen, D. J. Tildesley, *Computer Simulation of Liquids* Oxford, Oxford (1987).

W. J. Hehre, L. Radom, P. v.R. Schleyer, J. A. Pople, *Ab Initio Molecular Orbital Theory* John Wiley & Sons, New York (1986).

J. G. Csizmadia, R. Daudel, *Computational Theoretical Organic Chemistry* Reidel, Dordrecht (1981).

Computatons using parallel computers are discussed in

R. A. Kendall, R. J. Harrison, R. J. Littlefield, M. F. Guest, *Rev. Comput. Chem.* **6**, 209 (1995).

Reference for the Gaussian 98 software used to generate the data in Table 15.3

Gaussian 98, Revision A.7, M. J. Frisch, G. W. Trucks, H. B. Schlegel, G. E. Scuseria, M. A. Robb, J. R. Cheeseman, V. G. Zakrzewski, J. A. Montgomery, Jr., R. E. Stratmann, J. C. Burant, S. Dapprich, J. M. Millam, A. D. Daniels, K. N. Kudin, M. C. Strain, O. Farkas, J. Tomasi, V. Barone, M. Cossi, R. Cammi, B. Mennucci,

C. Pomelli, C. Adamo, S. Clifford, J. Ochterski, G. A. Petersson, P. Y. Ayala, Q. Cui, K. Morokuma, D. K. Malick, A. D. Rabuck, K. Raghavachari, J. B. Foresman, J. Cioslowski, J. V. Ortiz, A. G. Baboul, B. B. Stefanov, G. Liu, A. Liashenko, P. Piskorz, I. Komaromi, R. Gomperts, R. L. Martin, D. J. Fox, T. Keith, M. A. Al-Laham, C. Y. Peng, A. Nanayakkara, C. Gonzalez, M. Challacombe, P. M. W. Gill, B. Johnson, W. Chen, M. W. Wong, J. L. Andres, C. Gonzalez, M. Head-Gordon, E. S. Replogle, and J. A. Pople, Gaussian, Inc., Pittsburgh PA, 1998.

16 How to Conduct a Computational Research Project

When using computational chemistry to answer a chemical question, the obvious problem is that the researcher needs to know how to use the software. The difficulty sometimes overlooked is that one must estimate how accurate the answer will be in advance. The sections below provide a checklist to follow.

16.1 WHAT DO YOU WANT TO KNOW? HOW ACCURATELY? WHY?

If you cannot specifically answer these questions, then you have not formulated a proper research project. The choice of computational methods must be based on a clear understanding of both the chemical system and the information to be computed. Thus, all projects start by answering these fundamental questions in full. The statement "To see what computational techniques can do." is not a research project. However, it is a good reason to purchase this book.

16.2 HOW ACCURATE DO YOU PREDICT THE ANSWER WILL BE?

In analytical chemistry, a number of identical measurements are taken and then an error is estimated by computing the standard deviation. With computational experiments, repeating the same step should always give exactly the same result, with the exception of Monte Carlo techniques. An error is estimated by comparing a number of similar computations to the experimental answers or much more rigorous computations.

There are numerous articles and references on computational research studies. If none exist for the task at hand, the researcher may have to guess which method to use based on its assumptions. It is then prudent to perform a short study to verify the method's accuracy before applying it to an unknown. When an expert predicts an error or best method without the benefit of prior related research, he or she should have a fair amount of knowledge about available options: A savvy researcher must know the merits and drawbacks of various methods and software packages in order to make an informed choice. The bibliography at the end of this chapter lists sources for reviewing accuracy data. Appendix A of this book provides short reviews of many software packages.

There are many studies that examine the accuracy of computational techniques for modeling a particular compound or set of related compounds. It is far less often that a study attempts to quantify the accuracy of a method as a whole. This is usually done by giving some sort of average error for a large collection of molecules. Most studies of this type have focused on collections of organic and light main group compounds. Table 16.1 lists some of the accuracy data that have been published from such large studies. The sources of this data are listed in the bibliography for this chapter. Some of the online databases allow the user to specify individual molecules, or select a set of molecules and then retrieve accuracy data.

16.3 HOW LONG DO YOU PREDICT THE RESEARCH WILL TAKE?

If the world were perfect, a researcher could tell her PC to calculate the exact solution to the Schrödinger equation and continue with the rest of her work. However, the exact solution to the Schrödinger equation has not yet been found and *ab initio* calculations approaching it for moderate-size molecules would be so time-consuming that it might take a decade to do a single calculation, if a machine with enough memory and disk space were available. However, many methods exist because each is best for some situation. The trick is to determine which one is best for a given project. The first step is to predict which method will give an acceptable accuracy. If timing information is not available, use the scaling information as described in the previous chapter to predict how long the calculation will take.

16.4 WHAT APPROXIMATIONS ARE BEING MADE? WHICH ARE SIGNIFICANT?

This is a check on the reasonableness of the method chosen. For example, it would not be reasonable to select a method to investigate vibrational motions that are very anharmonic with a calculation that uses a harmonic oscillator approximation. To avoid such mistakes, it is important the researcher understand the method's underlying theory.

Once all these questions have been answered, the calculations can begin. Now the researcher must determine what software is available, what it costs, and how to properly use it. Note that two programs of the same type (i.e., *ab initio*) may calculate different properties so the user must make sure the program does exactly what is needed.

When learning how to use a program, dozens of calculations may fail because the input was constructed incorrectly. Do not use the project molecule to do this. Make mistakes with something inconsequential, like a water molecule.

TABLE 16.1 Accuracies of Computational Chemistry Methods Relative to Experimental Results

Method	Property	Accuracy
MM2	ΔH_f^0	0.5 kcal/mol std. dev.
	Bond length	0.01 Å std. dev.
	Bond angle	1.0° std. dev.
	Dihedral angle	8.0° std. dev.
	Dipole	0.1 D std. dev.
MM3	ΔH_f^0	0.6 kcal/mol std. dev.
	Bond length	0.01 Å std. dev.
	Bond angle	1.0° std. dev.
	Dihedral angle	5.0° std. dev.
	Dipole	0.07 D std. dev.
CFF	ΔH_f^0	2 kcal/mol std. dev.
	Bond length	0.01 Å std. dev.
	Bond angle	1.0° std. dev.
	Sorption energy	5 kcal/mol std. dev.
CNDO	ΔH_f^0	200 kcal/mol std. dev.
INDO/1	ΔH_f^0	100 kcal/mol std. dev.
MINDO/3	ΔH_f^0	5 kcal/mol std. dev.
MNDO	ΔH_f^0	11 kcal/mol std. dev.
	Bond angle	4.3° RMS error
	Bond length	0.048 Å RMS error
	Dipole	0.3 D std. dev.
	IP	0.8 eV std. dev.
MNDO/*d*	ΔH_f^0	5 kcal/mol std. dev.
	Dipole	0.4 D std. dev.
	IP	0.6 eV std. dev.
AM1	ΔH_f^0	8 kcal/mol std. dev.
	Total energy	18.8 kcal/mol mean abs. dev.
	Bond angle	3.3° RMS error
	Bond length	0.048 Å RMS error
	Dipole	0.5 D std. dev.
	IP	0.6 eV std. dev.
PM3	ΔH_f^0	8 kcal/mol std. dev.
	Total energy	17.2 kcal/mol mean abs. dev.
	Bond angle	3.9° RMS error
	Bond length	0.037 Å RMS error
	Dipole	0.6 D std. dev.
	IP	0.7 eV std. dev.
SAM1	ΔH_f^0	8 kcal/mol std. dev.
	IP	0.4 eV std. dev.
SVWN	Dipole	0.1 D std. dev.
SVWN/3–21G(*)	Bond angle	2.0° RMS error
	Bond length	0.033 Å RMS error
SVWN/6–31G*	Bond angle	1.4° RMS error
	Bond length	0.023 Å RMS error

TABLE 16.1 (Continued)

Method	Property	Accuracy
SVWN/6−311+G(2*d*,*p*)	Total energy	19.2 kcal/mol std. dev.
	Bond angle	1.4° RMS error
	Bond length	0.021 Å RMS error
SVWN/6−311+G(3*df*,2*p*)	IP	0.594 eV mean abs. dev.
	EA	0.697 eV mean abs. dev.
SVWN5/6−311+G(2*d*,*p*)	Total energy	18.1 kcal/mol mean abs. dev.
BLYP/6−31G**	Reaction energy	9.95 kcal/mol mean abs. dev.
BLYP/6−31+G(*d*,*p*)	Total energy	3.9 kcal/mol mean abs. dev.
BLYP/6−311+G(2*d*,*p*)	Total energy	3.9 kcal/mol mean abs. dev.
BLYP/6−311+G(3*df*,2*p*)	IP	0.260 eV mean abs. dev.
	EA	0.113 eV mean abs. dev.
BLYP/DZVP	Reaction energy	7.73 kcal/mol mean abs. dev.
BP86/6−311+G(3*df*,2*p*)	IP	0.198 eV mean abs. dev.
	EA	0.193 eV mean abs. dev.
BP91/6−31G**	Reaction energy	9.35 kcal/mol mean abs. dev.
BP91/DZVP	Reaction energy	6.91 kcal/mol mean abs. dev.
BPW91/6−311+G(3*df*,2*p*)	IP	0.220 eV mean abs. dev.
	EA	0.121 eV mean abs. dev.
B3LYP	ΔH_f^0	2 kcal/mol std. dev.
B3LYP/3−21G(*)	Bond angle	2.0° RMS error
	Bond length	0.035 Å RMS error
B3LYP/6−31G(*d*)	Total energy	7.9 kcal/mol mean abs. dev.
	Bond angle	1.4° RMS error
	Bond length	0.020 Å RMS error
B3LYP/6−31+G(*d*,*p*)	Total energy	3.9 kcal/mol mean abs. dev.
B3LYP/6−311+G(2*d*,*p*)	Total energy	3.1 kcal/mol mean abs. dev.
	Bond angle	1.4° RMS error
	Bond length	0.017 Å RMS error
B3LYP/6−311+G(3*df*,2*p*)	IP	0.177 eV mean abs. dev.
	EA	0.131 eV mean abs. dev.
B3PW91/6−311+G(3*df*,2*p*)	IP	0.191 eV mean abs. dev.
	EA	0.145 eV mean abs. dev.
HF/STO−3G	Dipole	0.5 D std. dev.
	Total energy	93.3 kcal/mol mean abs. dev.
	Bond angle	1.7° RMS error
	Bond length	0.055 Å RMS error
HF/3−21G	ΔH_f^0	7 kcal/mol std. dev.
	Dipole	0.4 D std. dev.
HF/3−21G(*d*)	Total energy	58.4 kcal/mol mean abs. dev.
	Bond angle	1.7° RMS error
	Bond length	0.032 Å RMS error
HF/6−31G*	ΔH_f^0	4 kcal/mol std. dev.
	Total energy	51.0 kcal/mol mean abs. dev.
	Dipole	0.2 D std. dev.
	Bond angle	1.4° RMS error
	Bond length	0.032 Å RMS error

TABLE 16.1 (Continued)

Method	Property	Accuracy
HF/6–31G**	Reaction energy	54.2 kcal/mol mean abs. dev.
HF/6–31+G(*d*,*p*)	Total energy	46.7 kcal/mol mean abs. dev.
HF/6–311+G(2*d*,*p*)	Bond angle	1.3° RMS error
	Bond length	0.035 Å RMS error
HF/aug–cc–pVDZ	Atomization energy	85 kcal/mol mean abs. dev.
	Proton affinity	3.5 kcal/mol mean abs. dev.
	EA	26 kcal/mol mean abs. dev.
	IP	20 kcal/mol mean abs. dev.
	Bond length	0.01, 0.03 Å mean abs. dev.
	Bond angle	1.2° mean abs. dev.
	Scaled frequencies	70, 90, 110 cm^{-1} mean abs. dev.
HF/aug–cc–pVTZ	Atomization energy	67 kcal/mol mean abs. dev.
	Proton affinity	2.5 kcal/mol mean abs. dev.
	EA	28 kcal/mol mean abs. dev.
	IP	22 kcal/mol mean abs. dev.
	Bond length	0.015, 0.03 Å mean abs. dev.
	Bond angle	1.6° mean abs. dev.
	Scaled frequencies	80, 50, 110 cm^{-1} mean abs. dev.
HF/aug–cc–pVQZ	Atomization energy	63 kcal/mol mean abs. dev.
	Proton affinity	3.5 kcal/mol mean abs. dev.
	EA	28 kcal/mol mean abs. dev.
	IP	21 kcal/mol mean abs. dev.
	Bond length	0.015, 0.035 Å mean abs. dev.
	Bond angle	1.7° mean abs. dev.
	Scaled frequencies	40, 110 cm^{-1} mean abs. dev.
MP2/3–21G(*)	Bond angle	2.2° RMS error
	Bond length	0.044 Å RMS error
MP2/6–31G*	Bond angle	1.5° RMS error
	Bond length	0.048 Å RMS error
MP2/6–31G**	Reaction energy	11.86 kcal/mol mean abs. dev.
MP2/6–31+G(*d*,*p*)	Total energy	11.4 kcal/mol mean abs. dev.
MP2/6–311+G(2*d*,*p*)	Total energy	8.9 kcal/mol mean abs. dev.
MP2/aug–cc–pVDZ	Atomization energy	15 kcal/mol mean abs. dev.
	Proton affinity	3.8 kcal/mol mean abs. dev.
	EA	4 kcal/mol mean abs. dev.
	IP	5.5 kcal/mol mean abs. dev.
	Bond length	0.012, 0.038 Å mean abs. dev.
	Bond angle	0.5° mean abs. dev.
	Frequencies	50, 40, 180 cm^{-1} mean abs. dev.
MP2/aug–cc–pVTZ	Atomization energy	5 kcal/mol mean abs. dev.
	Proton affinity	2 kcal/mol mean abs. dev.
	EA	3 kcal/mol mean abs. dev.
	IP	4 kcal/mol mean abs. dev.
	Bond length	0.012, 0.021 Å mean abs. dev.
	Bond angle	0.3° mean abs. dev.
	Frequencies	60, 30, 120 cm^{-1} mean abs. dev.

TABLE 16.1 (Continued)

Method	Property	Accuracy
MP2/aug–cc–pVQZ	atomization energy	5 kcal/mol mean abs. dev.
	Proton affinity	1.8 kcal/mol mean abs. dev.
	EA	3 kcal/mol mean abs. dev.
	IP	3 kcal/mol mean abs. dev.
	Bond length	0.008, 0.015 Å mean abs. dev.
	Bond angle	0.3° mean abs. dev.
	Frequencies	40, 110 cm^{-1} mean abs. dev.
MP4(STDQ)/aug–cc–pVDZ	Atomization energy	16 kcal/mol mean abs. dev.
	Proton affinity	2.1 kcal/mol mean abs. dev.
	EA	4 kcal/mol mean abs. dev.
	IP	5.5 kcal/mol mean abs. dev.
	Bond length	0.018, 0.042 Å mean abs. dev.
	Bond angle	0.4° mean abs. dev.
	Frequencies	60, 40, 110 cm^{-1} mean abs. dev.
MP4(STDQ)/aug–cc–pVTZ	Atomization energy	4 kcal/mol mean abs. dev.
	Proton affinity	1.2 kcal/mol mean abs. dev.
	EA	2 kcal/mol mean abs. dev.
	IP	2 kcal/mol mean abs. dev.
	Bond length	0.008, 0.03 Å mean abs. dev.
	Bond angle	0.3° mean abs. dev.
	Frequencies	40, 30, 110 cm^{-1} mean abs. dev.
MP4(STDQ)/aug–cc–pVQZ	Atomization energy	2 kcal/mol mean abs. dev.
	Proton affinity	1 kcal/mol mean abs. dev.
	EA	2 kcal/mol mean abs. dev.
	IP	1 kcal/mol mean abs. dev.
	Bond length	0.008, 0.015 Å mean abs. dev.
	Bond angle	0.3° mean abs. dev.
	Frequencies	30, 110 cm^{-1} mean abs. dev.
CCSD/aug–cc–pVDZ	Atomization energy	21 kcal/mol mean abs. dev.
	Proton affinity	1.1 kcal/mol mean abs. dev.
	EA	5.5 kcal/mol mean abs. dev.
	IP	5.8 kcal/mol mean abs. dev.
	Bond length	0.017, 0.033 Å mean abs. dev.
	Bond angle	0.5° mean abs. dev.
	Frequencies	60, 40, 80 cm^{-1} mean abs. dev.
CCSD/aug–cc–pVTZ	Atomization energy	11 kcal/mol mean abs. dev.
	Proton affinity	0.9 kcal/mol mean abs. dev.
	EA	4 kcal/mol mean abs. dev.
	IP	3 kcal/mol mean abs. dev.
	Bond length	0.008, 0.01 Å mean abs. dev.
	Bond angle	0.3° mean abs. dev.
	Frequencies	40, 30, 70 cm^{-1} mean abs. dev.

TABLE 16.1 (Continued)

Method	Property	Accuracy
CCSD/aug–cc–pVQZ	Atomization energy	7 kcal/mol mean abs. dev.
	Proton affinity	0.9 kcal/mol mean abs. dev.
	EA	3.3 kcal/mol mean abs. dev.
	IP	2 kcal/mol mean abs. dev.
	Bond length	0.008, 0.01 Å mean abs. dev.
	Bond angle	0.2° mean abs. dev.
	Frequencies	30, 60 cm^{-1} mean abs. dev.
CCSD(T)/aug–cc–pVDZ	Atomization energy	18 kcal/mol mean abs. dev.
	Proton affinity	1.6 kcal/mol mean abs. dev.
	EA	4.5 kcal/mol mean abs. dev.
	IP	5.5 kcal/mol mean abs. dev.
	Bond length	0.018, 0.03 Å mean abs. dev.
	Bond angle	0.5° mean abs. dev.
	Frequencies	55, 40, 70 cm^{-1} mean abs. dev.
CCSD(T)/aug–cc–pVTZ	Atomization energy	5 kcal/mol mean abs. dev.
	Proton affinity	1 kcal/mol mean abs. dev.
	EA	2 kcal/mol mean abs. dev.
	IP	2 kcal/mol mean abs. dev.
	Bond length	0.009, 0.015 Å mean abs. dev.
	Bond angle	0.3° mean abs. dev.
	Frequencies	40, 30, 30 cm^{-1} mean abs. dev.
CCSD(T)/aug–cc–pVQZ	Atomization energy	2 kcal/mol mean abs. dev.
	Proton affinity	0.8 kcal/mol mean abs. dev.
	EA	1 kcal/mol mean abs. dev.
	IP	1 kcal/mol mean abs. dev.
	Bond length	0.008, 0.009 Å mean abs. dev.
	Bond angle	0.2° mean abs. dev.
	Frequencies	25, 40 cm^{-1} mean abs. dev.
G1	Total energy	1.6 kcal/mol mean abs. dev.
G2	Total energy	1.2 kcal/mol mean abs. dev.
	IP	0.063 eV mean abs. dev.
	IP	1.8 kcal/mol mean abs. dev.
	EA	0.061 eV mean abs. dev.
	EA	1.4 kcal/mol mean abs. dev.
	Atomization energy	1.5 kcal/mol mean abs. dev.
	Proton affinity	1 kcal/mol mean abs. dev.
G2(MP2)	Total energy	1.5 kcal/mol mean abs. dev.
	IP	0.076 eV mean abs. dev.
	EA	0.084 eV mean abs. dev.
CBS-4	Total energy	2.0 kcal/mol mean abs. dev.
CBS-Q	Total energy	1.0 kcal/mol mean abs. dev.
CBS-APNO	Total energy	0.5 kcal/mol mean. abs. dev.

That way, enormous amounts of time are not wasted (both yours and the computer's).

BIBLIOGRAPHY

Tabulations of method accuracy or individual results:

One issue of the *Journal of Molecular Structure* each year contains a tabulation referencing all theoretical computations published the previous year. More recent years of this compilation are available online at http://qcldb.ims.ac.jp/

http://www.emsl.pnl.gov:2080/proj/crdb/

http://srdata.nist.gov/cccbdb/

http://www.chamotlabs.com/cl/Freebies/Methods.html

http://www.csc.fi/lul/rtam/rtamquery.html

J. J. P. Stewart, *MOPAC 2000 Manual* Schrödinger, Portland OR (1999).

F. Jensen, *Introduction to Computational Chemistry* John Wiley & Sons, New York (1999).

K. K. Irikura, *J. Phys. Chem. A* **102**, 9031 (1998).

D. Feller, K. A. Peterson, *J. Chem. Phys.* **108**, 154 (1998).

L. A. Curtiss, P. C. Redfern, K. Raghavachari, J. A. Pople, *J. Chem. Phys.* **109**, 42 (1998).

A. C. Scheiner, J. Baker, J. W. Andzelm, *J. Comput. Chem.* **18**, 775 (1997).

M. Head-Gordon, *J. Phys. Chem.* **100**, 13213 (1996).

J. B. Foresman, Æ. Frisch, *Exploring Chemistry with Electronic Structure Methods* Gaussian, Pittsburgh (1996).

W. J. Hehre, *Practical Strategies for Electronic Structure Calculations* Wavefunction, Irvine (1995).

G. Náray-Szabó, P. R. Surján, J. G. Ángyán, *Applied Quantum Chemistry* D. Reidel, Dordrecht (1987).

W. J. Hehre, L. Radom, P. v.R. Schleyer, J. A. Pople, *Ab Initio Molecular Orbital Theory* John Wiley & Sons, New York (1986).

T. Clark, *A Handbook of Computational Chemistry* John Wiley & Sons, New York (1985).

Comparison of Ab Initio Quantum Chemistry with Experiment for Small Molecules R. J. Bartlett, Ed., D. Reidel, Dordrecht (1985).

W. G. Richards, P. R. Scott, V. Sackwild, S. A. Robins, *Bibliography of ab initio Molecular Wave Functions Supplement for 1978–1980* Oxford, Oxford (1981).

W. G. Richards, P. R. Scott, E. A. Colburn, A. F. Marchington, *Bibliography of ab initio Molecular Wave Functions Supplement for 1974–1977* Oxford, Oxford (1978).

W. G. Richards, T. E. H. Walker, L. Farnell, P. R. Scott, *Bibliography of ab initio Molecular Wave Functions Supplement for 1970–1973* Oxford, Oxford (1974).

W. G. Richards, T. E. H. Walker, R. K. Hinkley, *A Bibliography of ab initio Molecular Wave Functions* Oxford, Oxford (1971).

A sicklist of known failings of various methods is at

http://srdata.nist.gov/sicklist/

Theoretical work is often reviewed in

Advances in Chemical Physics
Advances in Molecular Electronic Structure Theory
Advances in Molecular Modeling
Advances in Quantum Chemistry
Annual Review of Physical Chemistry
Recent Trends in Computational Chemistry
Reviews in Computational Chemistry

Reviews of computational work are sometimes found in

Chemical Reviews
Chemical Society Reviews
Structure and Bonding

Other useful sites on the internet

The computational chemistry list (CCL) consists of a list server and web site, which is at http://server.ccl.net/ The web site contains information about computational chemistry and the archives from the discussion list. Subscribing to the list results in receiving about twenty messages per day. This is a good way to watch discussions of current issues. The etiquette on the list is that you attempt to find an answer to your question in the library and the web archives before asking a question. Once you have asked a question, please post a summary of the reponses received.

http://www.rsc.org/lap/rsccom/dab/mmglinks.htm
http://ellington.pharm.arizona.edu/~bear/

PART II
Advanced Topics

17 Finding Transition Structures

17.1 INTRODUCTION

A transition structure is the molecular species that corresponds to the top of the potential energy curve in a simple, one-dimensional, reaction coordinate diagram. The energy of this species is needed in order to determine the energy barrier to reaction and thus the reaction rate. A general rule of thumb is that reactions with a barrier of 21 kcal/mol or less will proceed readily at room temperature. The geometry of a transition structure is also an important piece of information for describing the reaction mechanism.

Short of determining an entire reaction coordinate, there are a number of structures and their energies that are important to defining a reaction mechanism. For the simplest single-step reaction, there would be five such structures:

1. The reactants separated by large distances
2. The van der Waals complex between the reactants
3. The transition structure
4. The van der Waals complex between the products
5. The products separated by large distances

This is illustrated in Figure 17.1. The energies of the van der Waals complexes are a better description of the separated species for describing liquid-phase reactions. The energies of the products separated by large distances are generally more relevant to gas-phase reactions.

A transition structure is mathematically defined as the geometry that has a zero derivative of energy with respect to moving every one of the nuclei and has a positive second derivative of energy for all but one geometric movement, which has a negative curvature. Unfortunately, this description describes many structures other than a reaction transition. Other structures at energy maxima are an eclipsed conformation, the intermediate point in a ring flip, or any structure with a higher symmetry than the compound should have.

Predicting what a transition structure will look like (without the aid of a computer) is difficult for a number of reasons. Such a prediction might be based on a proposed mechanism that is incorrect. The potential energy surface around the transition structure is often much more flat than the surface around a stable geometry. Thus, there may be large differences in the transition-structure geometry between two seemingly very similar reactions with similar energy barriers.

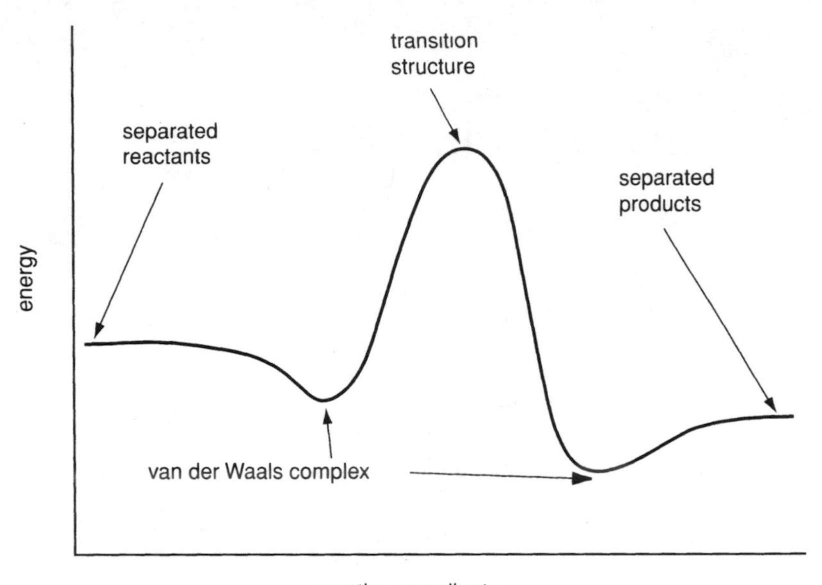

FIGURE 17.1 Points on a simple reaction coordinate.

It has been possible to determine transition structures computationally for many years, although not always easy. Experimentally, it has only recently become possible to examine reaction mechanisms directly using femtosecond pulsed laser spectroscopy. It will be some time before these techniques can be applied to all the compounds that are accessible computationally. Furthermore, these experimental techniques yield vibrational information rather than an actual geometry for the transition structure.

17.2 MOLECULAR MECHANICS PREDICTION

Traditionally, molecular mechanics has not been the method of choice for predicting transition structures. However, since it is the only method viable for many large molecules, some efforts have been made to predict transition structures. Since the bonds are explicitly defined in molecular mechanics methods, it is not possible to simply find a point that is an energy maximum, except for conformational intermediates.

Some force fields, such as MMX, have atom types designated as transition-structure atoms. When these are used, the user may have to define a fractional bond order, thus defining the transition structure to exist where there is a bond order of $\frac{1}{2}$ or $\frac{1}{3}$. Sometimes, parameters are available for common organic reactions. Other times, default values are available, based on general rules or assumptions. The geometry is then optimized to yield a bond length similar to that of the true transition structure. With the correct choice of parameters, this can give a reasonable transition-state geometry and differences in strain energy. This technique has only limited predictive value since it is dependent on an *a priori* knowledge of the transition structure.

The technique most often used (i.e., for an atom transfer) is to first plot the energy curve due to stretching a bond that is to be broken (without the new bond present) and then plot the energy curve due to stretching a bond that is to be formed (without the old bond present). The transition structure is next defined as the point at which these two curves cross. Since most molecular mechanics methods were not designed to describe bond breaking and other reaction mechanisms, these methods are most reliable when a class of reactions has been tested against experimental data to determine its applicability and perhaps a suitable correction factor.

Results using this technique are better for force fields made to describe geometries away from equilibrium. For example, it is better to use Morse potentials than harmonic potentials to describe bond stretching. Some researchers have created force fields for a specific reaction. These are made by fitting to the potential energy surface obtained from *ab initio* calculations. This is useful for examining dynamics on the surface, but it is much more work than simply using *ab initio* methods to find a transition structure.

This technique has been applied occasionally to orbital-based methods, where it is called *seam searching*. The rest of the techniques mentioned in this chapter are applicable to semiempirical, density functional theory (DFT), and *ab initio* techniques.

17.3 LEVEL OF THEORY

Transition structures are more difficult to describe than equilibrium geometries. As such, lower levels of theory such as semiempirical methods, DFT using a local density approximation (LDA), and *ab initio* methods with small basis sets do not generally describe transition structures as accurately as they describe equilibrium geometries. There are, of course, exceptions to this, but they must be identified on a case-by-case basis. As a general rule of thumb, methods that are empirically defined, such as semiempirical methods or the G1 and G2 methods, describe transition structures more poorly than completely *ab initio* methods do.

If the transition structure is the point where two electronic states cross, the single-determinant wave function approximation breaks down at that point. In this case, it may be impossible to find a transition structure using a single-determinant wave function, such as semiempiricals, HF, and single-determinant DFT calculations. If the two states have the same symmetry, single-determinant calculations will exhibit an avoided crossing, lowering the reaction barrier slightly as shown in Figure 17.2. If the two states do not have the same symmetry, a single-determinant calculation will often fail to find a transition structure. Multiple-determinant calculations with both states in the configuration space

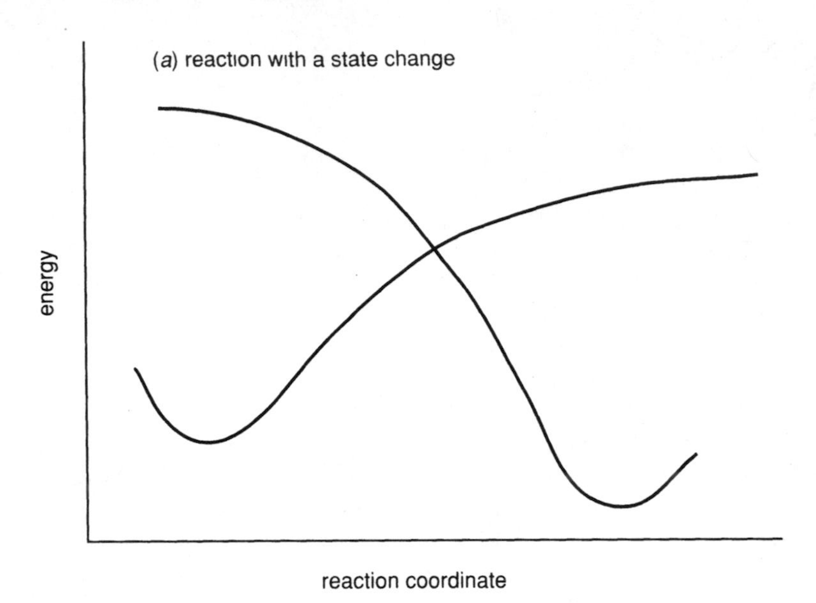

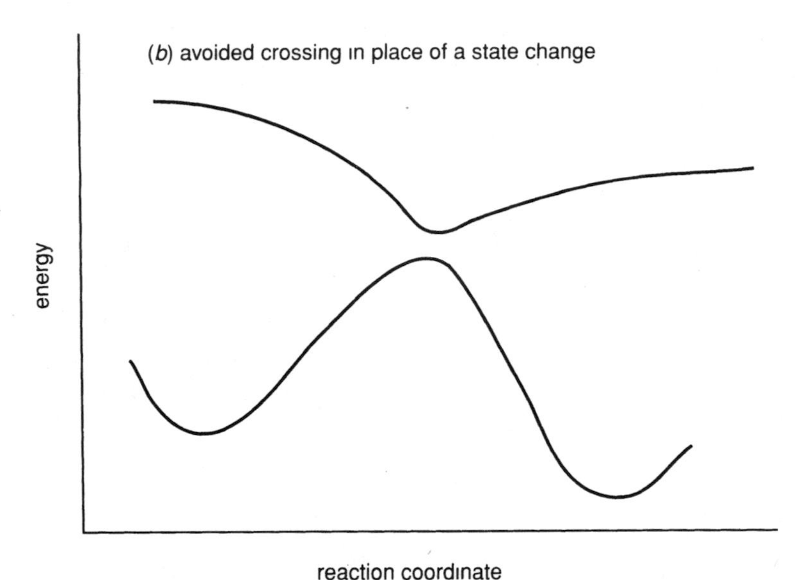

FIGURE 17.2 Illustration of the reaction coordinate for a reaction with a change in the electronic state. (*a*) Potential energy curves for the two electronic states of the system. (*b*) Avoided crossing that can be seen in single-determinant calculations.

tend to follow the reaction surface correctly. Hybrid and gradient-corrected DFT methods are unable to describe some transition structures due to their inability to adequately localize spin.

A few studies have found potential surfaces with a stable minimum at the transition point, with two very small barriers then going toward the reactants and products. This phenomenon is referred to as "Lake Eyring": Henry Eyring, one of the inventors of transition state theory, suggested that such a situation, analogous to a lake in a mountain cleft, could occur. In a study by Schlegel and coworkers, it was determined that this energy minimum can occur as an artifact of the MP2 wave function. This was found to be a mathematical quirk of the MP2 wave function, and to a lesser extent MP3, that does not correspond to reality. The same effect was not observed for MP4 or any other levels of theory.

The best way to predict how well a given level of theory will describe a transition structure is to look up results for similar classes of reactions. Tables of such data are provided by Hehre in the book referenced at the end of this chapter.

17.4 USE OF SYMMETRY

As mentioned above, a structure with a higher symmetry than is obtained for the ground state may satisfy the mathematical criteria defining a reaction structure. In a few rare (but happy) cases, the transition structure can be rigorously defined by the fact that it should have a higher symmetry. An example of this would be the symmetric S_N2 reaction:

$$F + CH_3F \rightarrow FCH_3 + F$$

In this case, the transition structure must have D_{3h} symmetry, with the two F atoms arranged axially and the H atoms being equatorial. In fact, the transition structure is the lowest energy compound that satisfies this symmetry criteria.

Thus, the transition structure can be found by forcing the structure to have the correct symmetry and then optimizing the geometry. This means geometry optimization rather than transition structure finding algorithms are used. This is a benefit because geometry optimization algorithms are generally more stable and reliable than transition structure optimization algorithms.

For systems where the transition structure is not defined by symmetry, it may be necessary to ensure that the starting geometry does not have any symmetry. This helps avoid converging to a solution that is an energy maximum of some other type.

17.5 OPTIMIZATION ALGORITHMS

If a program is given a molecular structure and told to find a transition structure, it will first compute the Hessian matrix (the matrix of second derivatives of energy with respect to nuclear motion). The nuclei are then moved in a manner that increases the energy in directions corresponding to negative values of the Hessian and decreases energy where there are positive values of the Hessian. This procedure has several implications, as follows.

This is most often done using a quasi-Newton technique, which implicitly assumes that the potential energy surface has a quadratic shape. Thus, the optimization will only be able to find the correct geometry if the starting geometry is sufficiently close to the transition structure geometry to make this a valid assumption. The starting geometry must also be closer to the reaction transition than to any other structure satisfying the same mathematical criteria, such as an eclipsed conformation. Quasi-Newton techniques are generally more sensitive to the starting geometry than the synchronous transit methods discussed below.

Simplex optimizations have been tried in the past. These do not assume a quadratic surface, but require far more computer time and thus are seldom incorporated in commercial software. Due to the unavailability of this method to most researchers, it will not be discussed further here.

The optimization of a transition structure will be much faster using methods for which the Hessian can be analytically calculated. For methods that incrementally compute the Hessian (i.e., the Berny algorithm), it is fastest to start with a Hessian from some simpler calculation, such as a semiempirical calculation. Occasionally, difficulties are encountered due to these simpler methods giving a poor description of the Hessian. An option to compute the initial Hessian at the desired level of theory is often available to circumvent this problem at the expense of additional CPU time.

When a transition structure is determined by starting from a single initial geometry, the calculation is very sensitive to the starting geometry. One excellent technique is to start with the optimized transition structure of another reaction that is expected to proceed by the same mechanism and then replace the functional groups to give the desired reactants without changing the arrangement of the atoms near the reaction site. This is sometimes called the template method.

If no known transition structure is available, try setting the lengths of bonds being formed or broken intermediate to their bonding and van der Waals lengths. Often, it is necessary for the starting geometry to have no symmetry. Ignoring wave function symmetry is usually not sufficient.

17.6 FROM STARTING AND ENDING STRUCTURES

Since transition-structure calculations are so sensitive to the starting geometry, a number of automated techniques for finding reasonable starting geometries have been proposed. One very useful technique is to start from the reactant and product structures.

The simplest way to guess the shape of a transition structure is to assume

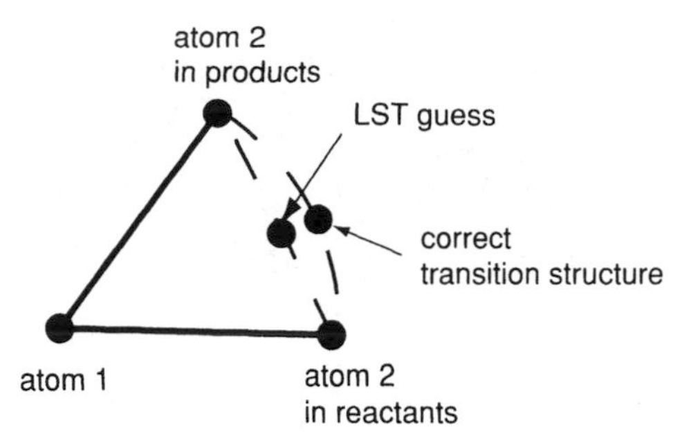

FIGURE 17.3 Illustration of the linear synchronous transit method for generating a starting point for a transition-structure optimization.

that each atom is directly between the position where it starts and the position where it ends. This linear motion approximation is called linear synchronous transit (LST). This is a good first approximation, but it has its failings. Consider the motion of an atom that is changing bond angle with respect to the rest of the molecule. The point half-way between its starting and ending positions on the line connecting those positions will give a shorter than expected bond length as shown in Figure 17.3 and thus be (perhaps significantly) higher in energy.

The logical extension of this technique is the quadratic synchronous transit method (QST). These methods assume that the coordinates of the atoms in the transition structure will lie along a parabola connecting the reactant and product geometries. QST generally gives some improvement over LST although the improvement may be very slight.

The QST3 technique requires the user to supply structures for the complex of reactants and the complex of products, and a guess of the transition state geometry. This helps assure that the desired path is examined, but the calculation is also biased by the user's predicted mechanism, which may be incorrect.

Many programs allow the user to input a weighting factor (i.e., to give a structure that is 70% of the way from reactants to products). This allows the application of the Hammond postulate: that the transition structure will look more like the reactants for an exothermic reaction and more like the products for an endothermic reaction.

These techniques have been very useful for simple reactions, but they do have limitations. The biggest limitation is that each of these is designed around the assumption that the reaction is a single step with a concerted motion of all atoms. For multistep reactions, these techniques can be used individually for each step. For a reaction that has only one transition structure but the motion is not concerted (i.e., breaking one bond and then forming another), it may be better to use starting geometries created by hand or to use eigenvalue following.

There are distinct differences in the way these methods are implemented in specific software packages. Some software packages will require the user to choose a transit method to obtain a starting geometry and then run a separate calculation with a quasi-Newton method. Other software packages will have an automated way of running the transit method calculation, followed by a quasi-Newton calculation. There have even been algorithms proposed for allowing the program to make decisions concerning which method to use at each step of the optimization.

17.7 REACTION COORDINATE TECHNIQUES

A transition structure is, of course, a maximum on the reaction pathway. One well-defined reaction path is the least energy or intrinsic reaction path (IRC). Quasi-Newton methods oscillate around the IRC path from one iteration to the next. Several researchers have proposed methods for obtaining the IRC path from the quasi-Newton optimization based on this observation.

Likewise, a transition structure can be obtained by following the reaction path from the equilibrium geometry to the transition structure. This technique is known as *eigenvalue following* because the user specifies which vibrational mode should lead to the desired reaction given sufficient kinetic energy. This is not the best way to obtain an IRC, nor is it the fastest or most reliable way to find a transition structure. However, it has the advantage of not making assumptions about concerted motions of atoms or what the transition structure will look like. When this algorithm fails, it is often because it began following a different motion on the potential energy surface or the potential surface for a different sate of the molecule. More information is provided in the chapter on reaction coordinates.

Another technique is to use a pseudo reaction coordinate. This can be quite a bit of work for the user and requires more computer time than most of the other techniques mentioned. However, it has the advantage of being very reliable and thus will work when all other techniques have failed. A pseudo reaction coordinate is calculated by first choosing a geometric parameter intimately involved in the reaction (such as the bond length for a bond that is being formed or broken). A series of calculations is then run in which this parameter is held fixed at various values, from the value in the reactants to the value in the products, and all other geometric parameters are optimized. This does not give a true reaction coordinate but an approximation to it, which matches the true intrinsic reaction coordinate perfectly only at the equilibrium geometries and transition structure. Typically, the highest-energy calculation from this set is used as the starting geometry for a quasi-Newton optimization. In a few rare cases involving very flat potential surfaces, the quasi-Newton optimization may still fail. In this case, the transition structure can be calculated to any desired accuracy (within the theoretical model) by varying the chosen geometric parameter in successively smaller increments to find the energy maximum. Some software packages have an automated algorithm for finding a pseudo reaction coordinate, called a *coordinate driving* algorithm.

17.8 RELAXATION METHODS

Algorithms have been devised for determining the reaction coordinate, transition structure, and optimized geometry all in a single calculation. The calculation simultaneously optimizes a whole set of geometries along the reaction coordinate. This has been found to be a reliable way of finding the transition structure and an overall improvement in the amount of computer time necessary for obtaining all this information. There are several algorithms following this general idea of computing the transition structure from the reaction coordinate, such as the chain method, locally updated planes method, and conjugate peak refinement method. These methods do require a significant amount of CPU time and are thus not used often.

17.9 POTENTIAL SURFACE SCANS

The reaction coordinate is one specific path along the complete potential energy surface associated with the nuclear positions. It is possible to do a series of calculations representing a grid of points on the potential energy surface. The saddle point can then be found by inspection or more accurately by using mathematical techniques to interpolate between the grid points.

This type of calculation does reliably find a transition structure. However, it requires far more computer time than any of the other techniques. As such, this is generally only done when the research requires obtaining a potential energy surface for reasons other than just finding the transition structure.

17.10 SOLVENT EFFECTS

It is well known that reaction rates can be affected by the choice of solvent. Solvent interactions can significantly affect the energy of the transition structure and generally only slightly change the transition-structure geometry. All the techniques for finding transition structures can be used when solvent effects are being included in the calculation. The presence of solvent interactions does not change the manner in which transition structures are found, although it might change the results. These methods are discussed in more detail in the Chapter 24.

17.11 VERIFYING THAT THE CORRECT GEOMETRY WAS OBTAINED

The primary means of verifying a transition structure is to compute the vibrational frequencies. A saddle point should have one negative frequency. The vibrational motion associated with this negative frequency is the motion going toward reactants in one direction and products in the other direction. Various programs may or may not print the six frequencies that are essentially zero for the three degrees of translational motion and three degrees of vibrational motion.

It is also important to always examine the transition structure geometry to make sure that it is the reaction transition and not the transition in the middle of a ring flip or some other unintended process. If it is not clear from the geometry that the transition structure is correct, displaying an animation of the transition vibrational mode should clarify this. If still unclear, a reaction coordinate can be computed.

It is possible that a transition structure calculation will give two negative frequencies (a second-order saddle point) or more. This gives a little bit of information about the potential energy surface, but it is extremely unlikely that such a structure has any significant bearing on how the reaction occurs. This type of structure will often be found if the starting geometry had a higher symmetry than the transition structure should have.

17.12 CHECKLIST OF METHODS FOR FINDING TRANSITION STRUCTURES

Many techniques for finding transition structures are discussed above. The following is a listing of each of these starting with those that are easiest to use and most often successful. In other words, start with number 1 and continue down the list until you find one that works.

1. If the system can only be modeled feasibly by molecular mechanics, use the potential energy curve-crossing technique or a force field with transition-structure atom types.
2. If the transition state can be defined by symmetry, do a normal geometry optimization calculation with the symmetry constrained.
3. If the structure of the intermediate for a very similar reaction is available, use that structure with a quasi-Newton optimization.
4. Quadratic synchronous transit followed by quasi-Newton.
5. Linear synchronous transit followed by quasi-Newton.
6. Try quasi-Newton calculations starting from structures that look like what you expect the transition structure to be and that have no symmetry. This is a skill that improves as you become more familiar with the mechanisms involved, but requires some trial-and-error work even for the most experienced researchers.
7. Eigenvalue-following.
8. Relaxation algorithms.

9. Use a pseudo reaction coordinate with one parameter constrained followed by a quasi-Newton optimization.
10. Use a pseudo reaction coordinate with one parameter constrained using successively smaller steps for the constrained parameter until the desired accuracy is reached.
11. Go back to options 9 and 10 and constrain a different parameter.
12. Consider the fact that some reactions have no barrier. You might also be making incorrect assumptions about the reaction mechanism. Consider these possibilities and start over.
13. Switch to a higher level of theory and start all over again.
14. Obtain the transition structure from the entire potential energy surface. It is questionable that there will be any case where this is the only option, but it should work as a desperate last resort.

Once you are experienced at finding transition structures for a particular class of reactions, you will probably go directly to the technique that has been most reliable for those reactions. Until that time, the checklist above is our best advice for finding a transition structure with the least amount of work for the researcher and the computer. Regardless of experience, it is common to experience quite a bit of trial and error in finding transition structures. Even experienced researchers find that the way they have been regarding a reaction is often much more simplistic than the molecular motions actually involved.

All the techniques discussed in this chapter are applicable to single-step reaction mechanism. For multiple-step mechanisms, it is necessary to work through this process for each step in the reaction.

BIBLIOGRAPHY

General discussion of transition structures are in

F. Jensen, *Introduction to Computational Chemistry* John Wiley & Sons, Chichester (1999).

J. B. Foresman, Æ. Frisch, *Exploring Chemistry with Electronic Structure Methods Second Edition* Gaussian, Pittsburgh (1996).

A. R. Leach *Molecular Modelling Principles and Applications* 240, Longman, Singapore, (1996).

W. H. Green, Jr., C. B. Moore, W. F. Polik, *Ann. Rev. Phys. Chem.* **43**, 591 (1992).

A discussion of the issues involved and tables of performance data can be found in

W. J. Hehre *Practical Strategies for Electronic Structure Calculations* Wavefunction, Irvine (1995).

C. W. Bauschlicker, Jr., S. R. Langhoff, P. R. Taylor, *Adv. Chem. Phys.* **77**, 103 (1990).

W. Chen, H. B. Schlegel, *J. Chem. Phys.*, **101**, 5957 (1994).

Review articles are

T. Bally, W. T. Borden, *Rev. Comput. Chem.* **13**, 1 (1999).

O. West, *Encycl. Comput. Chem.* **5**, 3104 (1998).

F. Jenson, *Encycl. Comput. Chem.* **5**, 3114 (1998).

M. L. McKee, M. Page, *Rev. Comput. Chem.* **4**, 35 (1993).

I. H. Williams, *Chem. Soc. Rev.* **22**, 277 (1993).

H. B. Schlegel, *New Theoretical Concepts for Understanding Organic Reactions* J. Bertrán, I. G. Csizmadia, Eds., 33, Kluwer, Dordrecht (1988).

A. E. Dorigo, K. N. Houk, *Advances in Molecular Modeling, Volume 1* 135, JAI, New York (1988)

F. Bernardi, M. A. Robb, *Adv. Chem. Phys.* **67**, 155 (1987).

H. B. Schlegel, *Adv. Chem. Phys.* **67**, 249 (1987).

S. Bell, J. S. Crighton, *J. Chem. Phys.* **80**, 2464 (1984).

C. E. Dykstra, *Ann. Rev. Phys. Chem.* **32**, 25 (1981).

For more information on synchronous transit methods see

C. Peng, H. B. Schlegel *Isr. J. Chem.* **33**, 449 (1993).

Eigenvalue following algorithms are described in

A. Bannerjee, N. Adams, J. Simons, R. Shepard, *J. Phys. Chem.* **89**, 52 (1985).

J. Simons, P. Jorgensen, H. Taylor, J. Ozment, *J Phys. Chem.* **87**, 2745 (1983).

C. J. Cerjan, W. H. Miller, *J. Chem. Phys.* **75**, 2800 (1981).

The relaxation method is described in

P. Y. Ayala, H. B. Schlegel, *J Chem. Phys.* **107**, 375 (1997).

Obtaining transition structures from molecular mechanics is discussed in

F. Jensen *J. Comp. Chem.* **15**, 1199 (1994).

J. E. Eksterowicz, K. N. Houk, *Chem. Rev.* **93**, 2439 (1993).

18 Reaction Coordinates

In order to define how the nuclei move as a reaction progresses from reactants to transition structure to products, one must choose a definition of how a reaction occurs. There are two such definitions in common use. One definition is the minimum energy path (MEP), which defines a reaction coordinate in which the absolute minimum amount of energy is necessary to reach each point on the coordinate. A second definition is a dynamical description of how molecules undergo intramolecular vibrational redistribution until the vibrational motion occurs in a direction that leads to a reaction. The MEP definition is an intuitive description of the reaction steps. The dynamical description more closely describes the true behavior molecules as seen with femtosecond spectroscopy.

18.1 MINIMUM ENERGY PATH

The MEP is defined as the path of steepest descent in mass-weighted Cartesian coordinates. This is also called intrinsic reaction coordinate (IRC). In reality, we know that many other paths close to the IRC path would also lead to a reaction and the percentage of the time each path is taken could be described by the Boltzmann distribution.

There are several algorithms for finding a MEP. The most reliable of these algorithms are the current generation of methods that start from the transition structure. Simply using a steepest-descent method does not give a good description of the MEP. This is because the points chosen by steepest-descent algorithms tend to oscillate around the reaction coordinate as shown in Figure 18.1. The algorithms incorporated in most software packages correct for this problem.

The reaction coordinate is calculated in a number of steps. If too few steps are used, then the points that are computed will follow the reaction coordinate less closely. Usually, the default number of points computed by software packages will give reasonable results. More points may be required for complex mechanisms. This algorithm is sometimes called the IRC algorithm, thus creating confusion over the definition of IRC.

Alternatively, an eigenvalue following algorithm can be used. This is a shallowest-ascent technique for following the motion of one of the vibrational modes. The primary advantage of this method is that it can follow the reaction coordinate energetically uphill from the reactants or products. This eliminates the need to compute a transition-structure geometry. The eigenvalue following path from a minimum to the transition structure gives an approximation to the reaction coordinate. The path from transition structure to a minimum found by

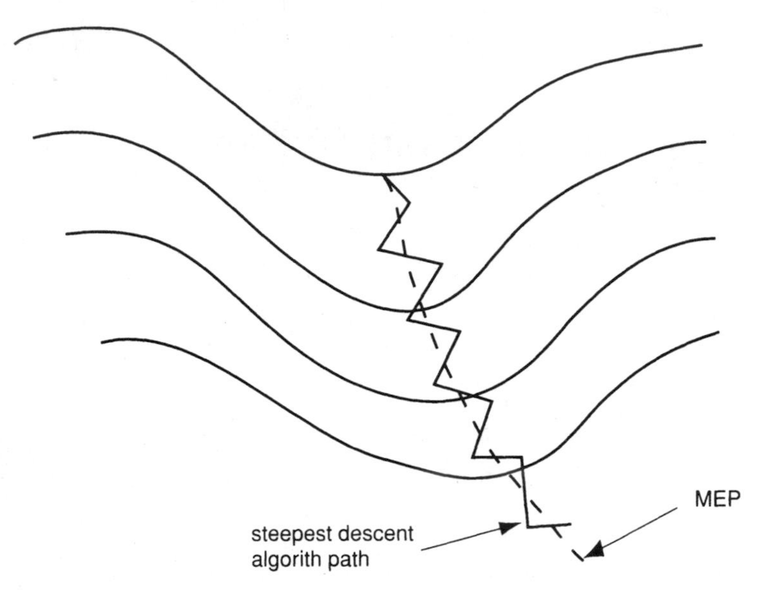

FIGURE 18.1 Illustration of how the steepest descent algorithm follows a path that oscillates around the minimum energy path.

this algorithm gives a better approximation of the MEP, but is not as accurate as the approximation of the IRC algorithms.

A pseudo reaction coordinate can be obtained by fixing one of the parameters and allowing the rest of the geometry to optimize. The fixed parameter is then stepped along the path and the rest of the structure reoptimized at each step. This is also called a trial reaction coordinate. This algorithm does not yield a rigorously correct MEP. This is because the chosen parameter may be the one moving the farthest at some points along the path, but not at others. The resulting MEP will look reasonable when the parameter gives a good description. When the parameter poorly describes the MEP, a very small change in the parameter value might result in a very large change in energy and geometry. Thus, the pseudo reaction coordinate will find energies on the MEP, but will locate them at an incorrect distance along reaction route. A poorly chosen reaction coordinate will sometimes result in a hysteresis in which a different IRC is obtained by increasing the reaction coordinate rather than by decreasing it. In this case, the geometries obtained will not be exactly those along the true IRC, although they are often close. Some software packages have an automated version of this algorithm called a *coordinate driving* algorithm.

18.2 LEVEL OF THEORY

The types of algorithms described above can be used with any *ab initio* or semiempirical Hamiltonian. Generally, the *ab initio* methods give better results than semiempirical calculations. HF and DFT calculations using a single determinant are not able to correctly describe mechanisms involving the uncoupling of electrons. The MP2 method, and occasionally MP3, are known to sometimes incorrectly predict a tiny energy minimum where the transition structure should be. Unrestricted DFT calculations give reasonable results as long as they are not combined with a spin projection method, in contrast to unrestricted HF that is improved by spin projection. The G1 and G2 methods also predict reaction coordinates more poorly than they predict equilibrium geometry. The reasonable list of methods to use, in order of ascending accuracy, is as follows:

1. Semiempirical calculations, particularly PM3 or AM1, for cases where *ab initio* calculations are too expensive
2. HF or DFT where no uncoupling of electrons is involved; which is the better method is still a matter of debate
3. MP4
4. CI
5. MCSCF, CASSCF
6. MRCI
7. CC, which is superior to CI because it is rigorously size consistent

The way in which the calculation is performed is also important. Unrestricted calculations will allow the system to shift from one spin state to another. It is also often necessary to run the calculation without using wave function symmetry. The calculation of geometries far from equilibrium tends to result in more SCF convergence problems, which are discussed in Chapter 22.

If the complete potential energy surface has already been computed, a reaction coordinate can be determined using an adaptation of the IRC algorithm. The IRC computation requires very little computer time, but obtaining the potential energy surface is far more computation-intensive than an *ab initio* IRC calculation. Thus, this is only done when the potential energy surface is being computed for another reason.

18.3 LEAST MOTION PATH

Some early studies were done by computing energies as a molecule was pulled apart into fragments without allowing the fragments to change geometry or orientation. The resulting reaction path is called a least motion path. These types of calculations are now generally assigned as introductory-level class exercises. The energetics of the least motion path are very poor. In some cases, the least motion path even shows energy barriers where none should exist.

18.4 RELAXATION METHODS

An algorithm has been proposed for determining the reaction coordinate, transition structure, and optimized geometry all in a single calculation. The calculation simultaneously optimizes a whole set of geometries along the reaction coordinate. This has been found to be a reliable way of finding the transition structure and represents an overall improvement in the amount of computer time necessary for obtaining all this information. The CHAIN algorithm is a relaxation method used with semiempirical calculations.

18.5 REACTION DYNAMICS

Both molecular dynamics studies and femtosecond laser spectroscopy results show that molecules with a sufficient amount of energy to react often vibrate until the nuclei follow a path that leads to the reaction coordinate. Dynamical calculations, called trajectory calculations, are an application of the molecular dynamics method that can be performed at semiempirical or *ab initio* levels of theory. See Chapter 19 for further details.

18.6 WHICH ALGORITHMS TO USE

Listing the available algorithms, starting with the best results and most reliable algorithm, and working to the lowest-quality results, we arrive at the following list:

1. If vibrational information is desired, use a trajectory calculation as described in Chapter 19.
2. If an entire potential energy surface has been computed, use an IRC algorithm with that surface.
3. Use an IRC algorithm starting from the transition structure.
4. Use an eigenvalue following algorithm.
5. Use a relaxation method.
6. Use a pseudo reaction coordinate algorithm.
7. As a last resort, compute the entire potential energy surface and then obtain a reaction coordinate from it.

An ensemble of trajectory calculations is rigorously the most correct description of how a reaction proceeds. However, the MEP is a much more understandable and useful description of the reaction mechanism. These calculations are expected to continue to be an important description of reaction mechanism in spite of the technical difficulties involved.

BIBLIOGRAPHY

Introductory discussions can be found in

J. B. Foresman, Æ. Frisch, *Exploring Chemistry with Electronic Structure Methods Second Edition* Gaussian, Pittsburgh (1996).

A. R. Leach *Molecular Modelling Principles and Applications* Longman, Essex (1996).

T. Clark, *A Handbook of Computational Chemistry* John Wiley & Sons, New York (1985).

Books on reaction coordinates are

The Reaction Path in chemistry: Current Approaches and Perspectives D. Heidrich, Ed., Kluwer, Dordrecht (1995).

D. Heidrich, W. Kliesch, W. Quapp, *Properties of Chemically Interesting Potential Energy Surfaces* Springer-Verlag, Berlin (1991).

Reviews articles are

H. B. Schlegel, *Encycl. Comput. Chem.* **4**, 2432 (1998).

E. Kraka, *Encycl. Comput. Chem.* **4**, 2437 (1998).

M. A. Collins, *Adv. Chem. Phys.* **93**, 389 (1996).

H. B. Schlegel, *Modern Electronic Structure Theory* D. R. Yarkony, Ed., 459, World Scientific, Singapore (1996).

M. L. McKee, M. Page, *Rev. Comput. Chem.* **4**, 35 (1993).

E. Kracka, T. H. Dunning, Jr., *Advances in Molecular Electronic Structure Theory; Calculation and Characterization of Molecular Potential Energy Surfaces* T. H. Dunning, Jr. Ed., 129, JAI, Greenwich (1990).

C. W. Bauschlicker, Jr., S. R. Langhoff, *Adv. Chem. Phys.* **77**, 103 (1990).

B. Friedrich, Z. Herman, R. Zahradnik, Z. Havlas, *Adv. Quantum Chem.* **19**, 247 (1988).

T. H. Dunning, Jr., L. B. Harding, *Theory of Chemical Reaction Dynamics Vol 1* M. Baer Ed. 1, CRC Boca Ratan, FL (1985).

P. J. Kuntz, *Theory of Chemical Reaction Dynamics Vol 1* M. Baer Ed. 71, CRC Boca Ratan, FL (1985).

M. Simonetta, *Chem. Soc. Rev.* **13**, 1 (1984).

Reviews of molecular mechanics trajectory calculations are listed in the bibliography to Chapter 19 as well as

R. M. Whitnell, K. R. Wilson, *Rev. Comput. Chem.* **4**, 67 (1993).

The IRC algorithm is described in

C. Gonzalez, H. B. Schlegel, *J. Phys. Chem.* **94**, 5523 (1990).

C. Gonzalez, H. B. Schlegel, *J. Chem. Phys.* **90**, 2154 (1989).

Eigenvalue following algorithms are described in

A. Bannerjee, N. Adams, J. Simons, R. Shepard, *J. Phys. Chem.* **89**, 52 (1985).

J. Simons, P. Jorgensen, H. Taylor, J. Ozment, *J. Phys. Chem.* **87**, 2745 (1983).

C. J. Cerjan, W. H. Miller, *J. Chem. Phys.* **75**, 2800 (1981).

19 Reaction Rates

The calculation of reaction rates has not seen as the widespread use as the calculation of molecular geometries. In recent years, it has become possible to compute reaction rates with reasonable accuracy. However, these calculations require some expertise on the part of the researcher. This is partly because of the difficulty in obtaining transition structures and partly because reaction rate algorithms have not been integrated into major computational chemistry programs and thus become automated.

19.1 ARRHENIUS EQUATION

The rate of a reaction r is dependent on the reactant concentrations. For example, a bimolecular reaction between the reactants B and C could have a rate expression, such as

$$r = k[B][C] \tag{19.1}$$

The simplest expression for the temperature dependence of the rate constant k is given by the Arrhenius equation

$$k = Ae^{-E_a/RT} \tag{19.2}$$

where E_a is the activation energy and A is called the pre-exponential factor. These can be obtained from experimentally determined reaction rates, *ab initio* calculations, trajectory calculations, or some simple theoretical method such as a hard-sphere collision description.

Although the Arrhenius equation does not predict rate constants without parameters obtained from another source, it does predict the temperature dependence of reaction rates. The Arrhenius parameters are often obtained from experimental kinetics results since these are an easy way to compare reaction kinetics. The Arrhenius equation is also often used to describe chemical kinetics in computational fluid dynamics programs for the purposes of designing chemical manufacturing equipment, such as flow reactors. Many computational predictions are based on computing the Arrhenius parameters.

Some reactions, such as ion-molecule association reactions, have no energy barrier. These reactions cannot be described well by the Arrhenius equation or the transition state theory discussed below. These reactions can be modeled as a trajectory for a capture process.

19.2 RELATIVE RATES

There are a few cases where the rate of one reaction relative to another is needed, but the absolute rate is not required. One such example is predicting the regioselectivity of reactions. Relative rates can be predicted from a ratio of Arrhenius equations if the relative activation energies are known. Reasonably accurate relative activation energies can often be computed with HF wave functions using moderate-size basis sets.

An example of this would be examining a reaction in which there are two possible products, such as ortho or para addition products. The activation energy is computed for each reaction by subtracting the energy of reactants from the transition-state energy. The difference in activation energies can then be computed. For this example, let us assume that the ortho product has an activation energy which is 2.6 kcal/mol larger than the activation energy for the para product. The ratio of Arrhenius equations would be

$$\frac{k_{\text{para}}}{k_{\text{ortho}}} = \frac{A_{\text{para}} e^{-(0\,\text{kcal/mol})/RT}}{A_{\text{ortho}} e^{-(2.6\,\text{kcal/mol})/RT}} \tag{19.3}$$

Since the reactions are very similar, we will assume that the pre-exponential factors A are the same, thus giving.

$$\frac{k_{\text{para}}}{k_{\text{ortho}}} = e^{(2.6\,\text{kcal/mol})/RT} \tag{19.4}$$

Substituting $T = 298$ K and the gas constant gives a ratio of about 81. Thus, we expect there will be 80 times as much para product as ortho product, assuming that the kinetic product is obtained.

19.3 HARD-SPHERE COLLISION THEORY

The simplest approach to computing the pre-exponential factor is to assume that molecules are hard spheres. It is also necessary to assume that a reaction will occur when two such spheres collide in order to obtain a rate constant k for the reactants B and C as follows:

$$k = N_A \pi (r_B + r_C)^2 \left[\frac{8kT}{\pi} \left(\frac{m_B + m_C}{m_B m_C} \right) \right]^{1/2} \exp\left(\frac{-E_a}{k_B T} \right) \tag{19.5}$$

where r_B and r_C are radii of molecules with masses m_B and m_C.

To a first approximation, the activation energy can be obtained by subtracting the energies of the reactants and transition structure. The hard-sphere theory gives an intuitive description of reaction mechanisms; however, the predicted rate constants are quite poor for many reactions.

19.4 TRANSITION STATE THEORY

Simply using the activation energy assumes that the only way a reaction occurs is along the minimum energy path (MEP). The transition structure is the maximum along this path, which is used to obtain the activation energy. It would be more correct to consider that reactions may occur by going through a geometry very similar to the transition structure. Transition state theory calculations (TST) take this into account. Eyring originally referred to this as the absolute rate theory.

Transition state theory is built on several mathematical assumptions. The theory assumes that Maxwell–Boltzmann statistics will predict how many molecular collisions should have an energy greater than or equal to the activation energy. This is called a quasi-equilibrium because it is equivalent to assuming that the molecules at the transition structure are in equilibrium with the reactant molecules, even though molecules do not stay at the transition structure long enough to achieve equilibrium. Furthermore, it assumes that the molecules reaching the transition point react irreversibly.

Taking into account paths near the saddle point in a statistically valid way requires an integral over the possible energies. This may be formulated for practical purposes as an integral over reactant partition functions and the density of states. These calculations require information about the shape of the potential energy surface around the transition structure, frequently using the reaction coordinate or an analytic function describing the entire potential energy surface. When making *ab initio* calculations, reactant and transition structure vibrational frequencies are often used.

19.5 VARIATIONAL TRANSITION STATE THEORY

Examining transition state theory, one notes that the assumptions of Maxwell–Boltzmann statistics are not completely correct because some of the molecules reaching the activation energy will react, lose excess vibrational energy, and not be able to go back to reactants. Also, some molecules that have reacted may go back to reactants again.

Variational transition state theory (VTST) is formulated around a variational theorem, which allows the optimization of a hypersurface (points on the potential energy surface) that is the effective point of no return for reactions. This hypersurface is not necessarily through the saddle point. Assuming that molecules react without a reverse reaction once they have passed this surface corrects for the statistical effects of molecules being drawn out of the ensemble and molecules going back to reactants. This is referred to as calculating the one-way equilibrium flux in the product direction. This results in VTST taking both energy and entropy into account, whereas TST is based on energy only.

Several VTST techniques exist. Canonical variational theory (CVT), improved canonical variational theory (ICVT), and microcanonical variational theory (μVT) are the most frequently used. The microcanonical theory tends to be the most accurate, and canonical theory the least accurate. All these techniques tend to lose accuracy at higher temperatures. At higher temperatures, excited states, which are more difficult to compute accurately, play an increasingly important role, as do trajectories far from the transition structure. For very small molecules, errors at room temperature are often less than 10%. At high temperatures, computed reaction rates could be in error by an order of magnitude.

For reactions between atoms, the computation needs to model only the translational energy of impact. For molecular reactions, there are internal energies to be included in the calculation. These internal energies are vibrational and rotational motions, which have quantized energy levels. Even with these corrections included, rate constant calculations tend to lose accuracy as the complexity of the molecular system and reaction mechanism increases.

These calculations can also take into account tunneling through the reaction barrier. This is most significant when very light atoms are involved (i.e., hydrogen transfer). Tunneling effects are often included via an approximation method called a semiclassical tunneling calculation. This is an effective one-dimensional description of tunneling. This approximation results in the calculation requiring less CPU time without introducing a significant amount of error compared to other ways of including tunneling.

The calculation must be given a description of the potential energy surface either as an analytic function or as the output from molecular orbital calculations. Analytic functions are generally used in order to compare the results of trajectory calculations and VTST calculations for the same surface. Information from molecular calculations might be either a potential energy surface scan or a series of points along the reaction coordinate with their associated gradient and Hessian matrices. Information about the reactants, products, and transition structure, such as geometries and vibrational and rotational excited states, must also be provided. Electronic excited-state information may be necessary if the reaction involves a state crossing. These energy surfaces must be very accurate, often requiring correlated methods with polarized basis sets.

19.6 TRAJECTORY CALCULATIONS

Molecular dynamics studies can be done to examine how the path and orientation of approaching reactants lead to a chemical reaction. These studies require an accurate potential energy surface, which is most often an analytic function fitted to results from *ab initio* calculations. Accurate potential energy surfaces have also been obtained from femtosecond spectroscopy results. The amount of work necessary to study a reaction with these techniques may be far more than the work required to obtain the potential energy surface, which was not a trivial task in itself.

A classical trajectory calculation will use this potential energy function in order to run a molecular dynamics simulation. The cross section for reaction can be computed by solving the equations of motion. The rate constants can then be obtained from many trajectories weighted by the appropriate distribution function. Classical trajectory calculations are most accurate for reactions involving heavy atoms at high temperatures. These calculations are sensitive to a number of technical details, such as the choice of the dynamics time step and the choice of numerical integration schemes (see the Karplus, Porter, and Sharma article in the bibliography). Technical details affecting molecular dynamics results are discussed further in Chapter 7.

Quasiclassical calculations are similar to classical trajectory calculations with the addition of terms to account for quantum effects. The inclusion of tunneling and quantized energy levels improves the accuracy of results for light atoms, such as hydrogen transfer, and lower-temperature reactions.

Ab initio trajectory calculations have now been performed. However, these calculations require such an enormous amount of computer time that they have only been done on the simplest systems. At the present time, these calculations are too expensive to be used for computing rate constants, which require many trajectories to be computed. Semiempirical methods have been designed specifically for dynamics calculations, which have given insight into vibrational motion, but they have not been the methods of choice for computing rate constants since they are generally inferior to analytic potential energy surfaces fitted from *ab initio* results.

19.7 STATISTICAL CALCULATIONS

Rather than using transition state theory or trajectory calculations, it is possible to use a statistical description of reactions to compute the rate constant. There are a number of techniques that can be considered variants of the statistical adiabatic channel model (SACM). This is, in essence, the examination of many possible reaction paths, none of which would necessarily be seen in a trajectory calculation. By examining paths that are easier to determine than the trajectory path and giving them statistical weights, the whole potential energy surface is accounted for and the rate constant can be computed.

This technique has not been used as widely as transition state theory or trajectory calculations. The accuracy of results is generally similar to that given by μTST. There are a few cases where SACM may be better, such as for the reactions of some polyatomic polar molecules.

19.8 ELECTRONIC STATE CROSSINGS

A simple method for predicting electronic state crossing transitions is Fermi's golden rule. It is based on the electromagnetic interaction between states and is derived from perturbation theory. Fermi's golden rule states that the reaction rate can be computed from the first-order transition matrix $H^{(1)}$ and the density of states at the transition frequency ρ as follows:

$$r = \frac{2\pi}{\hbar} |H^{(1)}|^2 \rho \tag{19.6}$$

The golden rule is a reasonable prediction of state-crossing transition rates when those rates are slow. Crossings with fast rates are predicted poorly due to the breakdown of the perturbation theory assumption of a small interaction.

There are reaction rates that depend on radiationless transitions between electronic states. For example, photochemically induced reactions often consist of an initial excitation to an excited electronic state, followed by a geometric rearrangement to lower the energy. In the course of this geometric rearrangement, there may be one or more radiationless transitions from one electronic state to another. The rate for these transitions can be obtained from a transition dipole moment calculation, analogous to the transition dipole calculations that give electronic spectrum intensities. For some reactions, spin-orbit coupling is a significant factor in determining the state crossing. A more empirical approach is to use an adiabatic coupling term. It is still a matter of debate which of these techniques is most accurate or most conceptually correct.

19.9 RECOMMENDATIONS

Computing reaction rates is not as simple as choosing one more option in an electronic structure program. Deciding to compute reaction rates will require a significant investment of the researchers time in order to understand the various input options. These calculations can give good results, but are very sensitive to subtle details like using a mass-scaled (isoinertial) coordinate system to specify the geometry. Most *ab initio* programs use center-of-mass or center-of-nuclear-charge coordinates. The computational requirements for completing a reaction-rate calculation are fairly modest. The typical calculation will require less than 20 MB of memory and only minutes of CPU time. The POLYRATE software program is the most widely used for performing variational transition state calculations.

For relative reaction rates, *ab initio* calculations with moderate-size basis sets usually give sufficient accuracy.

For the accurate, *a priori* calculation of reaction rates, variational transition state calculations are now the method of choice. These calculations are capable of giving the highest-accuracy results, but can be technically difficult to perform correctly. They can be done for moderate-size organic molecules. Even with the best of these methods, relative rates are more accurate than absolute rate constants. Absolute rate constants can be in error by as much as a factor of 10 even when the barrier height has been computed to within 1 kcal/mol.

Transition state theory calculations present slightly fewer technical difficulties. However, the accuracy of these calculations varies with the type of reaction. With the addition of an empirically determined correction factor, these calculations can be the most readily obtained for a given class of reactions.

Quasiclassical trajectory calculations are the method of choice for determining the dynamics of intramolecular vibrational energy redistribution leading to a chemical reaction. If this information is desired, an accurate reaction rate can be obtained at little extra expense.

BIBLIOGRAPHY

Introductory descriptions are in

F. Jensen, *Introduction to Computational Chemistry* John Wiley & Sons, New York (1999).

I. N. Levine, *Physical Chemistry Fourth Edition* McGraw Hill (1995).

W. H. Green, Jr., C. B. Moore, W. P. Polik, *Ann. Rev. Phys. Chem.* **43**, 591 (1992).

D. M. Hirst, *A Computational Approach to Chemistry* Blackwell Scientific, Oxford (1990).

R. Daudel, *Adv Quantum Chem.* **3**, 161 (1967).

D. G. Truhlar, B. C. Garret, *Acc. Chem. Res.* **13**, 440 (1980).

Introductory descriptions of trajectory calculations are

D. L. Bunker, *Methods Comput. Phys.* **10**, 287 (1971).

M. Karplus, R. N. Porter, R. D. Sharma, *J. Chem. Phys.* **43**, 3259 (1965).

Relative reaction rate comparisons are discussed in

W. J. Hehre, *Practical Strategies for Electronic Structure Calculations* Wavfunction, Irvine (1995).

The most hands on description of doing these calculations is in the software manuals, such as

R. Steckler, Y.-Y. Chuang, E. L. Coitino, W.-P. Hu, Y.-P. Lin, G. C. Lynch, K. A. Nguyen, C. F. Jackels, M. Z. Gu, I. Rossi, P. Fast, S. Clayton, V. S. Melissas, B. C. Garrett, A. D. Isaacson, D. G. Truhlar, POLYRATE Manual (1999).

Mathematical developments of reaction rate theory are given in

G. C. Schatz, M. A. Ratner, *Quantum Mechanics in Chemistry* Prentice Hall (1993).

D. G. Truhlar, A. D. Isaacson, B. C. Garrett, *Theory of Chemical Reaction Dynamics Vol. IV* M. Baer Ed., 65, CRC (1985).

A review of reactions in solution is

J. T. Hynes, *Theory of Chemical Reaction Dynamics Volume IV* B. Bauer, Ed., 171, CRC, Boca Raton (1985).

Reviews of transition state theory and variational transition state theory are

W. H. Miller, *Encycl. Comput. Chem.* **4**, 2375 (1998).

M. Quack, J. Troe, *Encycl. Comput. Chem.* **4**, 2708 (1998).

B. C. Garrett, D. G. Truhlar, *Encycl. Comput. Chem.* **5**, 3094 (1998).

D. G. Truhlar, B. C. Garrett, S. J. Klippenstein, *J. Phys. Chem.* **100**, 12771 (1996).

A. D. Isaacson, D. G. Truhlar, S. N. Rai, R. Steckler, G. C. Hancock, B. C. Garrett, M. J. Redmon, *Comp. Phys. Commun.* **47**, 91 (1987).

M. M. Kreevoy, D. G. Truhlar, *Investigation of Rates and Mechanisms of Reactions, Part 1, 4th Edition* C. F. Bernasconi Ed., 13, John Wiley, New York (1986).

T. Fonseca, J. A. N. F. Gomes, P. Grigolini, F. Marchesoni, *Adv. Chem. Phys.* **62**, 389 (1985).

K. J. Laidler, M. C. King, *J. Phys. Chem.* **87**, 2657 (1983).

D. G. Truhlar, W. L. Hase, J. T. Hynes, *J. Phys. Chem.* **87**, 2664 (1983).

P. Pechukas, *Ann. Rev Phys. Chem.* **32**, 159 (1981).

R. B. Walker, J. C. Light, *Ann. Rev. Phys. Chem.* **31**, 401 (1980).

T. F. George, J. Ross, *Ann. Rev. Phys. Chem.* **24**, 263 (1973).

J. C. Light, *Adv. Chem. Phys.* **19**, 1 (1971).

E. V. Waage, B. S. Rabinovitch, *Chem. Rev.* **70**, 377 (1970).

J. C. Keck, *Adv. Chem. Phys.* **13**, 85 (1967).

Reviews of trajectory calculations are

G. D. Billing, *Encycl. Comput. Chem.* **3**, 1587 (1998).

G. D. Billing, K. V. Mikkelsen, *Introduction to Molecular Dynamics and Chemical Kinetics* John Wiley & Sons, New York (1996).

J. M. Bowman, G. C. Schatz, *Ann. Rev. Phys. Chem.* **46**, 169 (1995).

Bimolecular Collisions M. N. R. Ashford, J. E. Battott, Eds., Royal Society of Chemistry, Herts (1989).

G. C. Schatz, *Ann. Rev. Phys. Chem.* **39**, 317 (1988).

D. G. Truhlar, *J. Phys. Chem.* **83**, 188 (1979).

D. G. Truhlar, J. T. Muckerman, *Atom-Molecule Collision Theory. A Guide for the Experimentalist* R. B. Bernstein, Ed., 505, Plenum (1979).

R. N. Porter, *Ann. Rev. Phys. Chem.* **25**, 317 (1974).

B. Widom, *Adv. Chem. Phys.* **5**, 353 (1963).

Tunneling in reactions is reviewed in

E. F. Caldin, *Chem Rev.* **69**, 135 (1969).

H. S. Johnston, *Adv. Chem. Phys* **3**, 131 (1961).

Radiationless transition calculations are described in

M. Klessinger, *Theoretical Organic Chemistry* C. Párkáni Ed., 581, Elsevier (1998).

J. Simons, J. Nichols, *Quantum Mechanics in Chemistry* 314, Oxford, New York (1997).

V. V. Kocharovsky, V. V. Kocharovsky, S. Tasaki, *Adv. Chem. Phys.* **99**, 333 (1997).

Adv. Chem. Phys. P. Gaspard, I. Burghardt, Eds., **101** (1997).

R. H. Landau, *Quantum Mechanics II; A Second Course in Quantum Theory Second Edition* 314, John Wiley & Sons, New York (1996).

F. Bernardi, M. Olivucci, M. A. Robb, *Chem. Soc. Rev.* **25**, 321 (1996).

L. Serrano-Andres, M. Merchan, I. Nebot-Gil, R. Lindh, B. O. Roos, *J. Chem. Phys.* **98**, 3151 (1993).

Adv. Chem. Phys. M. Baer, C. –Y. Ng, Eds., **82** (1992).

M. Baer, *Theory of Chemical Reaction Dynamics Vol. II* M. Baer Ed., 219, CRC (1985).

A. A. Ovchinnikov, M. Y. Ovchinnokova, *Adv. Quantum Chem.* **16**, 161 (1982).

C. Cohen-Tannoudji, B. Diu, F. Laloë, *Quantum Mechanics* 1299, John Wiley & Sons, New York (1977).

E. E. Nikitin, *Adv. Quantum Chem.* **5**, 135 (1970).

Statistical calculations are reviewed in

J. Troe, *Adv. Chem. Phys.* **101**, 819 (1997).

D. A. McQuarrie, *Adv. Chem. Phys.* **15**, 149 (1969).

20 Potential Energy Surfaces

In the chapter on reaction rates, it was pointed out that the perfect description of a reaction would be a statistical average of all possible paths rather than just the minimum energy path. Furthermore, femtosecond spectroscopy experiments show that molecules vibrate in many different directions until an energetically accessible reaction path is found. In order to examine these ideas computationally, the entire potential energy surface (PES) or an approximation to it must be computed. A PES is either a table of data or an analytic function, which gives the energy for any location of the nuclei comprising a chemical system.

20.1 PROPERTIES OF POTENTIAL ENERGY SURFACES

Once a PES has been computed, it can be analyzed to determine quite a bit of information about the chemical system. The PES is the most complete description of all the conformers, isomers, and energetically accessible motions of a system. Minima on this surface correspond to optimized geometries. The lowest-energy minimum is called the global minimum. There can be many local minima, such as higher-energy conformers or isomers. The transition structure between the reactants and products of a reaction is a saddle point on this surface. A PES can be used to find both saddle points and reaction coordinates. Figure 20.1 illustrates these topological features. One of the most common reasons for doing a PES computation is to subsequently study reaction dynamics as described in Chapter 19. The vibrational properties of the molecule can also be obtained from the PES.

In describing PES, the terms *adiabatic* and *diabatic* are used. In the older literature, these terms are used in confusing and sometimes conflicting ways. For the purposes of this discussion, we will follow the conventions described by Sidis; they are both succinct and reflective of the most common usage. The term adiabatic originated in thermodynamics, where it means no heat transfer ($d_q = 0$). An adiabatic PES is one in which a particular electronic state is followed; thus no transfer of electrons between electronic states occurs. Only in a few rare cases is this distinction noted by using the term *electronically adiabatic*. A diabatic surface is the lowest-energy state available for each set of nuclear positions, regardless of whether it is necessary to switch from one electronic state to another. These are illustrated in Figure 20.2.

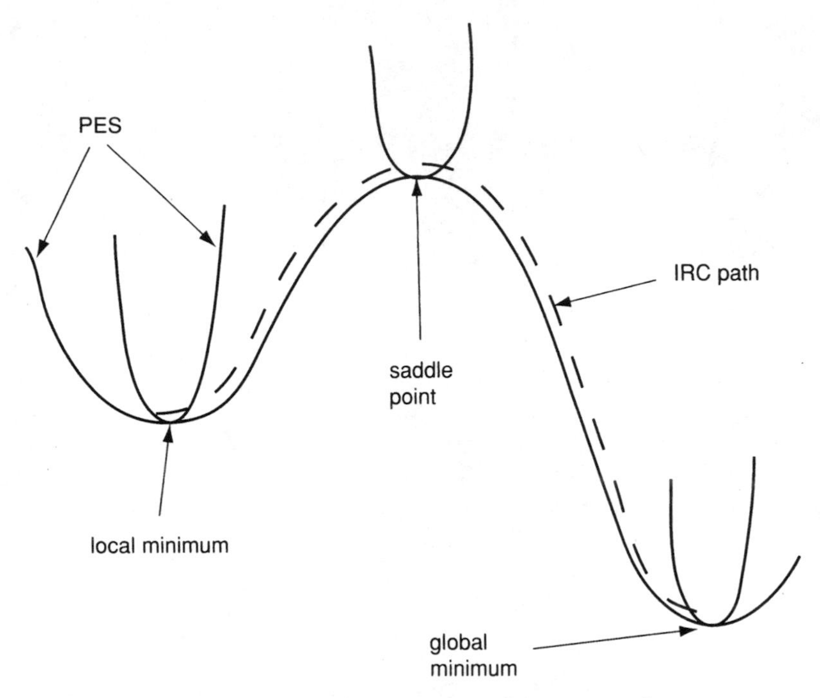

FIGURE 20.1 Points on a potential energy surface.

The mathematical definition of the Born–Oppenheimer approximation implies following adiabatic surfaces. However, software algorithms using this approximation do not necessarily do so. The approximation does not reflect physical reality when the molecule undergoes nonradiative transitions or two

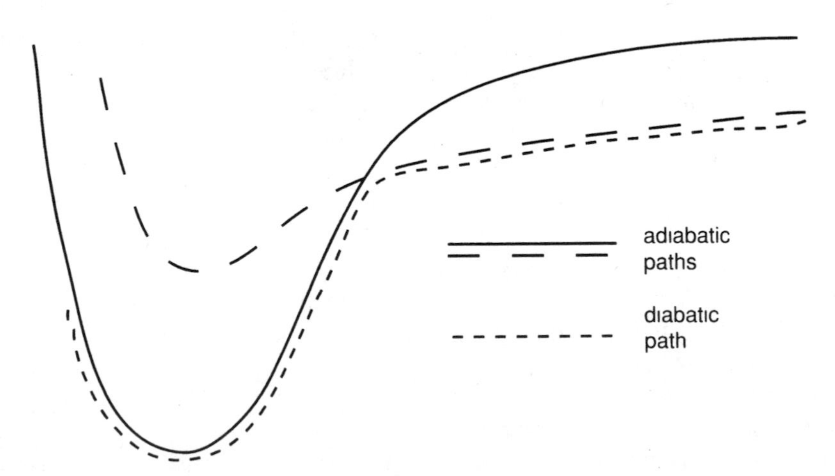

FIGURE 20.2 Adiabatic paths for bond dissociation in two different electronic states and the diabatic path.

electronic states are intimately involved in a process, such as Jahn–Teller distortions or vibronic coupling effects.

Depending on the information desired, the researcher might wish to have a calculation follow either an adiabatic path or a diabatic path, as shown in Figure 20.2. Which potential energy surface is followed depends on both the construction of the wave function and the means by which wave function symmetry is used by the software. A single-determinant wave function in which wave function symmetry is imposed will follow the adiabatic surface where two states of different symmetry cross. When two states of the same symmetry cross, a single-determinant wave function will exhibit an avoided crossing, as shown in Figure 17.2. A multiple-determinant wave function, in which both states are in the configuration space, will have the flexibility to follow the diabatic path. It is sometimes possible to get a multiple-determinant calculation (particularly MCSCF) to follow the adiabatic path by moving in small steps and using the optimized wave function from the previous point as the initial guess for each successive step.

20.2 COMPUTING POTENTIAL ENERGY SURFACES

Computing a complete PES for a molecule with N atoms requires computing energies for geometries on a grid of points in $3N$-6 dimensional space. This is extremely CPU-intensive because it requires computing a set of X points in each dimension, resulting in X^{3N-6} single point computations. Because of this, PES's are typically only computed for systems with a fairly small number of atoms.

Some software packages have an automated procedure for computing all the points on a PES, but a number of technical problems commonly arise. Some programs have a function that halts the execution when two nuclei are too close together; this function must be disabled. SCF procedures often exhibit convergence problems far from equilibrium. Methods for fixing SCF convergence problems are discussed in Chapter 22. At different points on the PES, the molecule may have different symmetries. This often results in errors when using software packages that use molecular symmetry to reduce computation time. The use of symmetry by a program can often be turned off. Computation of the PES for electronic excited states can lead to additional technical difficulties as described in Chapter 25.

The level of theory necessary for computing PES's depends on how those results are to be used. Molecular mechanics calculations are often used for examining possible conformers of a molecule. Semiempiricial calculations can give a qualitative picture of a reaction surface. *Ab initio* methods must often be used for quantitatively correct reaction surfaces. Note that size consistent methods must be used for the most accurate results. The specific recommendations given in Chapter 18 are equally applicable to PES calculations.

20.3 FITTING PES RESULTS TO ANALYTIC EQUATIONS

Once a PES has been computed, it is often fitted to an analytic function. This is done because there are many ways to analyze analytic functions that require much less computation time than working directly with *ab initio* calculations. For example, the reaction can be modeled as a molecular dynamics simulation showing the vibrational motion and reaction trajectories as described in Chapter 19. Another technique is to fit *ab initio* results to a semiempirical model designed for the purpose of describing PES's.

Of course, the analytic surface must be fairly close to the shape of the true potential in order to obtain physically relevant results. The criteria on fitting PES results to analytic equations have been broken down into a list of 10 specific items, all of which have been discussed by a number of authors. Below is the list as given by Schatz:

1. The analytic function should accurately characterize the asymptotic reactant and product molecules.
2. It should have the correct symmetry properties of the system.
3. It should represent the true potential accurately in the interaction regions for which experimental or nonempirical theoretical data are available.
4. It should behave in a physically reasonable manner in those parts of the interaction regions for which no experimental or theoretical data are available.
5. It should smoothly connect the asymptotic and interaction regions in a physically reasonable way.
6. The interpolating function and its derivatives should have as simple an algebraic form as possible consistent with the desired goodness of fit.
7. It should require as small a number of data points as possible to achieve an accurate fit.
8. It should converge to the true surface as more data become available.
9. It should indicate where it is most meaningful to compute the data points.
10. It should have a minimal amount of "ad hoc" or "patched up" character.

Criteria 1 through 5 must be obeyed in order to obtain reasonable results in subsequent calculations using the function. Criteria 6 through 10 are desirable for practical reasons. Finding a function that meets these criteria requires skill and experience, and no small amount of patience.

The analytic PES function is usually a summation of two- and three-body terms. Spline functions have also been used. Three-body terms are often polynomials. Some of the two-body terms used are Morse functions, Rydberg functions, Taylor series expansions, the Simons–Parr–Finlan (SPF) expansion, the Dunham expansion, and a number of trigonometric functions. Ermler & Hsieh, Hirst, Isaacson, and Schatz have all discussed the merits of these functions. Their work is referenced in the bibliography at the end of this chapter.

20.4 FITTING PES RESULTS TO SEMIEMPIRICAL MODELS

The most commonly used semiempirical for describing PES's is the diatomics-in-molecules (DIM) method. This method uses a Hamiltonian with parameters for describing atomic and diatomic fragments within a molecule. The functional form, which is covered in detail by Tully, allows it to be parameterized from either *ab initio* calculations or spectroscopic results. The parameters must be fitted carefully in order for the method to give a reasonable description of the entire PES. Most cases where DIM yielded completely unreasonable results can be attributed to a poor fitting of parameters. Other semiempirical methods for describing the PES, which are discussed in the reviews below, are LEPS, hyperbolic map functions, the method of Agmon and Levine, and the molecules-in-molecules (MIM) method.

BIBLIOGRAPHY

Introductory discussions can be found in

J. B. Foresman, Æ. Frisch, *Exploring Chemistry with Electronic Structure Methods Second Edition* Gaussian, Pittsburgh (1996).

A. R. Leach *Molecular Modelling Principles and Applications* Longman, Essex (1996).

I. N. Levine, *Physical Chemistry Fourth Edition* McGraw Hill, New York (1995).

P. W. Atkins, *Quanta* Oxford, Oxford (1991).

T. Clark, *A Handbook of Computational Chemistry* John Wiley & Sons, New York (1985).

Books on potential energy surfaces are

V. I. Minkin, B. Y. Simkin, R. M. Minyaev, *Quantum Chemistry of Organic Compounds; Mechanisms of Reactions* Springer-Verlag, Berlin (1990).

P. G. Mezey, *Potential Energy Hypersurfaces* Elsevier, Amsterdam (1987).

J. N. Murrell, S. Carter, S. C. Farantos, P. Huxley, A. J. C. Varandas, *Molecular Potential Energy Functions* John Wiley & Sons, New York (1984).

Potential Energy Surfaces and Dynamics Calculations D. G. Truhlar, Ed., Plunum, New York (1981).

Review articles are

M. L. McKee, M. Page, *Rev. Comput. Chem.* **4**, 35 (1993).

Adv. Chem. Phys. M. Baer, C.-Y. Ng, Eds., **82** (1992).

L. B. Harding, *Advances in Molecular Electronic Structure Theory* **1**, 45, T. H. Dunning, Jr., Ed., JAI, Greenwich (1990).

K. Balasubramanian, *Chem. Rev.* **90**, 93 (1990).

Advances in Molecular Electronic Structure Theory, Calculation and Characterization of Molecular Potential Energy Surfaces T. H. Dunning, Jr., Ed., JAI, Greenwich (1990).

New Theoretical Concepts for Understanding Organic Reactions J. Bertrán, I. G. Csizmadia, Eds., Kluwer, Dordrecht (1988).

D. G. Truhlar, R. Steckler, M. S. Gordon, *Chem. Rev.* **87**, 217 (1987).

P. J. Bruna, S. D. Peyerimhoff, *Adv. Chem. Phys.* **67**, 1 (1987).

T. H. Dunning, Jr., L. B. Harding, *Theory of Chemical Reaction Dynamics Vol 1* M. Baer Ed. 1, CRC (1985).

P. J. Kuntz, *Theory of Chemical Reaction Dynamics Vol 1* M. Baer Ed. 71, CRC (1985).

D. M. Hirst, *Adv. Chem. Phys.* **50**, 517 (1982).

Adv. Chem. Phys. K. P. Lawley, Ed., **42** (1980).

J. N. Murrell, *Gas Kinetics and Energy Transfer, Volume 3* P. G. Ashmore, R. J. Donovan (Eds.) 200, Chemical Society, London (1978).

W. A. Lester, Jr., *Adv. Quantum Chem.* **9**, 199 (1975).

G. G. Balint-Kurti, *Adv. Chem. Phys.* **30**, 137 (1975).

Reactions involving crossing between excited state surfaces are discussed in

M. Klessinger, *Theoretical Organic Chemistry* C. Párkáni Ed., 581, Elsevier (1998).

F. Bernardi, M. Olivucci, M. A. Robb, *Chem. Soc. Rev.* **25**, 321 (1996).

V. Sidis, *Adv. Chem. Phys.* **82**, 73 (1992).

J. J. Kaufman, *Adv. Chem. Phys.* **28**, 113 (1975).

Fitting to analytic functions is reviewed in

A. D. Isaacson, *J. Phys. Chem.* **96**, 531 (1992).

W. C. Ermler, H. C. Hsieh, *Advances in Molecular Electronic Structure Theory* T. H. Dunning, Jr., Ed., 1, JAI, Greenwich (1990).

G. C. Schatz, *Advances in Molecular Electronic Structure Theory* T. H. Dunning, Jr., Ed., 85, JAI, Greenwich (1990).

The DIM method is reviewed in

J. C. Tully, *Potential Energy Surfaces* K. P. Lawly, Ed., 63 John Wiley & Sons, New York (1980).

21 Conformation Searching

Geometry optimization methods start with an initial geometry and then change that geometry to find a lower-energy shape. This usually results in finding a local minimum of the energy as depicted in Figure 21.1. This local minimum corresponds to the conformer that is closest to the starting geometry. In order to find the most stable conformer (a global minimum of the energy), some type of algorithm must be used and then many different geometries tried to find the lowest-energy one.

For any given molecule, this is important because the lowest-energy conformers will have the largest weight in the ensemble of energetically accessible conformers. This is particularly important in biochemical research in order to determine a protein structure from its sequence. A protein sequence is much easier to determine than an experimental protein structure by X-ray diffraction or neutron diffraction. Immense amounts of work have gone into attempting to solve this protein-folding problem computationally. This is very difficult due to the excessively large number of conformers for such a big molecule. This problem is complicated by the fact that even molecular mechanics calculations require a finite amount of computer time, which may be too long to spend on each trial conformer of a large protein.

Which conformation is most important falls into one of three categories. First, simply asking what shape a molecule has corresponds to the lowest-energy conformer. Second, examining a reaction corresponds to asking about the conformer that is in the correct shape to undergo the reaction. Since the difference in energy between conformers is often only a few kcal/mole, it is not uncommon to find reactions in which the active conformer is not necessarily the lowest-energy one. Third, predicting an observable property of the system may require using the statistical weights of that property for all the energetically accessible conformers of the system.

Conformation search algorithms are an automated means for generating many different conformers and then comparing them based on their relative energies. Due to the immensely large number of possible conformers of a large molecule, it is desirable to do this with a minimum amount of CPU time. Quite often, all bond lengths are held fixed in the course of the search, which is a very reasonable approximation. Frequently, bond angles are held fixed also, which is a fairly reasonable approximation.

Algorithms that displace the Cartesian coordinates of atoms have also been

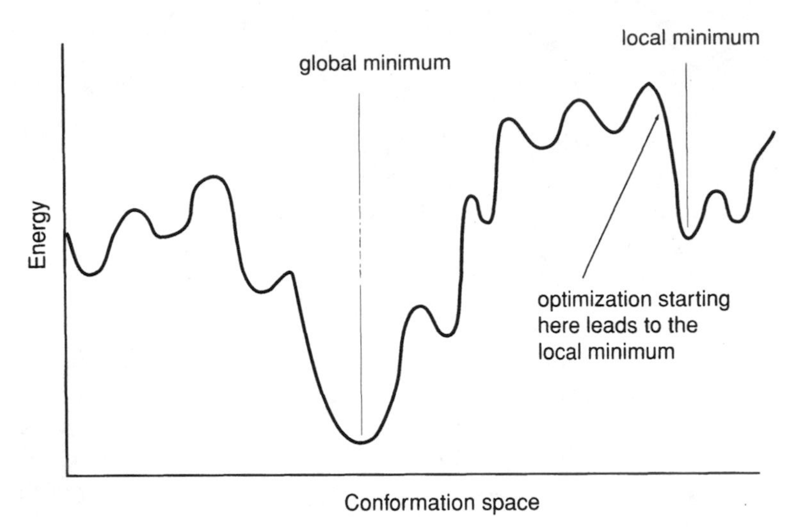

FIGURE 21.1 A one-dimensional representation of the energy of all possible conformers of a simple molecule.

devised. This is usually one of the less efficient methods for exploring conformers of flexible molecules. It can also have the unintended side effect of changing the stereochemistry of the compound. This can be an efficient way to find conformers of nonaromatic ring systems.

From geometry optimization studies, it is known that energy often decreases quite rapidly in the first few steps of the optimization. This is particularly common when using conjugate gradient algorithms, which are often employed with molecular mechanics methods. Because of this, geometry minimization is sometimes incorporated in conformation search algorithms. Some algorithms do one or two minimization steps after generating each trial conformer. Other algorithms will perform a minimization only for the lowest-energy conformers. It is advisable to do an energy minimization for a number of the lowest-energy structures found by a conformation search, not just the single lowest-energy structure.

The sections below describe the most commonly used techniques. There are many variations and permutations for all of these. The reader is referred to the software documentation and original literature for clarification of the details.

21.1 GRID SEARCHES

The simplest way to search the conformation space is by simply choosing a set of conformers, each of which is different by a set number of degrees from the previous one. If each bond is divided into M different angles and N bonds are

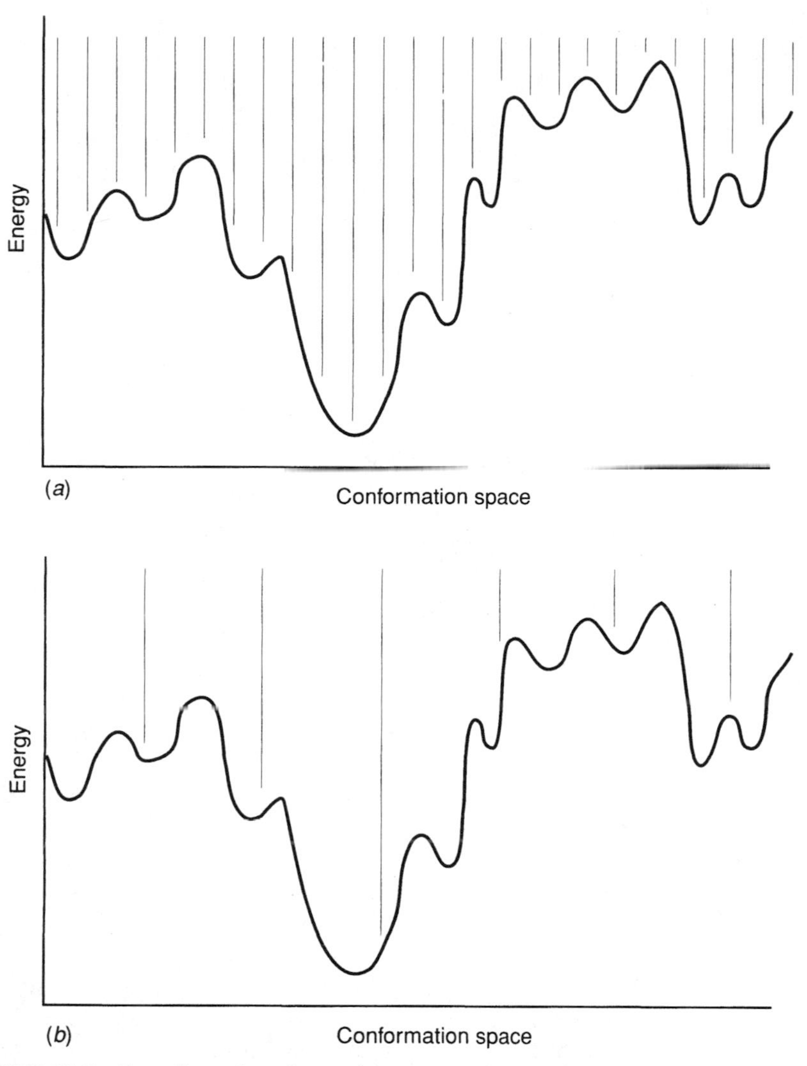

FIGURE 21.2 Sampling of conformation space using a grid search. (*a*) Sampling with a fine grid. (*b*) Sampling with a coarse grid.

rotated, this results in M^N possible conformers. In order to search the conformation space adequately, these points must be fairly close together as depicted in Figure 21.2. Although very slow, a grid search with a fairly small step size is the only way to be completely sure that the absolute global minimum has been found. The number of steps can be best limited by checking only the known staggered conformations for each bond, but this can still give a large number of conformers.

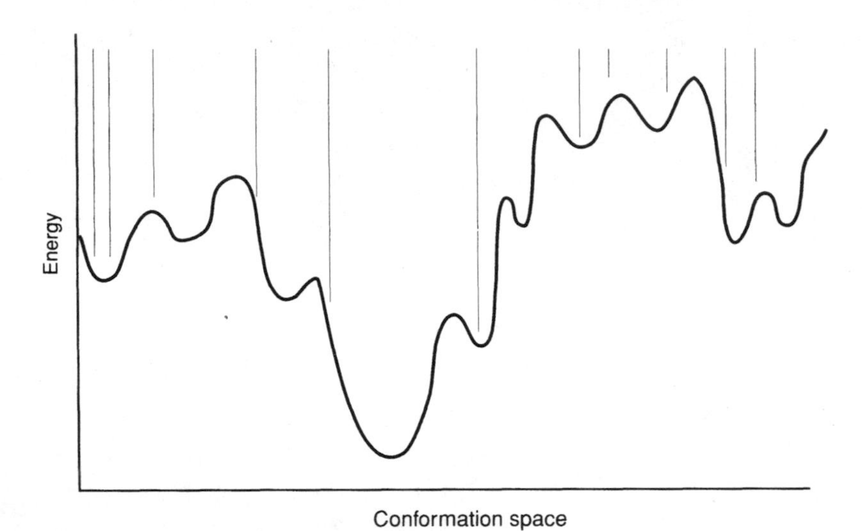

FIGURE 21.3 Sampling of conformation space using a Monte Carlo search (with a small number of iterations).

21.2 MONTE CARLO SEARCHES

Monte Carlo methods, which randomly choose the conformer angles, are a more efficient way to sample the entire conformation space. For the same amount of CPU time, a Monte Carlo method has a better chance of finding a lower-energy conformer than a grid search. However, unlike a grid search with a small step size, the results from a Monte Carlo optimization will not be guaranteed rigorously as the goal minimum, but rather as a near-optimum solution. This is because the Monte Carlo algorithm may search some regions more thoroughly than others, as shown in Figure 21.3. As illustrated in this figure, the lowest-energy conformer found may not lead to the global minimum even after an optimization is performed. Monte Carlo searches can be efficient in finding conformers that are very close in energy although much different in shape. This is because the entire conformation space is being searched. There is no way to predict how many iterations will be necessary to completely search the entire conformation space.

Monte Carlo searching becomes more difficult for large molecules. This is because a small change in the middle of the molecule can result in a large displacement of the atoms at the ends of the molecule. One solution to this problem is to hold bond lengths and angles fixed, thus changing conformations only, and to use a small maximum displacement.

A second option is to displace all atoms in Cartesian coordinates and then run an optimization. This second option works well for ring systems, but is not so efficient for long chains. This may also result in changing the stereochemistry of the molecule.

Several general rules of thumb can be given for doing very thorough Monte Carlo searches. The first is to run 1000 to 2000 Monte Carlo steps for each degree of freedom in the molecule. It is best to perform a couple of minimization steps at each point in the search. The lowest-energy structures should be saved and minimized fully. This should be done with anywhere from a few to a few hundred of the lowest-energy conformers, depending on the complexity of the molecule. Combine the results of several searches, keeping the unique structures. This process is continued until no more unique structures are found.

21.3 SIMULATED ANNEALING

A simulated annealing calculation has a greater level of sophistication than a Monte Carlo simulation. This algorithm is based on the observation that crystals are global energy minima for very complex systems. Crystals are typically formed by slowly decreasing the temperature. This is because the system will start out with enough energy to cross energy barriers from less favorable to more favorable configurations. As the temperature is lowered, the least favorable configurations are rendered energetically inaccessible and an increasingly larger number of molecules must populate the lowest-energy orientations.

A simulated annealing algorithm is a molecular dynamics simulation, in which the amount of kinetic energy in the molecule (the simulation temperature) slowly decreases over the course of the simulation. At the beginning of the simulation, many high-energy structures are being examined and high-energy barriers can be crossed. At the end of the calculation, only structures that are close to the best-known low-energy structures are examined, as depicted in Figure 21.4. This is generally an improvement over Monte Carlo methods in terms of the number of low-energy conformers found for a given amount of computation time.

There are similar algorithms, also called *simulated annealing*, that are Monte Carlo algorithms in which the choice conformations obey a Gaussian distribution centered on the lowest-energy value found thus far. The standard deviation of this distribution decreases over the course of the simulation.

In practice, simulated annealing is most effective for finding low-energy conformers that are similar in shape to the starting geometry. A simulated annealing algorithm starts from a given geometry and has the ability to cross barriers to other conformers. If the global minimum is reasonably similar to this geometry, a simulated annealing algorithm will have a good chance of finding it. If there are a number of high-energy barriers between the starting geometry and global minimum, a simulated annealing algorithm will not be the most efficient way to find the global minimum. This is because it will require a long computation time, starting from a high temperature and cooling slowly in order to adequately cross those barriers and search all the conformation space. Monte Carlo methods will be a better choice when there are a large number of nearly equivalent minima with significantly different structures. If the simula-

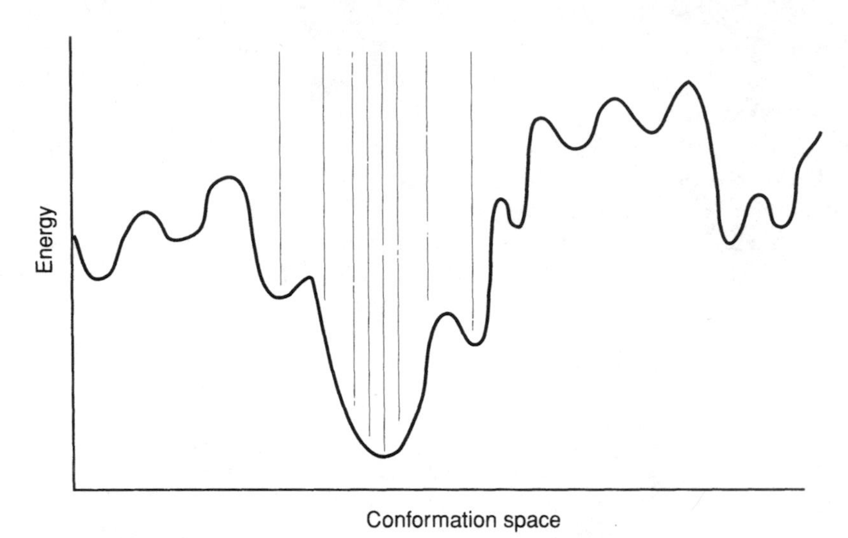

FIGURE 21.4 Sampling of conformation space using a simulated annealing algorithm: Looks like the Monte Carlo sampling in Figure 21.3 at the beginning of the simulation and like the sampling shown here at the end of the simulation.

tion is being run long enough with a sufficiently slow decrease in temperature, multiple runs starting from different geometries should give the same ending structure most of the time. Thus, the results of several identical simulations are examined to determine if the simulation was large enough to yield results that can be trusted.

21.4 GENETIC ALGORITHMS

Genetic algorithms stem from the observation that the evolution process tends to produce increasingly well-adapted populations. In keeping with this model, there must be some set of numbers representing the conformation of the molecule that are referred to collectively as a "chromosome." There must also be a criteria for measuring fitness that is generally the conformational energy. At the start of the simulation, a "population" consisting of many different conformers is chosen randomly. Over the course of the simulation, these are combined to produce new conformers using algorithms that represent reproduction, replacement, and mutation.

The reproduction process includes functions for selection, replacement, crossover, mutation, and elitism. Selection ensures that individuals in one generation have a higher probability of being included in the reproduction process if they have a higher fitness. Replacement is the copying of some individuals from one generation to another, thus resulting in some overlap between generations. Crossover is the process of individuals acquiring parts of their chromosomes from each parent. Mutation is accomplished by randomly changing a small percentage of the chromosomes in a generation. Elitism is the unnatural choice of copying the very best individual in the population into the next generation without mutation. All these operations can be defined as occurring on an array of numbers or a string of binary bits. As this process continues, the population becomes better adapted, as indicated by the fact that most of the population has the same value for one of its "genes," which has converged. Once most of the genes have converged, the simulation is stopped and the most fit individual is accepted as the global optimum.

There are a number of known problems with genetic algorithms, most of which can be somewhat corrected by more sophisticated variations on the basic algorithm. A few very fit individuals can force the population to a premature convergence on a local minimum. Premature convergence can be prevented by requiring a certain amount of diversity in the population. All the conformational space will be sampled even if some of it is physically unreasonable. A well-constructed fitness function can minimize this. The genetic algorithm only works well with multiple tests of many individuals and many generations; thus, these are inherently large simulations. The exact crossover algorithm can also have a significant bearing on the final results.

A genetic algorithm is generally an effective way of generating a large number of low-energy conformers. However, there is no guarantee that a global minimum will be found. A number of tests have shown genetic algorithms to be superior to simulated annealing and grid searches.

21.5 DISTANCE-GEOMETRY ALGORITHMS

The amount of computation necessary to try many conformers can be greatly reduced if a portion of the structure is known. One way to determine a portion of the structure experimentally is to obtain some of the internuclear distances from two-dimensional NMR experiments, as predicted by the nuclear Overhauser effect (NOE). Once a set of distances are determined, they can be used as constraints within a conformation search. This has been particularly effective for predicting protein structure since it is very difficult to obtain crystallographic structures of proteins. It is also possible to define distance constraints based on the average bond lengths and angles, if we assume these are fairly rigid while all conformations are accessible.

The experimentally determined correct distances are incorporated into the energy expression by defining a penalty function, which is zero for a reasonable range of correct distances and then increases outside of that range. Once this energy penalty has been defined, other search techniques such as Monte Carlo simulation and simulated annealing can be used. This technique has the added advantage of searching a space of conformers that are relevant to the experimental results, even when that might not be the global minimum.

If the molecular motion is faster than the NMR timescale, the distance predicted by an NOE measurement will be a time average only. In this case, it might be necessary to use a wider range of acceptable distances in order to incorporate all conformers consistent with the NOE data. Time-averaged constraint algorithms have been proposed, but these are difficult to use effectively.

21.6 THE FRAGMENT APPROACH

One way to reduce computation time is to optimize one part of a molecule at a time. For example, the conformation of a *t*-butyl group is well known; thus, there is no need to expend computation time searching conformations of that group. This concept of fragmentation can be automated by constructing a database of molecular fragments in their lowest-energy conformation. An algorithm can then be designed to use those fragments as they appear in the database, thus only optimizing the conformations between fragments.

Another variation on this technique is to first optimize the side chains and then keep the side chains fixed while optimizing the backbone. In an extreme case, representing these fixed side chains as large polygons with some net interaction potential can increase the calculation speed even more.

21.7 CHAIN GROWTH

The fragment approach has sometimes been combined with a chain growth or buildup algorithm. The chain growth algorithm is one in which the full molecule is built up one unit at a time. As each unit (monomer or functional group) is added, its conformation is searched without changing the rest of the chain. This results in a CPU time requirement that is directly proportional to the number of units and is thus much faster than some of the other algorithms.

This is a fairly reasonable way to describe man-made amorphous polymers, which had not been given time to anneal. For polymers that form very quickly, a quick Monte Carlo search on addition can insert an amount of nonoptimal randomness, as is expected in the physical system.

This is not a good way to describe polymers that have been annealed to give a crystalline form or some crystalline domains. The chain growth algorithm as described above is a fairly poor way to describe biomolecules. This is because the core regions of biomolecules will not be solvent-accessible while the outer regions are solvent-accessible, something which is not taken into account by this simple algorithm. More sophisticated chain growth algorithms have been applied to biomolecular systems.

21.8 RULE-BASED SYSTEMS

Rule-based systems try to identify certain subsequences of amino acids that tend to have a particular secondary structure, such as sheets, α-helices, β-strands, and so on. These sections can then be held rigid while the conformations of the connecting fragments are searched. The most successful way of doing this is by analyzing many known structures to determine which amino acids or sequences of amino acids have a statistical tendency to adopt each type of secondary structure. The known structures are also analyzed to determine which tend to initiate or terminate regions of a particular secondary structure. The drawback of this method is that there are exceptions to every rule, which can lead to false results.

21.9 USING HOMOLOGY MODELING

Homology algorithms are based on finding similar molecules. These are most often applied to proteins since it is easy to automate the step of comparing the sequence of a protein with the sequences of known proteins. For example, the same protein from several different species will often have a very similar sequence and conformation. Once the similarity algorithm has been used to identify the most similar sequence, the geometry of the previously determined structure (called the template) can be used with the necessary substitutions to generate a very reasonable starting geometry (called the model). From this starting geometry, minimization algorithms or very limited conformation searches can be used. Homology techniques can also be applied to nucleic acids.

There are several significant advantages to homology modeling compared to other conformation search techniques. Homology techniques are less computer-intensive than the alternatives (but still not trivial). They can give high-quality results. Homology is also good at obtaining the tertiary structure.

There are also some disadvantages to homology modeling. It is still a relatively new technique. Manual intervention is necessary. The result is never perfect. Side chains and loops are difficult to position. Worst of all is that homology cannot always be used since it depends on finding a known structure with a similar sequence. Overall, the advantages outweigh the disadvantages, making homology a very important technique in the pharmaceutical industry.

The degree of sequence similarity determines how much work is involved in homology modeling and how accurate the results can be expected to be. If similarity is greater than 90%, homology can match crystallography within the experimental error, with the biggest difference in side chain rotations. A similarity of 75 to 90% can still result in very good results. At this point, the accuracy of the final results is limited by the computer power available, mostly for model optimization. A similarity of 50 to 75% can be expected to result in an RMS error of about 1.5 Å, with large errors in some sections of the molecule. A case with a similarity of 25 to 50% can give adequate results with manual intervention. These cases are limited by the alignment algorithm. With a similarity of less than 25%, the homology is unreliable, and distance-geometry usually gives superior results.

The homology process consists of eight steps:

1. Template recognition
2. Alignment
3. Alignment correction
4. Backbone generation
5. Generation of loops
6. Side-chain generation
7. Model optimization
8. Model verification

Template recognition is the process of finding the most similar sequence. The researcher must choose how to compute similarity. It is possible to run a fast, approximate search of many sequences or a slow, accurate search of a few sequences. Sequences that should be analyzed more carefully are the same protein from a different species, proteins with a similar function or from the same metabolic pathway, or a library of commonly observed substructures if available.

The alignment algorithm scores how well pairs match: high scores for identical pairs, medium scores for similar pairs (i.e., both hydrophobic), low scores for dissimilar pairs. Alignment is difficult for low-similarity sequences, thus requiring manual alignment. Multiple-sequence alignment is the process of using sections from several template compounds.

Alignment correction is used to compensate for limitations of the alignment procedure. Alignment algorithms are based on primary structure only. Decisions between two options with medium scores are not completely reliable. Alignment correction reexamines these marginal cases by looking at the secondary and tertiary structure in that localized region. This may be completely manual, or at the very least it may require visual examination.

Backbone generation is the first step in building a three-dimensional model of the protein. First, it is necessary to find structurally conserved regions (SCR) in the backbone. Next, place them in space with an orientation and conformation best matching those of the template. Single amino acid exchanges are assumed not to affect the tertiary structure. This often results in having sections of the model compound that are unconnected.

The generation of loops is necessary because disconnected regions are often separated by a section where a few amino acids have been inserted or omitted. These are often extra loops that can be determined by several methods. One method is to perform a database search to find a similar loop and then use its geometric structure. Often, other conformation search methods are used. Manual structure building may be necessary in order to find a conformation that connects the segments. Visual inspection of the result is recommended in any case.

Side chain generation is often a source of error. It will be most reliable if certain rules of thumb are obeyed. Start with structurally conserved side chains and hold them fixed. Then look at the energy and entropy of rotamers for the remaining side chains. Conventional conformation search techniques are often used to place each side chain.

Model optimization is a further refinement of the secondary and tertiary structure. At a minimum, a molecular mechanics energy minimization is done. Often, molecular dynamics or simulated annealing are used. These are frequently chosen to search the region of conformational space relatively close to the starting structure. For marginal cases, this step is very important and larger simulations should be run.

Model verification provides a common-sense check of results. One quick check is to compare the minimized energy to that of similar proteins. It is also important to examine the structure to ensure that hydrophobic groups point inward and hydrophilic groups point outward.

An ideal case would be one with high similarity, in which most differences are single amino acid exchanges and the entire tertiary structure can be determined from the template. A good case is one in which sections of 10 to 30 pairs are conserved, and conserved regions are separated by short strings of inserted, omitted, or exchanged amino acids (to be examined in the loop generation phase). An acceptable case is one in which many small regions are identical and separated by single or double exchanges, or one in which larger sections are taken from different templates. Thus, alignment may be difficult for automated algorithms and should be examined by hand. A poor case would be one in which there is low similarity, causing alignment algorithms to fail and requiring extensive model optimization. For this poor case, results are expected to be marginal even with considerable manual intervention.

21.10 HANDLING RING SYSTEMS

Ring systems present a particular difficulty because many of the structures generated by a conformation search algorithm will correspond to a broken ring. Grid searches and Monte Carlo searches of ring systems are often done by first breaking the ring and then only accepting conformations that put the two ends within a reasonable distance. Monte Carlo searches are also sometimes done in Cartesian coordinates.

Simulated annealing searches will find conformations corresponding to ring changes, without unreasonably breaking the ring. There are also good algorithms for randomly displacing the atoms and then performing a minimization of the geometry. If the ring is one well understood, such as cyclohexane, it is sometimes most efficient to simply try all the known chair and boat forms of the base compound.

21.11 LEVEL OF THEORY

It is quite common to do the conformation search with a very fast method and to then optimize a collection of the lowest-energy conformers with a more accurate method. In some cases, single geometry calculations with more accurate methods are also performed. Solvent effects may also be important as discussed in Chapter 24.

The simplest and most quickly computed models are those based solely on steric hindrance. Unfortunately, these are often too inaccurate to be trusted. Molecular mechanics methods are often the method of choice due to the large amount of computation time necessary. Semiempirical methods are sometimes used when molecular mechanics does not properly represent the molecule. *Ab initio* methods are only viable for the very smallest molecules. These are discussed in more detail in the applicable chapters and the sources mentioned in the bibliography.

21.12 RECOMMENDED SEARCH ALGORITHMS

Since computation time is the most important bottleneck to conformation searching, the following list starts with the methods most amenable to the largest molecular systems:

1. Homology-based starting structures
2. Distance-geometry algorithms
3. Fragment-based algorithms
4. Chain growth algorithms where applicable
5. Rule-based systems
6. Genetic algorithms
7. Simulated annealing
8. Monte Carlo algorithms
9. Grid searches

BIBLIOGRAPHY

Sources, which compare conformation search algorithms are

F. Jensen, *Introduction to Computational Chemistry* John Wiley & Sons, New York (1999).

A. R. Leach *Molecular Modellıng Principles and Applications* Longman, Essex (1996).

M. Vásquez, G. Némethy, H. A. Scheraga, *Chem. Rev.* **94**, 2183 (1994).

H. A. Scheraga, *Computer Simulation of Biomolecular Systems Voume 2* W. F. v. Gunsteren, P. K. Weiner, A. J. Wilkinson, Eds., ESCOM, Leiden (1993).

H. A. Scheraga, *Rev. Comput. Chem.* **3**, 73 (1992).

L. Ingber, B. Rosen, *Mathl. Comput. Modelling* **16**, 87 (1992).

A. R. Leach, *Rev. Comput. Chem.* **2**, 1 (1991).

J. Skolnick, A. Kolinski, *Ann. Rev. Phys. Chem.* **40**, 207 (1989).

Sources, which compare the accuracy of various levels of theory for describing conformational energy differences, are

W. J. Hehre, *Practical Strategies for Electronic Structure Calculations* Wavfunction, Irvine (1995).

I. Pettersson, T. Liljefors, *Rev. Comput. Chem.* **9**, 167 (1996).

I. N. Levine, *Quantum Chemistry Fourth Edition* Prentice Hall, Englewood Cliffs (1991).

A. Golebiewski, A. Parczewski, *Chem. Rev.* **74**, 519 (1974).

Review articles are

I. Kolossváry, W. C. Guida, *Encycl. Comput. Chem.* **1**, 513 (1998).

S. Vajda, *Encycl. Comput. Chem.* **1**, 521 (1998).

E. L. Eliel, *Encycl. Comput. Chem.* **1**, 531 (1998).

G. M. Crippen, *Encycl. Comput. Chem.* **1**, 542 (1998).

F. Neumann, J. Michl, *Encycl. Comput. Chem.* **1**, 556 (1998).

R. L. Jernigan, I. Bahar, *Encycl. Comput. Chem.* **1**, 562 (1998).

J. E. Strab, *Encycl. Comput. Chem.* **3**, 2184 (1998).

M. Saunders, *Encycl. Comput. Chem.* **4**, 2948 (1998).

Monte Carlo algorithms are discussed in

M. H. Kalos, P. A. Witlock, *Monte Carlo Methods Volume I: Basics* Wiley Interscience, New York (1986).

N. Metropolis, A. W. Rosenbluth, M. N. Rosenbluth, A. H. Teller, E. Teller, *J. Chem Phys.* **21**, 1087 (1953).

Simulated annealing is discussed in

S. Kirkpatrick, C. D. Gelatt, Jr., M. P. Vecchi, *Science* **220**, 671 (1983).

Genetic algorithms are discussed in

R. Judson, *Rev. Comput. Chem* **10**, 1 (1997).

Genetic Algorithms in Molecular Modelling J. Devillers, Ed., Academic Press, San Diego (1996).

C. B. Lucasius, G. Kateman, *Chemometrics Intell. Lab. Syst* **19**, 1 (1993).

D. B. Hibbert, *Chemometrics Intell. Lab. Syst.* **19**, 277 (1993).

H. M. Cartwright, *Applications of Artificial Intelligence in Chemistry* Oxford, Oxford (1993).

T. Brodmeier, E. Pretsch, *J. Comp. Chem* **15**, 588 (1994).

Distance-geometry algorithms are reviewed in

A. E. Torda, W. F. v. Gunsteren, *Rev. Comput. Chem* **3**, 143 (1992).

M.-P. Williamson, J. P. Walto, *Chem. Soc. Rev.* **21**, 227 (1992).

G. M. Crippen, T. F. Havel, *Distance Geometry and Molecular Conformation* John Wiley & Sons, New York (1988).

Homology modeling is discussed in recent computational drug design texts and

http://swift.embl-heidelberg.de/course/

S. Mosimann, R. Meleshko, M. N. G. James *Proteins* **23**, 301 (1995).

G. D. Schuler, S. F. Altschul, D. J. Lipman, *Proteins* **9**, 190 (1991).

J. Greer, *Proteins*, **7**, 317 (1990).

P. S. Shenkin, D. L. Yarmush, R. M. Fine, H. Wang, C. Levinthal, *Biopolymers* **26**, 2053 (1987).

J. W. Ponder, F. M. Richards, *J. Mol. Biol.* **193**, 775 (1987).

T. A. Jones, S. Thirup, *EMBO J.* **5**, 819 (1986).

22 Fixing Self-Consistent Field Convergence Problems

The self-consistent field (SCF) procedure is in its simplest description an equation of the form

$$x = f(x) \tag{22.1}$$

This means that an initial set of orbitals x is used to generate a new set of orbitals following a prescribed mathematical procedure $f(x)$. The procedure is repeated until some convergence criteria are met. These criteria may be slightly different from one software package to another. They are usually based on several aspects of the calculation. Two of the most common criteria are the change in total energy and the change in density matrix.

A mathematician would classify the SCF equations as nonlinear equations. The term "nonlinear" has different meanings in different branches of mathematics. The branch of mathematics called *chaos theory* is the study of equations and systems of equations of this type.

22.1 POSSIBLE RESULTS OF AN SCF PROCEDURE

In an SCF calculation, the energies from one iteration to the next can follow one of several patterns:

1. After a number of iterations, the energy from one iteration may be the same as from the previous iteration. This is what chemists desire: a converged solution.
2. The energies from one iteration to the next may oscillate between two values, four values, or any other power of 2. (The author is not aware of any examples other than powers of 2.)
3. The values could be almost repeating but not quite so. In chaos theory, these are called Lorenz attractor systems.
4. The energies may be random within some fixed range. Random-number generators use this property intentionally.
5. The values produced may be random and not bounded within any upper or lower limits. This may happen if the boundary conditions on the total wave function are violated.

We have encountered oscillating and random behavior in the convergence of open-shell transition metal compounds, but have never tried to determine if the random values were bounded. A Lorenz attractor behavior has been observed in a hypervalent system. Which type of nonlinear behavior is observed depends on several factors: the SCF equations themselves, the constants in those equations, and the initial guess.

22.2 HOW TO SAFELY CHANGE THE SCF PROCEDURE

Changing the constants in the SCF equations can be done by using a different basis set. Since a particular basis set is often chosen for a desired accuracy and speed, this is not generally the most practical solution to a convergence problem. Plots of results vs. constant values are the bifurcation diagrams that are found in many explanations of chaos theory.

Another way of changing the constants in an SCF calculation is to alter the geometry. Often, making a bond length a bit shorter than expected is effective (say, adjusting the length to 90% of its expected value). Lengthening bond lengths a bit and avoiding eclipsed or gauche conformations are the second and third best options. Once a converged wave function is obtained, move the geometry back where it should be and use the converged wave function as the initial guess or just complete an optimization from that point.

The initial value of variables can be changed by using a different initial guess in an SCF calculation. The best initial guess is usually a converged SCF calculation for a different state of the same molecule or a slightly different geometry of the same molecule. This can be a very effective way to circumvent convergence problems. In the worst case, it may be necessary to construct an initial guess by hand in order to ensure that the nodal properties of all the orbitals are correct for the desired electronic state of the molecule. The construction of the virtual orbitals as well as the occupied orbitals can have a significant effect on convergence. Multiconfiguration self-consistent field (MCSCF) calculations can be particularly sensitive to the initial guess.

There are quite a number of ways to effectively change the equation in an SCF calculation. These include switching computation methods, using level shifting, and using forced convergence methods.

Switching between Hartree–Fock (HF), DFT, semiempirical, generalized valence bond (GVB), MCSCF, complete active-space self-consistent field (CASSCF), and Møller–Plesset calculations (MPn) will change the convergence properties. Configuration interaction (CI) and coupled-cluster (CC) calculations usually start with an SCF calculation and thus they will not circumvent problems with an SCF. In general, higher levels of theory tend to be harder to converge. Ease of convergence as well as calculation speed are why lower-level calculations are usually used to generate the initial guess for higher-level calculations.

Oscillating convergence in an SCF calculation is usually an oscillation between wave functions that are close to different states or a mixing of states. Thus, oscillating convergence can often be aided by using a level-shifting algorithm. This artificially raises the energies of the virtual orbitals. Level shifting may or may not help in cases of random convergence.

Most programs will stop trying to converge a problem after a certain number of iterations. In a few rare cases, the wave function will converge if given more than the default number of iterations.

Most SCF programs do not actually compute orbitals from the previous iteration orbitals in the way that is described in introductory descriptions of the SCF method. Most programs use a convergence acceleration method, which is designed to reduce the number of iterations necessary to converge to a solution. The method of choice is usually Pulay's direct inversion of the iterative subspace (DIIS) method. Some programs also give the user the capability to modify the DIIS method, such as adding a dampening factor. These modifications can be useful for fixing convergence problems, but a significant amount of experience is required to know how best to modify the procedure. Turning off the DIIS extrapolation can help a calculation converge, but usually requires many more iterations.

Some convergence problems are due to numerical accuracy problems. Many programs use reduced accuracy integrals at the beginning of the calculation to save CPU time. However, this can cause some convergence problems for difficult systems. A course DFT integration grid can also lead to accuracy problems, as can an incremental Fock matrix formation procedure.

Some programs contain alternative convergence methods that are designed to force even the most difficult problems to converge. These methods are called direct minimization or quadratic convergence methods. Although these methods almost always work, they often require a very large number of iterations and thus a significant amount of CPU time.

22.3 WHAT TO TRY FIRST

If you have an SCF calculation that failed to converge, which of the techniques outlined here should you try first? Here are our suggestions, with the preferred techniques listed first:

1. Try a different initial guess. Many programs have several different initial guess procedures, often based on semiempirical calculations.
2. For an open-shell system, try converging the closed-shell ion of the same molecule and then use that as an initial guess for the open-shell calculation. Adding electrons may give more reasonable virtual orbitals, but as a general rule, cations are easier to converge than anions.
3. Another initial guess method is to first run the calculation with a small basis set and then use that wave function as the initial guess for a larger basis set calculation.

4. Try level shifting. This will usually work with the default parameters or not at all.
5. If the SCF is approaching but not reaching the convergence criteria, relax or ignore the convergence criteria. This is usually done for geometry optimizations that do not converge at the initial geometry. Geometry optimizations often converge better as they approach the equilibrium geometry.
6. Some programs use reduced-accuracy integrals to speed the SCF. Using full integral accuracy may be necessary for systems with diffuse functions, long-range interactions, or low-energy excited states. Turning off incremental Fock matrix formation may also be necessary for these systems.
7. For DFT calculations, use a finer integration grid.
8. Try changing the geometry. First, slightly shorten a bond length. Then, slightly extend a bond length and next shift the conformation a bit.
9. Consider trying a different basis set.
10. Consider doing the calculation at a different level of theory. This is not always practical, but beyond this point the increased number of iterations may make the computation time as long as that occurring with a higher level of theory anyway.
11. Turn off the DIIS extrapolation. Give the calculation more iterations along with this.
12. Give the calculation (with DIIS) more SCF iterations. This seldom helps, but the next option often uses so many iterations that it is worth a try.
13. Use a forced convergence method. Give the calculation an extra thousand iterations or more along with this. The wave function obtained by these methods should be tested to make sure it is a minimum and not just a stationary point. This is called a stability test.
14. See if the software documentation suggests any other ways to change the DIIS method. You may have to run hundreds of calculations to become experienced enough with the method to know what works when and by how much to adjust it.

BIBLIOGRAPHY

The manuals accompanying many software packages contain discussions of how to handle convergence difficulties.

There are discussions of handling convergence problems in

G. Vacek, J. K. Perry, J.-M. Langlois, *Chem. Phys. Lett.* **310**, 189 (1999).

F. Jensen, *Introduction to Computational Chemistry* John Wiley & Sons, New York (1999).

T. Clark, *A Handbook of Computational Chemistry* Wiley-Interscience, New York (1985).

The DIIS algorithm is presented in

P. Pulay, *Chem. Phys. Lett.* **73**, 393 (1980).

P. Pulay, *J. Comp. Chem.* **3**, 556 (1982).

A good introduction to chaos theory is

J. Gleick, *Chaos: Making a New Science* Viking, New York (1987).

More mathematical treatmentsof chaos theory are

S. H. Strogatz, *Nonlinear Dynamics and Chaos With Applications to Physics, Biology, Chemistry and Engineering* Addison Wesley, Reading (1994).

L. E. Reichl, *The Transition to Chaos In Conservative Classical Systems: Quantum Manifestations* Springer-Verlag, New York (1992).

23 QM/MM

Various computational methods have strengths and weaknesses. Quantum mechanics (QM) can compute many properties and model chemical reactions. Molecular mechanics (MM) is able to model very large compounds quickly. It is possible to combine these two methods into one calculation, which models a very large compound using MM and one crucial section of the molecule with QM. This calculation is designed to give results that have very good speed when only one region needs to be modeled quantum mechanically. It can also be used to model a molecule surrounded by solvent molecules. This type of calculation is called a QM/MM calculation.

23.1 NONAUTOMATED PROCEDURES

The earliest combined calculations were done simply by modeling different parts of the system with different techniques. For example, some crucial part of the system could be modeled by using an *ab initio* geometry-optimized calculation. The complete system could then be modeled using MM, by holding the geometry of the initial region fixed and optimizing the rest of the molecule.

This procedure yields a geometry for the whole system, although there is no energy expression that reflects nonbonded interactions between the regions. One use is to compute the conformational strain in ligands around a metal atom, which is important in determining the possibility of binding. In order to do this, the metal atom is removed from the calculation, leaving just the ligands in the geometry from the complete system. Two energy calculations on these ligands are then performed: one without geometry optimization and one with geometry optimization. The difference between these two energies is the conformational strain that must be introduced into the ligands in order to form the compound.

Another technique is to use an *ab initio* method to parameterize force field terms specific to a single system. For example, an *ab initio* method can be used to compute the reaction coordinate for a model system. An analytic function can then be fitted to this reaction coordinate. A MM calculation can then be performed, with this analytic function describing the appropriate bonds, and so on.

23.2 PARTITIONING OF ENERGY

Quantitative energy values are one of the most useful results from computational techniques. In order to develop a reasonable energy expression when two calculations are combined, it is necessary to know not only the energy of the two regions, but also the energy of interaction between those regions. There have been a number of energy computation schemes proposed. Most of these schemes can be expressed generally as

$$E = E_{\mathrm{QM}} + E_{\mathrm{MM}} + E_{\mathrm{QM/MM}} + E_{\mathrm{pol}} + E_{\mathrm{boundary}} \tag{23.1}$$

The first two terms are the energies of the individual computations. The $E_{\mathrm{QM/MM}}$ term is the energy of interaction between these regions, if we assume that both regions remain fixed. It may include van der Waals terms, electrostatic interactions, or any term in the force field being used. E_{pol} is the effect of either region changing as a result of the presence of the other region, such as electron density polarization or solvent reorganization. E_{boundary} is a way of representing the effect of the rest of the surroundings, such as the bulk solvent. The individual terms in $E_{\mathrm{QM/MM}}$, E_{pol}, and E_{boundary} are discussed in more detail in the following sections.

23.2.1 van der Waals

Most of the methods proposed include a van der Waals term for describing nonbonded interactions between atoms in the two regions. This is usually represented by a Leonard–Jones 6–12 potential of the form

$$E_{\mathrm{vdW}} = \frac{A}{r^{12}} - \frac{B}{r^{6}} \tag{23.2}$$

The parameters A and B are those from the force field being used. A few studies have incorporated a hydrogen-bonding term also.

23.2.2 Charge

The other term that is very widely used is a Coulombic charge interaction of the form

$$E_{\mathrm{Coulomb}} = \frac{q_i q_j}{r_{ij}} \tag{23.3}$$

The subscripts i and j denote two nuclei: one in the QM region and one in the MM region. The atomic charges for the MM atoms are obtained by any of the techniques commonly used in MM calculations. The atomic charges for the QM atoms can be obtained by a population analysis scheme. Alternatively, there might be a sum of interactions with the QM nuclear charges plus the interaction with the electron density, which is an integral over the electron density.

23.2.3 Describing Bonds between Regions

If the QM and MM regions are separate molecules, having nonbonded interactions only might be sufficient. If the two regions are parts of the same molecule, it is necessary to describe the bond connecting the two sections. In most cases, this is done using the bonding terms in the MM method. This is usually done by keeping every bond, angle, or torsion term that incorporates one atom from the QM region. Alternatively, a few studies have been done in which a separate orbital-based calculation was used to describe each bond connecting the regions.

23.2.4 Polarization

The energy terms above allow the shape of one region to affect the shape of the other and include the energy of interaction between regions. However, these nonbonded energy terms assume that the electron density in each region is held fixed. This can be a reasonable approximation for covalent systems. It is a poor approximation when the QM region is being stabilized by its environment, as is the case with polar solvent effects.

Polarization is usually accounted for by computing the interaction between induced dipoles. The induced dipole is computed by multiplying the atomic polarizability by the electric field present at that nucleus. The electric field used is often only that due to the charges of the other region of the system. In a few calculations, the MM charges have been included in the orbital-based calculation itself as an interaction with point charges.

23.2.5 Solvent Reorientation

Many of the methods define an energy function and then use that function for the geometry optimization. However, there are some methods that use a minimal coupling between techniques for the geometry optimization and then add additional energy corrections to the single point energy. In the latter case, some researchers have included a correction for the effect of the solvent molecules reorienting in response to the solute. This is not a widespread technique mostly because there is not a completely rigorous way to know how to correct for solvent reorientation.

23.2.6 Boundary Terms

It is sometimes desirable to include the effect of the rest of the system, outside of the QM and MM regions. One way to do this is using periodic boundary conditions, as is done in liquid-state simulations. Some researchers have defined a potential that is intended to reproduce the effect of the bulk solvent. This solvent potential may be defined just for this type of calculation, or it may be a continuum solvation model as described in the next chapter. For solids, a set of point charges, called a Madelung potential, is often used.

23.3 ENERGY SUBTRACTION

An alternative formulation of QM/MM is the energy subtraction method. In this method, calculations are done on various regions of the molecule with various levels of theory. Then the energies are added and subtracted to give suitable corrections. This results in computing an energy for the correct number of atoms and bonds analogous to an isodesmic strategy.

Three such methods have been proposed by Morokuma and coworkers. The integrated MO + MM (IMOMM) method combines an orbital-based technique with an MM technique. The integrated MO + MO method (IMOMO) integrates two different orbital-based techniques. The *our own* n-*layered integrated MO and MM* method (ONIOM) allows for three or more different techniques to be used in successive layers. The acronym ONIOM is often used to refer to all three of these methods since it is a generalization of the technique.

This technique can be used to model a complete system as a small model system and the complete system. The complete system would be computed using only the lower level of theory. The model system would be computed with both levels of theory. The energy for the complete system, combining both levels of theory, would then be

$$E = E_{\text{low, complete}} + E_{\text{high, model}} - E_{\text{low, model}} \tag{23.4}$$

Likewise, a three-layer system could be broken down into small, medium, and large regions, to be computed with low, medium, and high levels of theory (L, M, and H respectively). The energy expression would then be

$$E = E_{H,\text{small}} + E_{M,\text{medium}} - E_{M,\text{small}} + E_{L,\text{large}} - E_{L,\text{medium}} \tag{23.5}$$

This method has the advantage of not requiring a parameterized expression to describe the interaction of various regions. Any systematic errors in the way that the lower levels of theory describe the inner regions will be canceled out. The geometry of one region will affect the geometry of the other because interaction between regions is not a systematic effect. If we assume transferability of parameters, this method avoids any overcounting of the nonbonded interactions.

One disadvantage is that the lower levels of theory must be able to describe all atoms in the inner regions of the molecule. Thus, this method *cannot* be used to incorporate a metal atom into a force field that is not parameterized for it. The effect of one region of the molecule causing polarization of the electron density in the other region of the molecule is incorporated only to the extent that the lower levels of theory describe polarization. This method requires more CPU time than most of the others mentioned. However, the extra time should be minimal since it is due to lower-level calculations on smaller sections of the system.

23.4 SELF-CONSISTENT METHOD

Bersuker and coworkers have proposed a technique whereby the atoms on the boundary between regions are included in both calculations. In this procedure, optimizations are done with each method, using the boundary atom charge from the other method, and this is repeated until the geometry is consistent between the levels of theory. They specify that the boundary atom cannot be part of a π bridge between regions.

23.5 TRUNCATION OF THE QM REGION

MM methods are defined atom by atom. Thus, having a carbon atom without all its bonds does not have a significant affect on other atoms in the system. In contrast, QM calculations use a wave function that can incorporate second atom effects. An atom with a nonfilled valence will behave differently than with the valence filled. Because of this, the researcher must consider the way in which the QM portion of the calculation is truncated.

A few of the earliest methods did truncate the atom on the dividing line between regions. Leaving this unfilled valence is reasonable only for a few of the very approximate semiempirical methods that were used at that time.

A number of methods fill the valence of the interface atoms with an extra orbital, sometimes centered on the connecting MM atom. This results in filling out the valence while requiring a minimum amount of additional CPU time. The concern, which is difficult to address, is that this might still affect the chemical behavior of the interface atom or even induce a second atom affect.

The other popular solution is to fill out the valence with atoms. Usually, H atoms are used as shown in Figure 23.1. Pseudohalide atoms have been used

(a)

H

H Ph

F — C — C — $CH_2-CH_2-CH_2-CH_2-CH-CH_2-CH_3$

H

H H

(b)

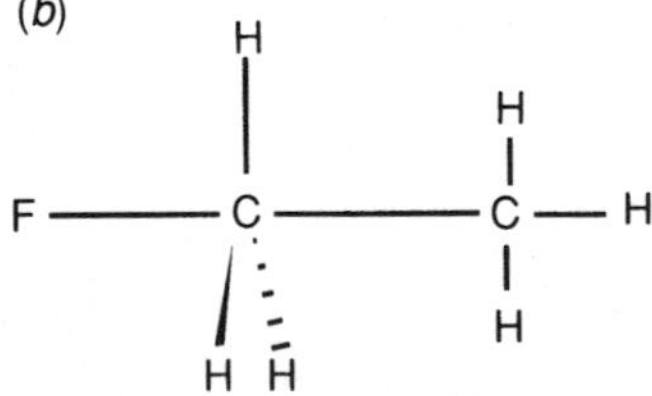

FIGURE 23.1 Example of a QM/MM region partitioning for a S_N2 reaction. (*a*) Entire molecule is shown with a dotted line denoting the QM region. (*b*) Molecule actually used for the QM calculation.

also. These pseudohalide atoms are parameterized to mimic the behavior of the MM atom for which they are substituted, such as a CH_2 group. These extra atoms are called *link atoms* or *junction dummy atoms*. The link atoms are not included in the energy expression used to describe the interaction between the regions of the system. The use of link atoms is somewhat questionable since they are often not subtracted from the final energy expression and may polarize the QM region incorrectly.

23.6 REGION PARTITIONING

The choice of where to locate the boundary between regions of the system is important. A number of studies have shown that very poor end results will be obtained if this is chosen improperly. There is no rigorous way to choose the best partitioning, but some general rules of thumb can be stated.

1. Any bonds that are being formed or broken must reside entirely in the QM region of the calculation.
2. Any section changing hybridization should be entirely in the QM region.
3. Keep conjugated or aromatic sections of the system completely in one region.
4. Where the second or third atom's effects are expected to be important, those atoms should be in the same region of the calculation.
5. QM/MM methods do not allow for charge transfer between different regions of the system. Thus, partitioning should not divide sections expected to have a charge separation.

Even with all these criteria met, researchers are advised to compare results from several different choices of boundary locations.

23.7 OPTIMIZATION

The more recently developed methods define an energy expression for the combined calculation and then use that expression to compute gradients for a geometry optimization. Some of the earlier methods would use a simpler level of theory for the geometry optimization and then add additional energy corrections to a final single point calculation. The current generation is considered to be the superior technique.

23.8 INCORPORATING QM TERMS IN FORCE FIELDS

Rather than doing several complete calculations with an additional interface, it is possible to incorporate orbital-based terms in a MM method. The first methods for doing this incorporated simple Hückel or PPP semiempirical models to help describe π system conjugation and aromaticity. There are also techniques for incorporating crystal field theory or ligand field theory descriptions of transition metals, which have proven difficult to model entirely with MM.

23.9 RECOMMENDATIONS

To date, there have not been any large-scale comparisons of QM/MM methods in which many different techniques were compared against experimental results for a large variety of chemical systems. There does tend to be some preference for the use of link atoms in order to ensure the correct chemical behavior of the QM region. Researchers are advised to consider the physical consequences of the effects that are included or excluded from various methods, as applied to their specific system. It is also prudent to verify results against experimental evidence when possible.

BIBLIOGRAPHY

Introductory descriptions are in

A. R. Leach *Molecular Modelling Principles and Applications* Longman, Essex (1996).

K. P. Eurenius, D. C. Chatfield, B. R. Brooks, M. Hodoscek, *Int. J. Quantum Chem.* **60**, 1189 (1996).

J. Gao, *Acc. Chem. Res* **29**, 298 (1996).

A book on QM/MM is

Combined Quantum and Molecular Mechanics Methods J. Gao, M. Thompson, Eds., American Chemical Society, Washington (1998).

Reviews of QM/MM are

P. Amara, M. J. Field, *Encycl. Comput. Chem.* **1**, 431 (1998).

R. D. J. Froese, K. Morokuma, *Encycl. Comput. Chem.* **2**, 1244 (1998).

J. Gao, *Encycl. Comput. Chem.* **2**, 1257 (1998).

K. M. Merz, Jr., R. V. Stanton, *Encycl. Comput. Chem.* **4**, 2330 (1998).

J. Tomasi, C. S. Pomelli, *Encycl. Comput. Chem.* **4**, 2343 (1998).

J. Gao, *Rev. Comput. Chem.* **7**, 119 (1992).

A. Warshel, *Computer Modeling of Chemical Reactions in Enzymes and Solutions* Wiley, New York (1991).

E. Clementi, *Computational Aspects for Large Chemical Systems* Springer, New York (1980).

Papers describing methods that incorporate modified orbitals are

J. Gao, P. Amara, C. Alhambra, J. J. Field, *J. Phys. Chem. A* **102**, 4714 (1998).

V. Théry, D. Rinaldi, J.-L. Rivail, B. Maigret, G. G. Ferency, *J. Comput Chem.* **15**, 269 (1994).

A. Warshel, M. Levitt, *J. Molec. Biol.* **103**, 227 (1976).

Methods describing methods using link atoms are

M. J. Field, P. A. Bash, M. Karplus, *J. Comput. Chem.* **11**, 700 (1990).

U. C. Singh, P. A. Kollman, *J. Comput. Chem.* **7**, 718 (1986).

Energy subtraction methods are described in

M. Svensson, S. Humbel, R. D. J. Froese, T. Matsubara, S. Sieber, K. Morokuma, *J. Phys. Chem.* **100**, 19357 (1996).

S. Humbel, S. Sieber, K. Morokuma, *J. Chem. Phys.* **105**, 1959 (1996).

F. Maseras, K. Morokuma, *J. Comput. Chem.* **16**, 1170 (1995).

The self consistent method is described in

I. B. Bersuker, M. K. Leong, J. E. Boggs, R. S. Pearlman, *Int. J. Quantum Chem.* **63**, 1051 (1997).

Methods incorporating orbital based terms in molecular mechanics force fields are described in

P. Comba, T. W. Hambley, *Molecular Modeling of Inorganic Compounds* VCH, New York (1995).

V. J. Burton, R. J. Deeth, *J. Chem Soc., Chem. Commun.* 573 (1995).

P. Comba, M. Zimmer, *Inorg. Chem.* **33**, 5368 (1994).

N. L. Allinger, J. T. Sprague, *J Am. Chem. Soc.* **95**, 3893 (1973).

24 Solvation

Most of the modeling methods discussed in this text model gas-phase molecular behavior, in which it is reasonable to assume that there is no interaction with other molecules. However, most laboratory chemistry is done in solution where the interaction between the species of interest and the solvent is not negligible.

The simulation of molecules in solution can be broken down into two categories. The first is a list of effects that are not defined for a single molecule, such as diffusion rates. These types of effects require modeling the bulk liquid as discussed in Chapters 7 and 39. The other type of effect is a solvation effect, which is a change in the molecular behavior due to the presence of a solvent. This chapter addresses this second type of effect.

24.1 PHYSICAL BASIS FOR SOLVATION EFFECTS

There is an energy of interaction between solute and solvent. Because of this, the solute properties dependent on energy, such as geometry, vibrational frequencies, total energy, and electronic spectrum, depend on the solvent. The presence of a solvent, particularly a polar solvent, can also stabilize charge separation within the molecule. This not only changes the energy, but also results in a shift in the electron density and associated properties. In reality, this is the result of the quantum mechanical interaction between solute and solvent, which must be averaged over all possible arrangements of solvent molecules according to the principles of statistical mechanics.

The energy of solvation can be further broken down into terms that are a function of the bulk solvent and terms that are specifically associated with the first solvation shell. The bulk solvent contribution is primarily the result of dielectric shielding of electrostatic charge interactions. In the simplest form, this can be included in electrostatic interactions by including a dielectric constant κ, as in the following Coulombic interaction equation:

$$E = \frac{q_i q_j}{\kappa r_{ij}} \tag{24.1}$$

This modification of the charge interaction is responsible for shifts in the electron density as permitted by the polarizability of the molecule.

There are several effects present in the region where the molecule meets the solvent shell. The first is referred to as a cavitation energy, which is the energy required to push aside the solvent molecules, thus making a cavity in which to place a solute molecule. The second is the force attracting the solute molecule to the solvent, which are the van der Waals, dispersion, and hydrogen-bonding interactions. Finally, the solvent molecules in the first shell can rearrange in order to maximize interactions with the solute. The largest amount of hydrogen-bonding energy is usually due to the solvent rearranging to the preferred hydrogen-bonding orientation. Methods modeling solvation energy based on breaking solvent–solvent "bonds" and forming solvent–solute "bonds" are called linear solvent energy relationships (LSER).

24.2 EXPLICIT SOLVENT SIMULATIONS

The most rigorously correct way of modeling chemistry in solution would be to insert all the solvent molecules explicitly and then run molecular dynamics or Monte Carlo calculations to give a time-averaged, ensemble average of the property of interest. This can be done using molecular-mechanics-style force fields, but even that is not a trivial amount of computational work. Furthermore, there are many properties that must be computed with orbital-based techniques. At present, there have been a few rare dynamics simulations at semiempirical or *ab initio* levels of theory, but most researchers do not have access to computing resources sufficient to complete quantum dynamics studies of molecules of interest. These calculations are often done using periodic boundary conditions so that long-range interactions will be accounted for.

In order to reduce the amount of computation time, some studies are conducted with a smaller number of solvent geometries, each optimized from a different starting geometry. The results can then be weighted by a Boltzmann distribution. This reduces computation time, but also can affect the accuracy of results.

In a few cases, where solvent effects are primarily due to the coordination of solute molecules with the solute, the lowest-energy solvent configuration is sufficient to predict the solvation effects. In general, this is a poor way to model solvation effects.

The primary problem with explicit solvent calculations is the significant amount of computer resources necessary. This may also require a significant amount of work for the researcher. One solution to this problem is to model the molecule of interest with quantum mechanics and the solvent with molecular mechanics as described in the previous chapter. Other ways to make the computational resource requirements tractable are to derive an analytic equation for the property of interest, use a group additivity method, or model the solvent as a continuum.

24.3 ANALYTIC EQUATIONS

It is reasonable to expect that the effect of a solvent on the solute molecule is, at least in part, dependent on the properties of the solute molecule, such as its size, dipole moment, polarizability, and so on. The earliest theoretical treatments were aimed at deriving analytic equations in which the molecular properties could be plugged in and the solvent effect computed. These equations are usually derived on theoretical grounds and by simply finding the parameters that give the best fit to experimental results. This type of scheme has been greatly improved and automated in recent years. Its current incarnation is quantitative structure property relationships (QSPR) and this method is discussed in Chapter 30.

Some molecular dynamics calculations use a potential of mean force (PMF). This requires reparameterization of the force field to give energetics reflecting the statistical weights found in solution. For example, the ratio of trans to gauche conformers in solution is actually due to both energetics and the dynamic effects of how the molecule can move most readily in solution, but PMF assumes that it is entirely energetic. These ratios can be determined either experimentally or from explicit solvent simulations. PMF calculations also incorporate a frictional drag term to simulate how the motion of the solute is dampened by the solvent. PMF calculations can provide dynamical information with a minimum amount of CPU time, but require that a new parameterization be done for each class of molecules and solvent.

24.4 GROUP ADDITIVITY METHODS

A similar technique is to derive a group additivity method. In this method, a contribution for each functional group must be determined. The contributions for the functional groups composing the molecule are then added. This is usually done from computations on a whole list of molecules using a fitting technique, similar to that employed in QSPR.

24.5 CONTINUUM METHODS

Another common approach is to do a calculation with the solvent included in some approximate manner. The simplest way to do this is to include the solvent as a continuum with a given dielectric constant. There are quite a few variations on this technique, only the most popular of which are included in the following sections.

24.5.1 Solvent Accessible Surface Area

The solvent accessible surface area (SASA) method is built around the assumption that the greatest amount of interaction with the solvent is in the area very close to the solute molecule. This is accounted for by determining a surface area for each atom or group of atoms that is in contact with the solvent. The free energy of solvation ΔG_s° is then computed by

$$\Delta G_S^{\circ} = \sum_i \sigma_i A_i \tag{24.2}$$

where σ_i is the surface tension associated with region i and A_i the surface area for that region.

This method does not attempt to distinguish between the various energy contributions. The surface tension parameter acts to include all interactions as much as possible. There are a number of algorithms for implementing this method, most of which differ in the means for determining the surface area associated with a particular group. This method is particularly popular for very large molecules, which can only be modeled by molecular mechanics.

24.5.2 Poisson Equation

The Poisson equation describes the electrostatic interaction between an arbitrary charge density $\rho(r)$ and a continuum dielectric. It states that the electrostatic potential ϕ is related to the charge density and the dielectric permitivity ε by

$$\nabla^2 \phi = \frac{-4\pi\rho(r)}{\varepsilon} \tag{24.3}$$

This can be solved analytically only for a few simplified systems. The Onsager model uses one of the known analytic solutions.

The Onsager model describes the system as a molecule with a multipole moment inside of a spherical cavity surrounded by a continuum dielectric. In some programs, only a dipole moment is used so the calculation fails for molecules with a zero dipole moment. Results with the Onsager model and HF calculations are usually qualitatively correct. The accuracy increases significantly with the use of MP2 or hybrid DFT functionals. This is not the most accurate method available, but it is stable and fast. This makes the Onsager model a viable alternative when PCM calculations fail.

The Poisson equation has been used for both molecular mechanics and quantum mechanical descriptions of solvation. It can be solved directly using numerical differential equation methods, such as the finite element or finite difference methods, but these calculations can be CPU-intensive. A more efficient quantum mechanical formulation is referred to as a self-consistent reaction field calculation (SCRF) as described below.

24.5.3 Poisson–Boltzmann Method

The Poisson equation assumes that the solvent is completely homogeneous. However, a solvent can have a significant amount of charge separation. An example of a heterogeneous solution would be a polar solute molecule surrounded by water with NaCl in solution. The positive sodium and negative chlorine ions will have a statistical tendency to migrate toward the negative and positive regions of the solute molecule.

The Poisson–Boltzmann equation is a modification of the Poisson equation. It has an additional term describing the solvent charge separation and can also be viewed mathematically as a generalization of Debye–Hückel theory.

24.5.4 Born Model

The Born model is based on electrostatic interactions, dielectric permitivity, and orbital overlaps. It has the advantage of being fairly straightforward and adaptable to computational methods. The free energy for the polarization of the solute is expressed as

$$G_p = -\frac{1}{2}\left(1 - \frac{1}{\varepsilon}\right) \sum_{ij} q_i q_j \gamma_{ij} \tag{24.4}$$

where q_i is the charge on center i and γ_{ij} the overlap between orbitals. Unlike the Poisson equation, this method is applicable to charged solutes also. It can be further simplified by approximating the overlap integrals. This allows it to be incorporated into molecular mechanics methods.

One very popular technique is an adaptation of the Born model for orbital-based calculations by Cramer and Truhlar, et. al. Their solvation methods (denoted SM1, SM2, and so on) are designed for use with the semiempirical and *ab initio* methods. Some of the most recent of these methods have a few parameters that can be adjusted by the user in order to customize the method for a specific solvent. Such methods are designed to predict ΔG_{solv} and the geometry in solution. They have been included in a number of popular software packages including the AMSOL program, which is a derivative of AMPAC created by Cramer and Truhlar.

The SM1–SM3 methods model solvation in water with various degrees of sophistication. The SM4 method models solvation in alkane solvents. The SM5 method is generalized to model any solvent. The SM5.42R method is designed to work with HF, DFT or hybrid HF/DFT calculations, as well as with AM1 or PM3. SM5.42R is implemented using a SCRF algorithm as described below. A description of the differences between these methods can be found in the manual accompanying the AMSOL program and in the reviews listed at the end of this chapter. Available Hamiltonians and solvents are summarized in Table 24.1.

The accuracy of these methods is tested by finding the mean absolute error between the computed and experimental free energies of solvation. The SM4 method does well for neutral molecules in alkane solvents with a mean absolute error of 0.3 kcal/mol. For neutral molecules, the SM5 methods do very well with mean absolute errors in the 0.3 to 0.6 kcal/mol range, depending on the method and solvent. For ions, the SM1 method seems to be most accurate with

TABLE 24.1 Cramer and Truhlar Semiempirical Solvation Methods

	Hamiltonian		Solvant			
Method	AM1	PM3	Water	Alkane	Organic	Custom
SM1	X		X			
SM1a	X		X			
SM2	X		X			
SM2.1	X		X			
SM2.2	X		X			
SM2.2PDA	X		X			
SM3		X	X			
SM3.1		X	X			
SM4	X	X		X		
SM5.0R	X		X	X	X	
SM5.05R	X		X		X	
SM5.2PDA	X		X			
SM5.2PDP		X	X			
SM5.2R[a]	X	X	X	X	X	
SM5.4PDA	X		X			
SM5.4PDP		X	X			
SM5.4PDU	X	X	X			
SM5.4U	X	X	X			
SM5.4A	X		X	X	X	X
SM5.4P		X	X	X	X	X
SM5.42R[b]	X	X	X	X	X	

[a] MNDO also.

[b] Also for HF, DFT, or hybrid DFT functionals.

a mean absolute error of 2.9 kcal/mol. Accuracy test results are summarized in more detail in the manual for the AMSOL program and in the review articles listed in the bibliography of this chapter.

24.5.5 GB/SA

The generalized Born/surface area (GB/SA) model is a combination of the Born and SASA models. This method has been effective in describing the solvation of biomolecular molecules. It is incorporated in the MacroModel software package.

24.5.6 Self-consistent Reaction Field

The self-consistent reaction field (SCRF) method is an adaptation of the Poisson method for *ab initio* calculations. There are quite a number of variations on this method. One point of difference is the shape of the solvent cavity. Various models use spherical cavities, spheres for each atom, or an isosurface of electron density. The second difference is the description of the solute, which could be a dipole, multipole expansion, or numerical integration over the charge density.

There are many technical details involved in SCRF calculations, many of which the user can control. Readers of this book are advised to use the default values as much as possible unless they have carefully examined the original literature and tested their modifications. PCM methods are generally more accurate than the Onsager and COSMO methods.

The most popular of the SCRF methods is the polarized continuum method (PCM) developed by Tomasi and coworkers. This technique uses a numerical integration over the solute charge density. There are several variations, each of which uses a nonspherical cavity. The generally good results and ability to describe the arbitrary solute make this a widely used method. However, it is sensitive to the choice of a basis set. Some software implementations of this method may fail for more complex molecules.

The original PCM method uses a cavity made of spherical regions around each atom. The isodensity PCM model (IPCM) uses a cavity that is defined by an isosurface of the electron density. This is defined iteratively by running SCF calculations with the cavity until a convergence is reached. The self-consistent isodensity PCM model (SCI-PCM) is similar to IPCM in theory, but different in implementation. SCI-PCM calculations embed the cavity calculation in the SCF procedure to account for coupling between the two parts of the calculation.

24.5.7 COSMO

The conductor-like screening model (COSMO) is a continuum method designed to be fast and robust. This method uses a simpler, more approximate equation for the electrostatic interaction between the solvent and solute. Line the SMx methods, it is based on a solvent accessible surface. Because of this, COSMO calculations require less CPU time than PCM calculations and are less likely to fail to converge. COSMO can be used with a variety of semiempirical, *ab initio*, and DFT methods. There is also some loss of accuracy as a result of this approximation.

An improved version, called COSMO for realistic solvents (COSMO-RS), has also been created. This method has an improved scheme for modeling nonelectrostatic effects. It can be adapted for modeling the behavior of molecules in any solvent and, gives increased accuracy of results as compared to COSMO.

24.6 RECOMMENDATIONS

The most accurate calculations are those that use a layer of explicit solvent molecules surrounded, in turn, by a continuum model. This adds the additional complexity of having to try various configurations of solvent molecules in order to obtain a statistical average. In some cases, biomolecules fold with solvent molecules locked into the folded structure, which is also represented well by this technique. The drawback of this technique is the large amount of work required.

There is no one best method for describing solvent effects. The choice of method is dependent on the size of the molecule, type of solvent effects being examined, and required accuracy of results. Many of the continuum solvation methods predict solvation energy more accurately for neutral molecules than for ions. The following is a list of preferred methods, with those resulting in the highest accuracy and the least amount of computational effort appearing first:

1. A layer of explicit solvent molecules surrounded by a continuum description for the highest possible accuracy.
2. SMx semiempirical methods for very modest computational demands.
3. PCM when quantum mechanics is necessary, but explicit solvent simulations are too CPU-intensive.
4. COSMO.
5. Onsager.
6. SASA or GB/SA for very large molecules.
7. Other continuum methods.
8. Analytic equations or group additivity techniques when applicable.
9. Potential of mean force for dynamics simulations.
10. Explicit solvent methods. Monte Carlo methods are somewhat more popular than molecular dynamics methods.

BIBLIOGRAPHY

Books discussing solvation methods are

F. Jensen, *Introduction to Computational Chemistry* John Wiley & Sons, New York (1999).

M. F. Schlecht, *Molecular Modeling on the PC* Wiley-VCH, New York (1998).

J. B. Foresman, Æ. Frisch, *Exploring Chemistry with Electronic Structure Methods Second Edition* Gaussian, Pittsburgh (1996).

G. H. Grant, W. G. Richards, *Computational Chemistry* Oxford, Oxford (1995).

C. Reichardt, *Solvents and Solvent Effects in Organic Chemistry* VCH, Weinheim (1988).

A. Ben-Naim, *Solvation Thermodynamics* Plenum, New York (1987).

Molecular Interactions Volume 3 H. Ratujczak, W. J. Orville-Thomas, M. Redshaw, Eds., John Wiley & Sons, New York (1982).

Review articles are

C. J. Cramer, D. G. Truhlar, *Chem. Rev.* **99**, 2161 (1999).
L. Gorb, J. Leszczynski, *Computational Chemistry Reviews of Current Trends Volume 3* 179, J. Leszczynski, Ed., World Scientific, Singapore (1999).
K. A. Sharp, *Encycl. Comput. Chem.* **1**, 571 (1998).
A. Klamt, *Encycl. Comput. Chem.* **1**, 604 (1998).
W. L. Jorgensen, *Encycl. Comput. Chem.* **2**, 1061 (1998).
R. B. Hermann, *Encycl. Comput. Chem.* **3**, 1703 (1998).
J. Tomasi, B. Mennucci, *Encycl. Comput. Chem.* **4**, 2547 (1998).
V. Gogonea, *Encycl. Comput. Chem.* **4**, 2560 (1998).
G. B. Bacskay, J. R. Reimers, *Encycl. Comput. Chem.* **4**, 2620 (1998).
M. Karelson, *Adv. Quantum Chem.* **28**, 141 (1997).
K. Coutinho, S. Canuto, *Adv. Quantum Chem.* **28**, 90 (1997).
C. Cativiela, J. I. García, J. A. Mayoral, L. Salvatella, *Chem. Soc. Rev.* **25**, 209 (1996).
C. J. Cramer, D. G. Truhlar, *Rev. Comput. Chem.* **6**, 1 (1995).
M. Vásquez, G. Némethy, H. A. Scheraga, *Chem. Rev.* **94**, 2183 (1994).
I. Alkorta, H. O. Villar, J. J. Perez, *J. Comput. Chem.* **14**, 620 (1993).
J. Tomasi, R. Bonaccorsi, R. Cammi, F. J. Olivares del Valle, *Theochem* **234**, 401 (1991).
A. Warshel, *Computer Modelling of Chemical Reactions in Enzymes and Solutions* John Wiley & Sons, New York (1991).
The Chemical Physics of Solvation Part A Theory of Solvation R. R. Dogonadze, E. Kálmán, a. A. Kornyshev, J. Ulstrup, Eds., Elsevier, Amsterdam (1985).
P. Schuster, W. Jakubetz, W. Marius, S. A. Rice, *Structure of Liquids* Springer-Verlag, Berlin (1975).
A. T. Amos, B. L. Burrows, *Adv. Quantum Chem.* **7**, 289 (1973).
S. Basu, *Adv. Quantum Chem.* **1**, 145 (1964).

Reviews of molecular dynamics simulations of reactions in solution are

R. M. Whitnell, K. B. Wilson, *Rev. Comput. Chem.* **1**, 67 (1993).
J. T. Hynes, *Theory of Chemical Reaction Dynamics Volume IV* B. Baer, Ed., 171, CRC, Boca Ratan (1985).

A review of QM/MM methods as applied to solvent-solute systems is

J. Gao, *Rev. Comput. Chem.* **7**, 119 (1996).

An example of using analytic equations is

M. H. Abraham, *Chem. Soc. Rev.* **22**, 73 (1993).

Group additivity methods are discussed in

G. P. Ford, J. D. Scribner, *J. Org. Chem.* **48**, 2226 (1983).
J. Hine, P. K. Mookerjee, *J. Org. Chem.* **40**, 292 (1975).

A review of modeling vibrational relaxation in liquids is

P. W. Oxtoby, *Ann. Rev. Phys. Chem.* **32**, 77 (1981).

A more mathematical description is in

A. H. De Vries, P. T. van Duijnen, A. H. Jaffer, J. A. Rullmann, J. P. Dijkman, H. Merenga, B. T. Thole, *J. Comput. Chem.* **16**, 37 (1995).

25 Electronic Excited States

This is an introduction to the techniques used for the calculation of electronic excited states of molecules (sometimes called *eximers*). Specifically, these are methods for obtaining wave functions for the excited states of a molecule from which energies and other molecular properties can be calculated. These calculations are an important tool for the analysis of spectroscopy, reaction mechanisms, and other excited-state phenomena.

These same techniques may also be necessary to find the ground-state wave function. Although most software packages attempt to compute the ground-state wave function, there is no way to guarantee that the algorithm will find the ground state even if the calculation does converge. Thus, it is sometimes necessary to attempt to find the first few states of a molecule just to ensure that the ground state has been found. Determining the ground-state electron configuration can be particularly difficult for compounds with low-energy excited states (i.e., transition metal systems).

Depending on the needs of the researcher, either vertical or adiabatic excitation energies may be desired. Vertical excitation energies are those in which the ground-state geometry is used, thus assuming that a fast process is being modeled. This is appropriate for electronic processes such as UV absorption or photo-electron spectroscopy. Adiabatic excitation energies are those in which the excited-state geometry has been optimized. Adiabatic excitations are more likely to reflect experimental results when the excited state is long lived relative to the time required for nuclear motion.

25.1 SPIN STATES

Ab initio programs attempt to compute the lowest-energy state of a specified multiplicity. Thus, calculations for different spin states will give the lowest-energy state and a few of the excited states. This is most often done to determine singlet-triplet gaps in organic molecules.

25.2 CIS

A single-excitation configuration interaction (CIS) calculation is probably the most common way to obtain excited-state energies. This is because it is one of the easiest calculations to perform.

A configuration interaction calculation uses molecular orbitals that have been optimized typically with a Hartree–Fock (HF) calculation. Generalized valence bond (GVB) and multi-configuration self-consistent field (MCSCF) calculations can also be used as a starting point for a configuration interaction calculation.

A CIS calculation starts with this initial set of orbitals and moves a single electron to one of the virtual orbitals from the original calculation. This gives a description of one of the excited states of the molecule, but does not change the quality of the description of the ground state as do double-excitation CIs. This gives a wave function of somewhat lesser quality than the original calculation since the orbitals have been optimized for the ground state. Often, this results in the ground-state energy being a bit low relative to the other states. The inclusion of diffuse basis functions can improve the accuracy somewhat.

The extended CIS method (XCIS) is a version of CIS for examining states that are doubly excited from the reference state. It does not include correlation and is thus similar in accuracy to CIS.

The CIS(D) method is designed to include some correlation in excited states. Initial results with this method show that it is stable and reliable and gives excitation energies significantly more accurate than those of CIS.

25.3 INITIAL GUESS

If the initial guess for a calculation is very close to an excited-state wave function, the calculation may converge to that excited state. This is typically done by doing an initial calculation and then using its wave function, with some of the orbitals switched as the initial guess for another calculation. This works best with HF calculations. It can work with MCSCF calculations too, but will not work with CI, CC, or MPn calculations. This is a very good way to find the ground state, or to at least verify that the state found was indeed the ground state.

The advantage of this method is that the orbitals have been optimized for the excited state. The disadvantage is that there is no guarantee it will work. If there is no energy barrier between the initial guess and the ground-state wave function, the entire calculation will converge back to the ground state. The convergence path may take the calculation to an undesired state in any case.

A second disadvantage of this technique applies if the excited state has the same wave function symmetry as a lower-energy state. There is no guarantee that the state obtained is completely orthogonal to the ground state. This means that the wave function obtained may be some mix of the lower-energy state and a higher-energy state. In practice, this type of calculation only converges to a higher state if a fairly reasonable description of the excited-state wave function is obtained. Mixing tends to be a significant concern if the orbital energies are very close together or the system is very sensitive to correlation effects.

25.4 BLOCK DIAGONAL HAMILTONIANS

Most *ab initio* calculations use symmetry-adapted molecular orbitals. Under this scheme, the Hamiltonian matrix is block diagonal. This means that every molecular orbital will have the symmetry properties of one of the irreducible representations of the point group. No orbitals will be described by mixing different irreducible representations.

Some programs such as COLUMBUS, DMOL, and GAMESS actually use a separate matrix for each irreducible representation and solve them separately. Such programs give the user the option of defining how many electrons are of each irreducible representation. This defines the symmetry of the total wave function. In this case, the resulting wave function is the lowest-energy wave function of a particular symmetry. This is a very good way to calculate excited states that differ in symmetry from the ground state and are the lowest-energy state within that symmetry.

25.5 HIGHER ROOTS OF A CI

For configuration interaction calculations of double excitations or higher, it is possible to solve the CI super-matrix for the 2nd root, 3rd root, 4th root, and so on. This is a very reliable way to obtain a high-quality wave function for the first few excited states. For higher excited states, CPU times become very large since more iterations are generally needed to converge the CI calculation. This can be done also with MCSCF calculations.

25.6 NEGLECTING A BASIS FUNCTION

Some programs, such as COLUMBUS, allow a calculation to be done with some orbitals completely neglected from the calculation. For example, in a transition metal compound, four *d* functions could be used so that the calculation would have no way to occupy the function that was left out.

This is a reliable way to obtain an excited-state wave function even when it is not the lowest-energy wave function of that symmetry. However, it might take a bit of work to construct the input.

25.7 IMPOSING ORTHOGONALITY: DFT TECHNIQUES

Traditionally, excited states have not been one of the strong points of DFT. This is due to the difficulty of ensuring orthogonality in the ground-state wave function when no wave functions are being used in the calculation.

The easiest excited states to find using DFT techniques are those that are the lowest state of a given symmetry, thus using a ground-state calculation. A promising technique is one that uses a variational bound for the average of the first M states of a molecule. A few other options have also been examined. However, there is not yet a large enough volume of work applying DFT to excited states to predict the reliability of any of these techniques.

25.8 IMPOSING ORTHOGONALITY: QMC TECHNIQUES

Quantum Monte Carlo (QMC) methods are computations that use a statistical integration to calculate integrals which could not be evaluated analytically. These calculations can be extremely accurate, but often at the expense of enormous CPU times. There are a number of methods for obtaining excited-state energies from QMC calculations. These methods will only be mentioned here and are explained more fully in the text by Hammond, Lester, and Reynolds.

Computations done in imaginary time can yield an excited-state energy by a transformation of the energy decay curve. If an accurate description of the ground state is already available, an excited-state description can be obtained by forcing the wave function to be orthogonal to the ground-state wave function.

Diffusion and Green's function QMC calculations are often done using a fixed-node approximation. Within this scheme, the nodal surfaces used define the state that is obtained as well as ensuring an antisymmetric wave function.

Matrix QMC procedures, similar to configuration interaction treatments, have been devised in an attempt to calculate many states concurrently. These methods are not yet well developed, as evidenced by oscillatory behavior in the excited-state energies.

25.9 PATH INTEGRAL METHODS

There has been some initial success at computing excited-state energies using the path integral formulation of quantum mechanics (Feynman's method). In this formulation, the energies are computed using perturbation theory. There has not yet been enough work in this area to give any general understanding of the reliability of results or relative difficulty of performing the calculations. However, the research that has been done indicates this may in time be a viable alternative to the other methods mentioned here.

25.10 TIME-DEPENDENT METHODS

Time-dependent calculations often result in obtaining a wave function that oscillates between the ground and first excited states. From this solution, it is possible to extract both these states.

25.11 SEMIEMPIRICAL METHODS

Most of the semiempirical methods are not designed to correctly predict the electronic excited state. Although excited-state calculations are possible, particularly using a CIS formulation, the energetics are not very accurate. However, the HOMO–LUMO gap is reasonably reproduced by some of the methods.

The one exception to this is the INDO/S method, which is also called ZINDO. This method was designed to describe electronic transitions, particularly those involving transition metal atoms. ZINDO is used to describe electronic excited-state energies and often transition probabilities as well.

25.12 STATE AVERAGING

State averaging gives a wave function that describes the first few electronic states equally well. This is done by computing several states at once with the same orbitals. It also keeps the wave functions strictly orthogonal. This is necessary to accurately compute the transition dipole moments.

25.13 ELECTRONIC SPECTRAL INTENSITIES

Intensities for electronic transitions are computed as transition dipole moments between states. This is most accurate if the states are orthogonal. Some of the best results are obtained from the CIS, MCSCF, and ZINDO methods. The CASPT2 method can be very accurate, but it often requires some manual manipulation in order to obtain the correct configurations in the reference space.

25.14 RECOMMENDATIONS

Methods for obtaining electronic excited-state energies could be classified by their accuracy, ease of use, and computational resource requirements. Such a list, in order of preferred method, would be as follows:

1. Spin-state transitions
2. CIS, XCIS, CIS(D)
3. Block diagonal Hamiltonians
4. ZINDO
5. Higher roots of a CI
6. Time-dependent calculations
7. Choice of initial guess
8. Neglecting basis functions

9. DFT with orthogonality
10. Path integral techniques
11. QMC methods

Note that trying different initial guesses is usually best for verifying that the correct ground state has been found.

Regardless of the choice of method, excited-state modeling usually requires a multistep process. The typical sequence of steps is:

1. Find which excited states exist and which are of interest.
2. Do a geometry optimization for the excited state.
3. Complete a frequency calculation to verify that the geometry is correct.
4. Compute excited-state properties.

BIBLIOGRAPHY

For introductory descriptions see

J. B. Foresman, Æ. Frisch, *Exploring Chemistry with Electronic Structure Methods Second Edition* Gaussian, Pittsburgh (1996).

I. N. Levine, *Quantum Chemistry Fourth Edition* Prentice Hall, Englewood Cliffs (1991).

D. A. McQuarrie, *Quantum Chemistry* University Science Books, Mill Valley CA (1983).

M. Karplus, R. N. Porter, *Atoms & Molecules: An Introduction For Students of Physical Chemistry* W. A. Benjamin, Reading (1970).

Books addressing excited states are

M. Klessinger, J. Michl, *Excited States and Photochemistry of Organic Molecules* VCH, New York (1995).

C. E. Dykstra, *Quantum Chemistry and Molecular Spectroscopy* Prentice Hall, Englewood Cliffs (1992).

N. Mataga, T. Kubota, *Molecular Interactions and Electronic Spectra* Marcel Dekker, New York (1970).

Review articles are

Excited States, a series of collections of review articles edited by E. C. Lim, Academic Press, New York 1974–1982.

T. Bally, W. T. Borden, *Rev. Comput. Chem.* **13**, 1 (1999).

M. Quack, *Encycl. Comput. Chem.* **3**, 1775 (1998).

S. D. Peyerimhoff, *Encycl. Comput. Chem.* **4**, 2646 (1998).

P. J. Bruna, *Computational Chemistry; Structure, Interactions and Reactivity Part A* S. Fraga, Ed., 379, Elsevier, Amsterdam (1992).

R. Moccia, P. Spizzo, *Computational Chemistry; Structure, Interactions and Reactivity Part B* S. Fraga, Ed., 76, Elsevier, Amsterdam (1992).

Adv. Chem. Phys. J. W. McGowan, Ed. vol. **45** (1981).

S. D. Peyerimhoff, R. J. Buenker, *Adv. Quantum Chem.* **9**, 69 (1975).

Excited states of solids and liquids are discussed in

W. L. Greer, S. A. Rice, *Adv. Chem. Phys.* **17**, 229 (1970).

Excited states of polymers are discussed in

S. Basu, *Adv. Quantum Chem.* **11**, 33 (1978).

For quantum Monte Carlo methods see

B. L. Hammond, W. A. Lester, Jr., P. J. Reynolds, *Monte Carlo Methods in Ab Initio Quantum Chemistry* World Scientific, Singapore (1994).

For density functional theory see

R. G. Parr, W. Yang, *Density-Functional Theory of Atoms and Molecules* Oxford, Oxford (1989).

For an introduction to excited states via path integral methods see

T. E. Sorensen, W. B. England, *Mol. Phys.* **89**, 1577 (1996).

26 Size-Consistency

It is a well-known fact that the Hartree–Fock model does not describe bond dissociation correctly. For example, the H_2 molecule will dissociate to an H^+ and an H^- atom rather than two H atoms as the bond length is increased. Other methods will dissociate to the correct products; however, the difference in energy between the molecule and its dissociated parts will not be correct. There are several different reasons for these problems: size-consistency, size-extensivity, wave function construction, and basis set superposition error.

The above problem with H_2 dissociation is a matter of wave function construction. The functional form of a restricted single-determinant wave function will not allow a pair of electrons in an orbital to separate into two different orbitals. Wave function construction issues were addressed in greater detail in Chapters 3 through 6.

An even more severe example of a limitation of the method is the energy of molecular mechanics calculations that use harmonic potentials. Although harmonic potentials are very reasonable near the equilibrium geometry, they are not even qualitatively correct for bond dissociation. These methods are very reasonable for comparing the energies of conformers but not for bond dissociation processes. Molecular mechanics methods using Morse potentials will give reasonable dissociation energies only if the method was parameterized to describe dissociation.

The literature contains some conflicting terminology regarding size-consistency and size-extensivity. A size-consistent method is one in which the energy obtained for two fragments at a sufficiently large separation will be equal to the sum of the energies of those fragments computed separately. A size–extensive method is one that gives an energy that is a linear function of the number of electrons. Some authors use the term size-consistent to refer to both criteria. Another error in the energy of separated fragments is basis set superposition error (BSSE), which is discussed in Chapter 28.

A mathematical analysis can be done to show whether a particular method is size-consistent. Strictly speaking, this analysis is only applicable to the behavior at infinite separation. However, methods that are size-consistent tend to give reasonable energetics at any separation. Some methods are approximately size-consistent, meaning that they fail the mathematical test but are accurate enough to only exhibit very small errors. Size-consistency is of primary importance for correctly describing the energetics of a system relative to the separated pieces of the system (i.e., bond dissociation or van der Waals bonding in a given system).

TABLE 26.1 Size Properties of Methods

Method	Size-consistent	Size-extensive
HF	Y D	Y
Full CI	Y	Y
Limited CI	N	N
CISDTQ	A	A
MPn	Y D	Y
Other MBPT	—	N
CC	Y D	Y
SAC	Y D	Y
MRCI	A D	A
MRPT	A D	A
Semiempiricals	A D	A

Y = yes.

N = no.

D = true only if the reference space dissociates to the correct state for that system.

A = approximately.

Size-extensivity is of importance when one wishes to compare several similar systems with different numbers of atoms (i.e., methanol, ethanol, etc.). In all cases, the amount of correlation energy will increase as the number of atoms increases. However, methods that are not size-extensive will give less correlation energy for the larger system when considered in proportion to the number of electrons. Size-extensive methods should be used in order to compare the results of calculations on different-size systems. Methods can be approximately size–extensive. The size-extensivity and size-consistency of various methods are summarized in Table 26.1.

26.1 CORRECTION METHODS

It is possible to make a method approximately size–extensive by adding a correction to the final energy. This has been most widely used for correcting CISD energies. This is a valuable technique because a simple energy correction formula is easier to work with than full CI calculations, which require an immense amount of computational resources. The most widely used correction is the Davidson correction:

$$\Delta E_{\mathrm{DC}} = E_{\mathrm{SD}}(1 - C_0^2) \tag{26.1}$$

where ΔE_{DC} is the energy lowering, E_{SD} the CISD energy, and C_0 the weight of the HF reference determinant in the CI expansion. This was designed to give the additional energy lowering from a CISD energy to a CISDTQ energy. This results in both a more accurate energy and in making the energy approximately size-extensive. The method is based on perturbation theory and thus may perform poorly if the HF wave function has a low weight in the CI expansion.

A slightly improved form of this equation is the renormalized Davidson correction, which is also called the Brueckner correction:

$$\Delta E_{\mathrm{RD}} = E_{\mathrm{SD}} \frac{1 - C_0^2}{C_0^2} \tag{26.2}$$

This generally gives a slight improvement over the Davidson correction, although it does not reach the full CI limit.

A more detailed perturbation theory analysis leads to an improved correction formula. This method, known as the Davidson and Silver or Siegbahn correction, is

$$\Delta E_{\mathrm{DS}} = E_{\mathrm{SD}} \frac{1 - C_0^2}{2C_0^2 - 1} \tag{26.3}$$

This correction does approximate the full CI energy, although it may overcorrect the energy.

There are many more error correction methods, which are reviewed in detail by Duch and Diercksen. They also discuss the correction of other wave functions, such as multireference methods. In their tests with various numbers of Be atoms, the correction most closely reproducing the full CI energy is

$$\Delta E_{\mathrm{PC}} = E_{\mathrm{SD}} \frac{1 - \sqrt{1 - 4C_0^2(1 + C_0^2)(1 - 2/N)}}{(2C_0^2 - 1) + \sqrt{1 - 4C_0^2(1 - C_0^2)(1 - 2/N)}} \tag{26.4}$$

where N is the number of electrons in the system.

Another method for making a method size-extensive is called the self-consistent dressing of the determinant energies. This is a technique for modifying the CI superdeterminant in order to make a size-extensive limited CI. The accuracy of this technique is generally comparable to the Davidson correction. It performs better than the Davidson correction for calculations in which the HF wave function has a low weight in the CI expansion.

26.2 RECOMMENDATIONS

Size-consistency and size-extensivity are issues that should be considered at the outset of any study involving multiple molecules or dissociated fragments. As always, the choice of a computational method is dependent on the accuracy desired and computational resource requirements. Correction formulas are so simple to use that several of them can readily be tried to see which does best for the system of interest. The Davidson, Brueckner, and Siegbahn corrections are commonly compared. There is not a large enough collection of results to make any general comments on the merits of the other correction methods.

BIBLIOGRAPHY

Introductory discussions can be found in

P. W. Atkins, R. S. Friedman, *Molecular Quantum Mechanics Third Edition* Oxford, Oxford (1997).

I. N. Levine, *Quantum Chemistry Fourth Edition* Prentice Hall, Englewood Cliffs (1991).

W. J. Hehre, L. Radom, P. v. R. Schleyer, J. A. Pople, *Ab Initio Molecular Orbital Theory* John Wiley & Sons, New York (1986).

More detailed reviews are

W. Duch, G. H. F. Dieckson, *J. Chem. Phys.* **101**, 3018 (1994).

A. Povill, R. Caballol, J. Rubio, J. P. Malrieu, *Chem. Phys. Lett.* **209**, 126 (1993).

I. Shavitt, *Advanced Theories and Computational Approaches to the Electronic Structure of Molecules* C. E. Dykstra, Ed., 185, Reidel, Dordrecht (1984).

27 Spin Contamination

Introductory descriptions of Hartree–Fock calculations [often using Rootaan's self-consistent field (SCF) method] focus on singlet systems for which all electron spins are paired. By assuming that the calculation is restricted to two electrons per occupied orbital, the computation can be done more efficiently. This is often referred to as a spin-restricted Hartree–Fock calculation or RHF.

For systems with unpaired electrons, it is not possible to use the RHF method as is. Often, an unrestricted SCF calculation (UHF) is performed. In an unrestricted calculation, there are two complete sets of orbitals: one for the alpha electrons and one for the beta electrons. These two sets of orbitals use the same set of basis functions but different molecular orbital coefficients.

The advantage of unrestricted calculations is that they can be performed very efficiently. The alpha and beta orbitals should be slightly different, an effect called spin polarization. The disadvantage is that the wave function is no longer an eigenfunction of the total spin $\langle S^2 \rangle$. Thus, some error may be introduced into the calculation. This error is called spin contamination and it can be considered as having too much spin polarization.

27.1 HOW DOES SPIN CONTAMINATION AFFECT RESULTS?

Spin contamination results in a wave function that appears to be the desired spin state, but is a mixture of some other spin states. This occasionally results in slightly lowering the computed total energy because of greater variational freedom. More often, the result is to slightly raise the total energy since a higher-energy state is mixed in. However, this change is an artifact of an incorrect wave function. Since this is not a systematic error, the difference in energy between states will be adversely affected. A high spin contamination can affect the geometry and population analysis and significantly affect the spin density. Exactly how these results are changed depends on the nature of the state being mixed with the ground state. Spin contamination can also result in the slower convergence of MPn calculations. Transition states and high-spin transition metal compounds tend to be particularly susceptible to spin contamination.

As a check for the presence of spin contamination, most *ab initio* programs will print out the expectation value of the total spin $\langle S^2 \rangle$. If there is no spin contamination, this should equal $s(s+1)$, where s equals $\frac{1}{2}$ times the number of unpaired electrons. One rule of thumb, which was derived from experience with

TABLE 27.1 Spin Eigenfunctions

Number of unpaired electrons	s	$\langle S^2 \rangle$
0	0	0
1	0.5	0.75
2	1	2.0
3	1.5	3.75
4	2	6.0
5	2.5	8.75

organic molecule calculations, is that the spin contamination is negligible if the value of $\langle S^2 \rangle$ differs from $s(s+1)$ by less than 10%. Although this provides a quick test, it is always advisable to doublecheck the results against experimental evidence or more rigorous calculations. Expected values of $\langle S^2 \rangle$ are listed in Table 27.1.

Spin contamination is often seen in unrestricted Hartree–Fock (UHF) calculations and unrestricted Møller–Plesset (UMP2, UMP3, UMP4) calculations. UMP2 is often the most sensitive to spin contamination, followed by UHF, UMP3, and UMP4. Local MP2 (LMP2) usually has less spin contamination than MP2. It is less common to find any significant spin contamination in DFT calculations, even when unrestricted Kohn–Sham orbitals are used. Spin contamination has little effect on CC and CI calculations, in which the variational principle will result in correcting for spin contamination in the reference wave function.

Unrestricted calculations often incorporate a spin annihilation step, which removes a large percentage of the spin contamination from the wave function. This helps minimize spin contamination but does not completely prevent it. The final value of $\langle S^2 \rangle$ is always the best check on the amount of spin contamination present. In the Gaussian program, the option "iop(5/14=2)" tells the program to use the annihilated wave function to produce the population analysis.

27.2 RESTRICTED OPEN-SHELL CALCULATIONS

It is possible to run spin-restricted open-shell calculations (ROHF). The advantage of this is that there is no spin contamination. The disadvantage is that there is an additional cost in the form of CPU time required in order to correctly handle both singly occupied and doubly occupied orbitals and the interaction between them. As a result of the mathematical method used, ROHF calculations give good total energies and wave functions but the singly occupied orbital energies do not rigorously obey Koopman's theorem.

ROHF does not include spin polarization. Thus, it is not useful for some purposes, such as predicting EPR spectra. Also because of this, it cannot reliably predict spin densities.

ROHF calculations can also exhibit symmetry breaking. Symmetry breaking is due to the calculation converging to one of the resonance structures instead of the correct ground state, which is a superposition of possible resonance structures. This causes the molecule to distort to a lower-symmetry geometry. How much symmetry breaking affects the geometry of a molecule is difficult to determine because there are many examples in which a low symmetry is the correct shape of a molecule. In cases where the symmetry breaking is a non-physical artifact of the calculation, the wave function often exhibits an abnormally pronounced localization of the spin density to one atom or bond.

Within some programs, the ROMPn methods do not support analytic gradients. Thus, the fastest way to run the calculation is as a single point energy calculation with a geometry from another method. If a geometry optimization must be done at this level of theory, a non-gradient-based method such as the Fletcher–Powell optimization should be used.

When it has been shown that the errors introduced by spin contamination are unacceptable, restricted open-shell calculations are often the best way to obtain a reliable wave function.

27.3 SPIN PROJECTION METHODS

Another approach is to run an unrestricted calculation and then project out the spin contamination after the wave function has been obtained (PUHF, PMP2). This gives a correction to the energy but does not affect the wave function. Spin projection nearly always improves *ab initio* results, but may seriously harm the accuracy of DFT results.

A spin projected result does not give the energy obtained by using a restricted open-shell calculation. This is because the unrestricted orbitals were optimized to describe the contaminated state, rather than the spin-projected state. In cases of very-high-spin contamination, the spin projection may fail, resulting in an increase in spin contamination.

A similar effect is obtained by using the spin-constrained UHF method (SUHF). In this method, the spin contamination error in a UHF wave function is constrained by the use of a Lagrangian multiplier. This removes the spin contamination completely as the multiplier goes to infinity. In practice, small positive values remove most of the spin contamination.

27.4 HALF-ELECTRON APPROXIMATION

Semiempirical programs often use the half-electron approximation for radical calculations. The half-electron method is a mathematical technique for treating a singly occupied orbital in an RHF calculation. This results in consistent total energy at the expense of having an approximate wave function and orbital energies. Since a single-determinant calculation is used, there is no spin contamination.

The consistent total energy makes it possible to compute singlet-triplet gaps using RHF for the singlet and the half-electron calculation for the triplet. Koopman's theorem is not followed for half-electron calculations. Also, no spin densities can be obtained. The Mulliken population analysis is usually fairly reasonable.

27.5 RECOMMENDATIONS

If spin contamination is small, continue to use unrestricted methods, preferably with spin-annihilated wave functions and spin projected energies. Do not use spin projection with DFT methods. When the amount of spin contamination is more significant, use restricted open-shell methods. If all else fails, use highly correlated methods.

BIBLIOGRAPHY

Some discussion and results are in

F. Jensen, *Introduction to Computational Chemistry* John Wiley & Sons, New York (1999).

W. J. Hehre, L. Radom, P. v.R. Schleyer, J. A. Pople, *Ab Initio Molecular Orbital Theory* John Wiley & Sons, New York (1986).

T. Clark, *A Handbook of Computational Chemistry* Wiley, New York (1985).

Review articles are

T. Bally, W. T. Borden, *Rev. Comput. Chem.* **13**, 1 (1999).

H. B. Schlegel, *Encycl. Comput. Chem.* **4**, 2665 (1998).

A more mathematical treatment can be found in the paper

J. S. Andrews, D. Jayatilake, R. G. A. Bone, N. C. Handy, R. D. Amos, *Chem. Phys. Lett.* **183**, 423 (1991).

Articles that compare unrestricted, restricted and projected results are

M. W. Wong, L. Radom, *J. Phys. Chem.* **99**, 8582 (1995).

J. Baker, A. Scheiner, J. Andzelm, *Chem. Phys Lett.* **216**, 380 (1993).

SUHF results are examined in

P. K. Nandi, T. Kar, A. B. Sannigrahi, *Journal of Molecular Structure (Theochem)* **362**, 69 (1996).

Spin projection is addressed in

J. M. Wittbrodt, H. B. Schlegel, *J. Chem. Phys.* **105**, 6574 (1996).

I. Mayer, *Adv. Quantum Chem.* **12**, 189 (1990).

28 Basis Set Customization

Chapter 10 represented a wave function as a linear combination of Gaussian basis functions. Today, there are so many basis sets available that many researchers will never need to modify a basis set. However, there are occasionally times when it is desirable to extend an existing basis set in order to obtain more accurate results. The savvy researcher also needs to be able to understand the older literature, in which basis sets were customized routinely.

28.1 WHAT BASIS FUNCTIONS DO

The tight functions in the basis set are those having large Gaussian exponents. These functions described the shape of the electron density near the nucleus; they are responsible for a very large amount of total energy due to the high kinetic and potential energy of electrons near the nucleus. However, the tight functions have very little effect on how well the calculation describes chemical bonding. Altering the tight basis functions may result in slightly shifting the atomic size. Although it is, of course, possible to add additional tight functions to an existing basis, this is very seldom done because it is difficult to do correctly and it makes very little difference in the computed chemical properties. It is advisable to completely switch basis sets if the description of the core region is of concern.

Diffuse functions are those functions with small Gaussian exponents, thus describing the wave function far from the nucleus. It is common to add additional diffuse functions to a basis. The most frequent reason for doing this is to describe orbitals with a large spatial extent, such as the HOMO of an anion or Rydberg orbitals. Adding diffuse functions can also result in a greater tendency to develop basis set superposition error (BSSE), as described later in this chapter.

Polarization functions are functions of a higher angular momentum than the occupied orbitals, such as adding *d* orbitals to carbon or *f* orbitals to iron. These orbitals help the wave function better span the function space. This results in little additional energy, but more accurate geometries and vibrational frequencies.

28.2 CREATING BASIS SETS FROM SCRATCH

Creating completely new basis sets is best left to professionals because it requires a very large amount of technical expertise. To be more correct, anyone could create a new basis set, but it is extremely difficult to create one better than the existing sets. Creating a basis set requires first obtaining an initial set of functions and then optimizing both coefficients and exponents in order to get the best variational energy. An initial set of functions can be created by using a known approximate mathematical relationship between basis functions or fitting to high-accuracy results, such as numerical basis set calculations. Even once a low variational energy has been obtained, that does not mean it is a good basis set. A good description of the core electrons will yield a low energy. It is also necessary to give a good description of the valence region, where bonding occurs. In order to describe the valence region well, the basis must have both basis functions in the correct region of space and enough flexibility (uncontracted functions) to describe the shift in electron density as bonds are being formed. Then the issues of polarization and diffuse functions must be addressed.

28.3 COMBINING EXISTING BASIS SETS

It is possible to mix basis sets by choosing the functions for different elements from different existing basis sets. This can be necessary if the desired basis set does not have functions for a particular element. It can result in very good results as long as basis sets being combined are of comparable accuracy. Mixing very large and very small basis sets can result in inaccurate calculations due to uneven spanning of the function space, called an unbalanced basis.

How does one tell if two basis sets have the same accuracy? If the two were defined for the same atom, then they would be expected to have a similar number of Gaussian primitives, similar number of valence contractions, and similar total energy's for the atom-only calculation. If they are not defined for the same elements, then a similar number of valence contractions is a crude indicator. In this case, the number of primitives should follow typical trends. For example, examining a basis set that has been defined for a large number of elements will show general trends, such as having more primitives for the heavier elements and more primitives for the lower-angular-momentum orbitals. These patterns have been defined based on studies in which the variational energy lowering for each additional primitive was examined. A well-balanced basis set is one in which the energy contribution due to adding the final primitive to each orbital is approximately equivalent.

What if an unbalanced basis is used? This is tantamount to asking what would happen if the calculation contains two nearby atoms: one of which is described by a large basis and the other by a small basis. In this case, the energy of the atom with a small basis can be variationally lowered if the basis functions of the other atom are weighted in order to describe the electron density around the first atom. This leads to an extreme case of basis set superposition error. The energy of the atom with the smaller basis can be lowered even more if it moves closer to the atom with the large basis. Thus, this leads to errors in both energy and geometry. An unbalanced basis can give results of a poorer accuracy than a small but balanced basis.

```
N 0
 S    1 1.00
  6.7871900000E+02  1.0000000000E+00
 S    1 1.00
  1.0226600000E+02  1.0000000000E+00
 S    1 1.00
  2.2906600000E+01  1.0000000000E+00
 S    1 1.00
  6 1064900000E+00  1.0000000000E+00
 S    1 1.00
  8.3954000000E-01  1.0000000000E+00
 S    1 1.00
  2.5953000000E-01  1 0000000000E+00
 P    1 1.00
  2.3379500000E+00  1 0000000000E+00
 P    1 1.00
  4.1543000000E-01  1.0000000000E+00
 ****
```

FIGURE 28.1 Uncontracted basis input.

28.4 CUSTOMIZING A BASIS SET

The most difficult part of creating a basis set from scratch is optimizing the exponents. In customizing a basis, one must start with a set of exponents at the very least. The steps involved then consist of creating contractions, adding additional functions to the valence contractions, adding polarization functions, and adding diffuse functions. It is possible to add outer primitives (low exponents) that are nearly optimal. It is more difficult to add near-optimal inner functions (large exponents). Thus, the description of the core region provided by the starting exponents must be acceptable.

Determining contraction coefficients is something that occasionally must be done. Early calculations did not use contracted basis functions and so older basis sets do not include contraction coefficients. The user must determine the contractions in order to make optimal use of these basis sets with the current generation of software, which is designed to be used with contracted basis sets. If no contraction coefficients are available, the first step is to run a calculation on the atom with the basis set completely uncontracted. For example, Figure 28.1 shows a $6s2p$ nitrogen atom basis created by Duijneveldt and formatted for input in the Gaussian program.

Figure 28.2 shows the molecular orbitals from the uncontracted atom calculation. A reasonable set of contractions will be obtained by using the largest-weight primitives from each of these orbitals. There is not a specific value that is considered a large weight. However, some guidance can be obtained by examining established basis sets. In this example, the basis will be partitioned into two s and one p contracted functions. Our choice of coefficients are those printed in bold in Figure 28.2. The p_x and p_z orbitals show some mixing to give an arbitrary orientation in the xz plane, which does not change the energy of a single atom. The p_y orbitals were used since they mix very little with the x and

			1	2	3	4	5
			O	O	O	O	O
EIGENVALUES --			-15.67268	-1.15465	-0.49434	-0.49434	-0.49434
1 1	N	1S	**0.01725**	-0.00401	0.00000	0.00000	0.00000
2		2S	**0.12052**	-0.02888	0.00000	0.00000	0.00000
3		3S	**0.42594**	-0.11729	0.00000	0.00000	0.00000
4		4S	**0.54719**	-0.23494	0.00000	0.00000	0.00000
5		5S	0.05956	**0.47481**	0.00000	0.00000	0.00000
6		6S	-0.01658	**0.64243**	0.00000	0.00000	0.00000
7		7PX	0.00000	0.00000	0.25850	0.21951	-0.06145
8		7PY	0.00000	0.00000	0.00343	0.08916	**0.33289**
9		7PZ	0.00000	0.00000	-0.22792	0.25029	-0.06469
10		8PX	0.00000	0.00000	0.60087	0.51024	-0.14284
11		8PY	0.00000	0.00000	0.00796	0.20725	**0.77381**
12		8PZ	0.00000	0.00000	-0.52980	0.58180	-0.15037

FIGURE 28.2 Uncontracted orbitals.

z. If all the *p* orbitals were mixed, it would be necessary to orthogonalize the orbitals or enforce some symmetry in the wave function.

The contracted basis set created from the procedure above is listed in Figure 28.3. Note that the contraction coefficients are not normalized. This is not usually a problem since nearly all software packages will renormalize the coefficients automatically. The atom calculation rerun with contracted orbitals is expected to run much faster and have a slightly higher energy.

Likewise, a basis set can be improved by uncontracting some of the outer basis function primitives (individual GTO orbitals). This will always lower the total energy slightly. It will improve the accuracy of chemical predictions if the primitives being uncontracted are those describing the wave function in the middle of a chemical bond. The distance from the nucleus at which a basis function has the most significant effect on the wave function is the distance at which there is a peak in the radial distribution function for that GTO primitive. The formula for a normalized radial GTO primitive in atomic units is

```
 N 0
S    4 1.00
 6.7871900000E+02  1.7250000000E-02
 1.0226600000E+02  1.2052000000E-01
 2.2906600000E+01  4.2594000000E-01
 6.1064900000E+00  5.4719000000E-01
S    2 1 00
 8.3954000000E-01  4.7481000000E-01
 2 5953000000E-01  6.4243000000E-01
P    2 1.00
 2.3379500000E+00  3 3289000000E-01
 4.1543000000E-01  7.7381000000E-01
****
```

FIGURE 28.3 Contracted basis.

$$R(r) = \left[\frac{2^7\zeta^3}{\pi}\right]^{1/4} e^{-\zeta r^2} \tag{28.1}$$

where ζ is the Gaussian exponent. The formula for the radial distribution function of a GTO primitive is

$$\text{RDF} = r^2 R^2(r) = r^2 \left[\frac{2^7\zeta^3}{\pi}\right]^{1/2} e^{-2\zeta r^2} \tag{28.2}$$

The maximum in this function occurs at

$$r = \frac{1}{\sqrt{2\zeta}} \tag{28.3}$$

Thus, a primitive with an exponent of 0.2 best describes the wave function at a distance of 0.79 Bohrs.

The contracted basis in Figure 28.3 is called a minimal basis set because there is one contraction per occupied orbital. The valence region, and thus chemical bonding, could be described better if an additional primitive were added to each of the valence orbitals. This is almost always done using the "even-tempered" method. This method comes from the observation that energy-optimized exponents tend to nearly follow an exponential pattern given by

$$\zeta_i = \alpha\beta^i \tag{28.4}$$

where ζ_i is the ith exponent and α and β are fitted parameters. This equation can be used to generate additional primitive from the two outer primitives. An additional s primitive would be generated from the last two primitives as follows. First, divide to obtain β:

$$\frac{\zeta_6}{\zeta_5} = \frac{\alpha\beta^6}{\alpha\beta^5} = \beta = \frac{0.25957}{0.83954} = 0.309 \tag{28.5}$$

Rearrange and substitute to get α:

$$\alpha = \frac{\zeta_6}{\beta^6} = \frac{0.25957}{0.309^6} = 298 \tag{28.6}$$

Then use the original formula to get ζ_7

$$\zeta_7 = \alpha\beta^7 = 298 \cdot 0.309^7 = 0.0802 \tag{28.7}$$

```
N 0
S   4 1.00
 6.7871900000E+02  1.7250000000E-02
 1.0226600000E+02  1.2052000000E-01
 2.2906600000E+01  4.2594000000E-01
 6.1064900000E+00  5.4719000000E-01
S   2 1.00
 8.3954000000E-01  4.7481000000E-01
 2.5953000000E-01  6.4243000000E-01
S   1 1.00
 8.0200000000E-02  1.0000000000E+00
P   2 1.00
 2.3379500000E+00  3.3289000000E-01
 4.1543000000E-01  7.7381000000E-01
P   1 1.00
 7.3880000000E-02  1 0000000000E+00
D   1 1.00
 8.0000000000E-01  1.0000000000E+00
****
```

FIGURE 28.4 Final basis.

Note that the answers have been rounded to three significant digits. Since the even-tempered formula is only an approximation, this does not introduce any significant additional error.

Although the even tempered function scheme is fairly reasonable far from the nucleus, each function added is slightly further from the energy-optimized value. Generally, two or three additional functions at the most will be added to a basis set. Beyond this point, it is most efficient to switch to a different, larger basis.

A different scheme must be used for determining polarization functions and very diffuse functions (Rydberg functions). It is reasonable to use functions from another basis set for the same element. Another option is to use functions that will depict the electron density distribution at the desired distance from the nucleus as described above.

Having polarization functions of higher angular momentum than the highest occupied orbitals is usually the most polarization that will benefit HF or DFT results. Higher-angular-momentum functions are important for very-high-accuracy configuration interaction and coupled-cluster calculations. As a general rule of thumb, uncontracting a valence primitive generally lowers the variational energy by about as much as adding a set of polarization functions.

A new basis for the element can now be created by combining these techniques. The basis in Figure 28.4 was created from the contracted set illustrated in Figure 28.3. Additional even-tempered exponents have been added to both the s and p functions. A polarization function of d symmetry was obtained from the 6−31G(d) basis set. In a realistic scenario, a certain amount of trial-and-error work, based on obtaining low variational energies and stronger chemical bonds, would be involved in this process. This nitrogen example is somewhat artificial because there are many high-quality basis functions available for nitrogen that would be preferable to customizing a basis set.

The final step is to check the performance of the basis set. This can be done by first doing a single-atom calculation to check the energy and virial theorem value. The UHF calculation for this basis gave a virial theorem check of −1.9802, which is in reasonable agreement with the correct value of −2. The UHF atom energy is −54.10814 Hartrees for this example. This is really not a very good total energy for nitrogen due to the fact that the example started with a fairly small basis set. The 6−31G(d) basis gives a total energy of −54.38544 Hartrees for nitrogen. The basis in this example should probably not be extended any more than has been done here, since it would lead to having a disproportionately well-described valence region and poorly described core.

The final test of the basis quality, particularly in the valence region, is the result of molecular calculations. This basis gave an N_2 bond length of 1.1409 Å at the HF level of theory and 1.1870 at the CCSD level of theory, in only moderate agreement with the experimental value of 1.0975 Å. The larger 6−31G(d) basis set gives a bond length of 1.0783 Å at the HF level of theory. The experimental bond energy for N_2 is 225.9 kcal/mol. The HF calculation with this example basis yields 89.9 kcal/mol, compared to the HF 6−31G(d) bond energy of 108.6 kcal/mol. At the CCSD level of theory, the sample basis gives a bond energy of 170.3 kcal/mol.

28.5 BASIS SET SUPERPOSITION ERROR

Basis set superposition error (BSSE) is an energy lowering of a complex of two molecules with respect to the sum of the individual molecule energies. This results in obtaining van der Waals and hydrogen bond energies that are too large because the basis functions on one molecule act to describe the electron density of the other molecule. In the limit of an exact basis set, there would be no superposition error. The error is also small for minimal basis sets, which do not have functions diffuse enough to describe an adjacent atom. The largest errors occurred when using moderate-size basis sets.

The procedure for correcting for BSSE is called a counterpoise correction. In this procedure, the complex of molecules is first computed. The individual molecule calculations are then performed using all the basis functions from the complex. For this purpose, many *ab initio* software programs contain a mechanism for defining basis functions that are centered at a location which is not on one of the nuclei. The interaction energy is expressed as the energy for the complex minus the individual molecule energies computed in this way. In equation form, this is given as

$$E_{\text{interaction}} = E_{AB}(AB) - E_{AB}(A) - E_{AB}(B) \tag{28.8}$$

where the subscripts denote the basis functions being used and the letters in parentheses denote the molecules included in each calculation.

Counterpoise correction should, in theory, be unnecessary for large basis sets. However, practical applications have shown that it yields a significant improvement in results even for very large basis sets. The use of a counterpoise correction is recommended for the accurate computation of molecular interaction energies by *ab initio* methods.

BIBLIOGRAPHY

Text books containing detailed basis set discussions are

J. B. Foresman, A. Frisch, *Exploring Chemistry with Electronic Structure Methods Second Edition* Gaussian, Pittsburgh (1996).

A. R. Leach *Molecular Modelling Principles and Applications* Longman, Essex (1996).

I. N. Levine, *Quantum Chemistry Fourth Edition* Prentice Hall, Englewood Cliffs (1991).

W. J. Hehre, L. Radom, P. v. R. Schleyer, J. A. Pople, *Ab Initio Molecular Orbital Theory* John Wiley & Sons, New York (1986).

Review articles and more detailed sources are

T. H. Dunning, K. A. Peterson, D. E. Woon, *Encycl. Comput. Chem.* **1**, 88 (1998).

F. B. van Duijneveldt, J. G. C. M. van Duijneveldt-van de Rijdt, J. H. van Lenthe, *Chem. Rev.* **94**, 1873 (1994).

A. D. Buckingham, P. W. Fowler, J. M. Hutson, *Chem. Rev.* **88**, 963 (1988).

S. Wilson, *Ab Initio Methods in Quantum Chemistry-I* 439 K. P. Lawley, Ed., John Wiley & Sons, New York (1987).

E. R. Davidson, D. Feller, *Chem. Rev.* **86**, 681 (1986).

J. Andzelm, M. Kobukowski, E. Radzio-Andzelm, Y. Sakai, H. Tatewaki, *Gaussian Basis Sets for Molecular Calculations* S. Huzinaga, Ed., Elsevier, Amsterdam (1984).

Basis set superposition error is reviewed in

N. R. Kestner, *Rev. Comput. Chem.* **13**, 99 (1999).

F. B. van Duijneveldt, *Molecular Interactions* S. Scheiner, Ed., 81, John Wiley & Sons, New York (1997).

29 Force Field Customization

It is occasionally desirable to add new parameters to a molecular mechanics force field. This might mean adding an element that is not in the parameterization set or correctly describing a particular atom in a specific class of molecules.

29.1 POTENTIAL PITFALLS

It is tempting to take parameters from some other force field. However, unlike *ab initio* basis sets, this is not generally a viable method. Force fields are set up with different lists of energy terms. For example, one force field might use stretch, bend, and stretch–bend terms, whereas another uses stretch and bend terms only. Using the stretch and bend parameters from the first without the accompanying stretch–bend term would result in incorrectly describing both bond stretching and bending.

From one force field to the next, the balance of energy terms may be different. For example, one force field might use a strong van der Waals potential and no electrostatic interaction, while another force field uses a weaker van der Waals potential plus a charge term. Even when the same terms are present, different charge-assignment algorithms yield systematic differences in results and the van der Waals term may be different to account for this.

When the same energy terms are used in two force fields, it may be acceptable to transfer bond-stretching and angle-bending terms. These are fairly stiff motions that do not change excessively. The force constants for these terms vary between force fields, much more than the unstrained lengths and angles.

Transferring torsional and nonbonded terms between force fields is much less reliable. These are lower-energy terms that are much more interdependent. It is quite common to find force fields with significantly different parameters for these contributions, even when the exact same equations are used.

Atoms with unusual hybridizations can be particularly difficult to include. Most organic force fields describe atoms with hybridizations whose bond angles are all equivalent (i.e., *sp*, sp^2, and sp^3 hybridizations with bond angles of 180, 120, and 109.5°, respectively). In contrast to this, a square planar atom will have some bond angles of 90° and some angles of 180°. In this case, it may be necessary to define the bond and angle terms manually, modify the software, or hold the bond angles fixed in the calculation.

29.2 ORIGINAL PARAMETERIZATION

Understanding how the force field was originally parameterized will aid in knowing how to create new parameters consistent with that force field. The original parameterization of a force field is, in essence, a massive curve fit of many parameters from different compounds in order to obtain the lowest standard deviation between computed and experimental results for the entire set of molecules. In some simple cases, this is done by using the average of the values from the experimental results. More often, this is a very complex iterative process.

The first step in creating a force field is to decide which energy terms will be used. This determines, to some extent, the ability of the force field to predict various types of chemistry. This also determines how difficult the parameterization will be. For example, more information is needed to parameterize anharmonic bond-stretching terms than to parameterize harmonic terms.

The parameters in the original parameterization are adjusted in order to reproduce the correct results. These results are generally molecular geometries and energy differences. They may be obtained from various types of experimental results or *ab initio* calculations. The sources of these "correct" results can also be a source of error. *Ab initio* results are only correct to some degree of accuracy. Likewise, crystal structures are influenced by crystal-packing forces.

Many parameterizations are merely a massive fitting procedure to determine which parameters will best reproduce these results. This procedure may be done with automated software or through the work and understanding of the designers. Often, a combination of both gives the best results. In recent years, global search techniques, such as genetic algorithms, have been used. This is usually an iterative procedure as parameters are adjusted and results computed for the test set of molecules.

A second procedure is called a rule-based parameterization. This is a way of using some simple relationship to predict a large number of parameters. For example, bond lengths might be determined as the geometric mean of covalent bond radii multiplied by a correction factor. In this case, determining one correction factor is tantamount to determining all needed bond lengths. This procedure has the advantage of being able to create a force field describing a large variety of compounds. The disadvantage is that the accuracy of results for a specific compound is not usually as good as that obtained with a force field parameterized specifically for that class of compounds.

29.3 ADDING NEW PARAMETERS

A measure of sophistication is necessary in order to obtain a reasonable set of parameters. The following steps are recommended in order to address the concerns above. They are ranked approximately best to worst, but it is advisable to use all techniques for the sake of doublechecking your work. Step 9 should always be included in the process. There are utilities available to help ease the amount of work involved, but even with these the researcher should still pay close attention to the steps being taken.

1. Examine the literature describing the original parameterization of the force field being used. Following this procedure as much as possible is advisable. This literature also gives insights into the strengths and limitations of a given force field.
2. Find articles describing how new parameters were added to the exact same force field. The procedure will probably be similar for your case.
3. If the atom being added has an unusual hybridization, examine the literature in which parameters were derived for that same hybridization.
4. If considering transferring parameters from one force field to another, examine the parameters for an atom that is in both force fields already. If the two sets of parameters are not fairly similar, do not use parameters from that force field.
5. First determine what parameters will be used for describing bond lengths and angles. Then determine torsional, inversion, and nonbonded interaction parameters.
6. Try using obvious values for the parameters, such as bond lengths directly from crystal structures. This assumes that no interdependence exists between parameters, but it is a starting point.
7. Use values from *ab initio* calculations.
8. Look for a very similar atom that has been parameterized for the force field and trying scaling its parameters by a suitable correction factor. Even if one of the steps above was used, this provides a quick check on the reasonableness of your parameterization.
9. Run test calculations with the new parameters. Then adjust the parameters as necessary to reproduce experimental results before using them to describe an unknown compound.

BIBLIOGRAPHY

Books discussing force field parameterization are

MacroModel Technical Manual Schrödinger, Portland OR (1999).

M. F. Schlecht, *Molecular Modeling on the PC* Wiley-VCH, New York (1998).

A. K. Rappé, C. J. Casewit, *Molecular Mechanics across Chemistry* University Science Books, Sausalito (1997).

A. R. Leach *Molecular Modelling Principles and Applications* Longman, Essex (1996).

P. Comba, T. W. Hambley, *Molecular Modeling of Inorganic Compounds* VCH, Weinheim (1995).

G. H. Grant, W. G. Richards, *Computational Chemistry* Oxford, Oxford (1995).

U. Burkert, N. L. Allinger, *Molecular Mechanics* American Chemical Society, Washington (1982).

Some journal articles with general discussions are

M. Zimmer, *Chem. Rev.* **95**, 2629 (1995).

B. P. Hay, *Coord. Chem. Rev.* **126**, 177 (1993).

J. P. Bowen, N. L. Allinger, *Rev. Comput. Chem.* **2**, 81 (1991).

J. R. Maple, U. Dinur, A. T. Hagler, *Proc. Natl. Acad Sci. USA* **85**, 5350 (1988).

A. J. Hopfinger, R. A. Pearlstein, *J. Comput. Chem.* **5**, 486 (1984).

A comprehensive listing of all published force field parameters is

M. Jalaie, K. B. Lipkowitz, *Rev. Comput. Chem.* **14**, 441 (2000).

E. Osawa, K. B. Lipkowitz, *Rev. Comput. Chem.* **6**, 355 (1995).

30 Structure–Property Relationships

Structure–property relationships are qualitative or quantitative empirically defined relationships between molecular structure and observed properties. In some cases, this may seem to duplicate statistical mechanical or quantum mechanical results. However, structure-property relationships need not be based on any rigorous theoretical principles.

The simplest case of structure-property relationships are qualitative rules of thumb. For example, the statement that branched polymers are generally more biodegradable than straight-chain polymers is a qualitative structure–property relationship.

When structure-property relationships are mentioned in the current literature, it usually implies a quantitative mathematical relationship. Such relationships are most often derived by using curve-fitting software to find the linear combination of molecular properties that best predicts the property for a set of known compounds. This prediction equation can be used for either the interpolation or extrapolation of test set results. Interpolation is usually more accurate than extrapolation.

When the property being described is a physical property, such as the boiling point, this is referred to as a quantitative structure–property relationship (QSPR). When the property being described is a type of biological activity, such as drug activity, this is referred to as a quantitative structure–activity relationship (QSAR). Our discussion will first address QSPR. All the points covered in the QSPR section are also applicable to QSAR, which is discussed next.

30.1 QSPR

The first step in developing a QSPR equation is to compile a list of compounds for which the experimentally determined property is known. Ideally, this list should be very large. Often, thousands of compounds are used in a QSPR study. If there are fewer compounds on the list than parameters to be fitted in the equation, then the curve fit will fail. If the same number exists for both, then an exact fit will be obtained. This exact fit is misleading because it fits the equation to all the anomalies in the data, it does not necessarily reflect all the correct trends necessary for a predictive method. In order to ensure that the method will be predictive, there should ideally be 10 times as many test compounds as fitted parameters. The choice of compounds is also important. For example, if the equation is only fitted with hydrocarbon data, it will only be reliable for predicting hydrocarbon properties.

The next step is to obtain geometries for the molecules. Crystal structure geometries can be used; however, it is better to use theoretically optimized geometries. By using the theoretical geometries, any systematic errors in the computation will cancel out. Furthermore, the method will predict as yet unsynthesized compounds using theoretical geometries. Some of the simpler methods require connectivity only.

Molecular descriptors must then be computed. Any numerical value that describes the molecule could be used. Many descriptors are obtained from molecular mechanics or semiempirical calculations. Energies, population analysis, and vibrational frequency analysis with its associated thermodynamic quantities are often obtained this way. *Ab initio* results can be used reliably, but are often avoided due to the large amount of computation necessary. The largest percentage of descriptors are easily determined values, such as molecular weights, topological indexes, moments of inertia, and so on. Table 30.1 lists some of the descriptors that have been found to be useful in previous studies. These are discussed in more detail in the review articles listed in the bibliography.

Once the descriptors have been computed, is necessary to decide which ones will be used. This is usually done by computing correlation coefficients. Correlation coefficients are a measure of how closely two values (descriptor and property) are related to one another by a linear relationship. If a descriptor has a correlation coefficient of 1, it describes the property exactly. A correlation coefficient of zero means the descriptor has no relevance. The descriptors with the largest correlation coefficients are used in the curve fit to create a property prediction equation. There is no rigorous way to determine how large a correlation coefficient is acceptable.

Intercorrelation coefficients are then computed. These tell when one descriptor is redundant with another. Using redundant descriptors increases the amount of fitting work to be done, does not improve the results, and results in unstable fitting calculations that can fail completely (due to dividing by zero or some other mathematical error). Usually, the descriptor with the lowest correlation coefficient is discarded from a pair of redundant descriptors.

A curve fit is then done to create a linear equation, such as

$$\text{Property} = c_0 + c_1 d_1 + c_2 d_2 + \cdots \tag{30.1}$$

where c_i are the fitted parameters and d_i the descriptors. Most often, the equation being fitted is a linear equation like the one above. This is because the use of correlation coefficients and linear equations together is an easily automated process. Introductory descriptions cite linear regression as the algorithm for determining coefficients of best fit, but the mathematically equivalent matrix least-squares method is actually more efficient and easier to implement. Occasionally, a nonlinear parameter, such as the square root or log of a quantity, is used. This is done when a researcher is aware of such nonlinear relationships in advance.

TABLE 30.1 Common Molecular Descriptors

Constitutional Descriptors
Molecular weight
Number of atoms of various elements
Number of bonds of various orders
Number of rings
Topological Descriptors
Weiner index
Randic indices
Kier and Hall indices
Information content
Connectivity index
Balaban index
Electrostatic Descriptors
Partial charges
Polarity indices
Topological electronic index
Multipoles
Charged partial surface areas
Polarizability
Anisotropy of polarizability
Geometrical Descriptors
Moments of inertia
Molecular volume
Molecular surface areas
Shadow indices
Taft steric constant
Length, width, and height parameters
Shape factor
Quantum Chemical Descriptors
Net atomic charges
Bond orders
HOMO and LUMO energies
FMO reactivity indices
Refractivity
Total energy
Ionization potential
Electron affinity
Energy of protonation
Orbital populations
Frontier orbital densities
Superdelocalizabilities

TABLE 30.1 (Continued)

Quantum Chemical Descriptors
Sum of the squared atomic charge densities
Sum of the absolute values of charges
Absolute hardness
Statistical Mechanical Descriptors
Vibrational frequencies
Rotational enthalpy and entropy
Vibrational enthalpy and entropy
Translational enthalpy and entropy

The process described in the preceding paragraphs has seen widespread use. This is partly because it has been automated very well in the more sophisticated QSPR programs.

It is possible to use nonlinear curve fitting (i.e., exponents of best fit). Nonlinear fitting is done by using a steepest-descent algorithm to minimize the deviation between the fitted and correct values. The drawback is possibly falling into a local minima, thus necessitating the use of global optimization algorithms. Automated algorithms for determining which descriptors to include in a nonlinear fit are possible, but there is not yet a consensus as to what technique is best. This approach can yield a closer fit to the data than multiple linear techniques. However, it is less often used due to the large amount of manual trial-and-error work necessary. Automated nonlinear fitting algorithms are expected to be included in future versions of QSPR software packages.

The validation of the prediction equation is its performance in predicting properties of molecules that were not included in the parameterization set. Equations that do well on the parameterization set may perform poorly for other molecules for several different reasons. One mistake is using a limited selection of molecules in the parameterization set. For example, an equation parameterized with organic molecules may perform very poorly when predicting the properties of inorganic molecules. Another mistake is having nearly as many fitted parameters as molecules in the test set, thus fitting to anomalies in the data rather than physical trends.

The development of group additivity methods is very similar to the development of a QSPR method. Group additivity methods can be useful for properties that are additive by nature, such as the molecular volume. For most properties, QSPR is superior to group additivity techniques.

Other algorithms for predicting properties have been developed. Both neural network and genetic algorithm-based programs are available. Some arguments can be made for the use of each. However, none has yet seen widespread use. This may be partially due to the greater difficulty in interpreting the chemical information that can be gained in addition to numerical predictions. Neural networks are generally known to provide a good interpolation of data, but rather poor extrapolation.

30.2 QSAR

QSAR is also called *traditional QSAR* or *Hansch QSAR* to distinguish it from the 3D QSAR method described below. This is the application of the technique described above to biological activities, such as environmental toxicology or drug activity. The discussion above is applicable but a number of other caveats apply; which are addressed in this section. The following discussion is oriented toward drug design, although the same points may be applicable to other areas of research as well.

In order to parameterize a QSAR equation, a quantified activity for a set of compounds must be known. These are called lead compounds, at least in the pharmaceutical industry. Typically, test results are available for only a small number of compounds. Because of this, it can be difficult to choose a number of descriptors that will give useful results without fitting to anomalies in the test set. Three to five lead compounds per descriptor in the QSAR equation are normally considered an adequate number. If two descriptors are nearly collinear with one another, then one should be omitted even though it may have a large correlation coefficient.

In the case of drug design, it may be desirable to use parabolic functions in place of linear functions. The descriptor for an ideal drug candidate often has an optimum value. Drug activity will decrease when the value is either larger or smaller than optimum. This functional form is described by a parabola, not a linear relationship.

The advantage of using QSAR over other modeling techniques is that it takes into account the full complexity of the biological system without requiring any information about the binding site. The disadvantage is that the method will not distinguish between the contribution of binding and transport properties in determining drug activity. QSAR is very useful for determining general criteria for activity, but it does not readily yield detailed structural predictions.

30.3 3D QSAR

For drug design purposes, it is desirable to construct a method that will predict the molecular structures of candidate compounds without requiring knowledge of the binding-site geometry. 3D QSAR has been fairly successful in fulfilling these criteria. It is similar to QSAR in that property descriptors, statistical analysis, and fitting techniques are used. Beyond that, the two computations are significantly different.

Like QSAR, molecular structures must be available for compounds that have known quantitatively defined activities. The first step is then to align the molecular structures. This alignment is based on the fact that all have a drug activity due to docking at a particular site. Alignment algorithms rotate and translate a molecule within the Cartesian coordinate space until it matches the location and rotation of another molecule as well as possible. This can be as simple as aligning the backbones of similar molecules or as complex as a sophisticated search and optimization scheme. For conformationally flexible compounds, both alignment and conformation must be addressed. Typically, the most rigid molecule in the set is the one to which the others are aligned. There are automated routines for finding the conformer of best alignment, or this can be done manually.

Once the molecules are aligned, a molecular field is computed on a grid of points in space around the molecule. This field must provide a description of how each molecule will tend to bind in the active site. Field descriptors typically consist of a sum of one or more spatial properties, such as steric factors, van der Waals parameters, or the electrostatic potential. The choice of grid points will also affect the quality of the final results.

The field points must then be fitted to predict the activity. There are generally far more field points than known compound activities to be fitted. The least-squares algorithms used in QSAR studies do not function for such an underdetermined system. A partial least squares (PLS) algorithm is used for this type of fitting. This method starts with matrices of field data and activity data. These matrices are then used to derive two new matrices containing a description of the system and the residual noise in the data. Earlier studies used a similar technique, called principal component analysis (PCA). PLS is generally considered to be superior.

The model obtained from the PLS algorithm gives two pieces of information on various regions of space. The first is how well the activity correlates to that region in space. The second is whether the functional group at that point should be electron-donating, electron-withdrawing, bulky, and so forth according to the choice of field parameters. This site description is called a pharmacophore in drug design work.

An examination of the plotted data reveals significant structural information, such as the fact that an electron-donating group should be a certain distance from a withdrawing group, and so on. Further examination of relative magnitudes can give an indication as to precisely which group might be best. Unknown compounds may then be run through the same analysis to obtain a quantitative prediction of their drug activities.

Ideally, the results should be validated somehow. One of the best methods for doing this is to make predictions for compounds known to be active that were not included in the training set. It is also desirable to eliminate compounds that are statistical outliers in the training set. Unfortunately, some studies, such as drug activity prediction, may not have enough known active compounds to make this step feasible. In this case, the estimated error in prediction should be increased accordingly.

30.4 COMPARATIVE QSAR

Comparative QSAR is a field currently under development by several groups. Large databases of known QSAR and 3D QSAR results have been compiled. Such a database can be used for more than simply obtaining literature citations. The analysis of multiple results for the same or similar systems can yield a general understanding of the related chemistry as well as providing a good comparison of techniques.

30.5 RECOMMENDATIONS

Floppy molecules present some additional difficulty in applying QSAR/QSPR. They are also much more difficult to work with in 3D QSAR. With QSAR/QSPR, this problem can be avoided by using only descriptors that do not depend on the conformation, but the accuracy of results may suffer. For more accurate QSPR, the lowest-energy conformation is usually what should be used. For QSAR or 3D QSAR, the conformation most closely matching a rigid molecule in the test set should be used. If all the molecules are floppy, finding the lowest-energy conformer for all and looking for some commonality in the majority might be the best option.

QSPR and QSAR are useful techniques for predicting properties that would be very difficult to predict by any other method. This is a somewhat empirical or indirect calculation that ultimately limits the accuracy and amount of information which can be obtained. When other means of computational prediction are not available, these techniques are recommended for use. There are a variety of algorithms in use that are not equivalent. An examination of published results and tests of several techniques are recommended.

BIBLIOGRAPHY

Introductory descriptions are in

A. K. Rappé, C. J. Casewit, *Molecular Mechanics across Chemistry* University Science Books, Sausalito (1997).

A. R. Leach *Molecular Modelling Principles and Applications* Longman, Essex (1996).

G. H. Grant, W. G. Richards, *Computational Chemistry* Oxford, Oxford (1995).

Books about QSAR/QSPR are

L. B. Kier, L. H. Hall, *Molecular Structure Description· The Electrotopological State* Academic Press, San Diego (1999).

Topological Indices and Related Descriptors in QSAR and QSPR J. Devillers, A. T. Balaban, Eds., Gordon and Breach, Reading (1999).

3D QSAR in Drug Design H. Kubinyi, Y. C. Martin, G. Folker, Eds., Kluwer, Norwell MA (1998). (3 volumes)

J. Devillers, *Neural Networks in QSAR and Drug Design* Academic Press, San Diego (1996).

C. Hansch, A. Leo, *Exploring QSAR* American Chemical Society, Washington (1995).

L. B. Kier, L. H. Hall, *Molecular Connectivity in Structure-Activity Analysis* Research Studies Press, Chichester (1986).

L. B. Kier, L. H. Hall, *Molecular Connectivity in Chemistry and Drug Research* Academic Press, San Diego (1976).

Review articles are

D. Ivanciuc, *Encycl. Comput. Chem.* **1**, 167 (1998).

V. Venkatasubramanian, a. Sundaram, *Encycl. Comput. Chem.* **2**, 1115 (1998).

G. Jones, *Encycl. Comput. Chem.* **2**, 1127 (1998).

D. Ivanciuc, A. T. Balaban, *Encycl. Comput. Chem* **2**, 1169 (1998).

J. Shorter, *Encycl. Comput. Chem.* **4**, 1487 (1998).

P. C. Jurs, *Encycl. Comput. Chem.* **1**, 2320 (1998).

M. Randic, *Encycl. Comput. Chem.* **5**, 3018 (1998).

S. Profeta, Jr., *Kirk-Othmer Encyclopedia of Chemical Technology Supplement* J. I. Kroschwitz (Ed.) 315, John Wiley & Sons, New York (1998).

G. A. Arteca, *Rev. Comput. Chem.* **9**, 191 (1996).

M. Karelson, V. S. Lobanov, A. R. Katritzky, *Chem. Rev.* **96**, 1027 (1996).

A. R. Katritzky, V. S. Lobanov, M. Karelson, *Chem. Soc Rev.* **24**, 279 (1995).

B. W. Clare, *Theor. Chim. Acta* **87**, 415 (1994).

L. H. Hall, L. B. Kerr, *Rev. Comput. Chem.* **2**, 367 (1991).

I. B. Bersuker, A. S. Dimoglo, *Rev. Comput. Chem.* **2**, 423 (1991).

S. P. Gupta, *Chem. Rev.* **87**, 1183 (1987).

3D QSAR reviews are

H. Kubinyi, *Encycl. Comput. Chem.* **1**, 448 (1998).

T. I. Oprea, C. L. Waller, *Rev. Comput. Chem.* **11**, 127 (1997).

G. Greco, E. Novellino, Y. C. Martin, *Rev. Comput. Chem.* **11**, 183 (1997).

Comparative QSAR reviews are

H. Gao, J. A. Katzenellenbogen, R. Garg, C. Hansch, *Chem. Rev.* **99**, 723 (1999).

C. Hansch, G. Gao, *Chem. Rev.* **97,** 2995 (1997).

C. Hansch, D. Hoekmen, H. Gao, *Chem. Rev.* **96**, 1045 (1996).

Many resources are listed at the web site of The QSAR and Modelling Society

http://www.pharma.ethz.ch/qsar

QSAR applications in various fields

J. Devillers, *Encycl. Comput. Chem.* **2**, 930 (1998).

H. Kubinyi, *Encycl. Comput. Chem.* **4**, 2309 (1998).

F. Leclerc, R. Cedergren, *Encycl. Comput. Chem.* **4**, 2756 (1998).

QSAR in Environmental Toxicology-IV Elsevier, Amsterdam (1991).

Practical Applications of Quantitative Structure-Activity Relationships (QSAR) in Environmental Chemistry and Toxicology W. Karcher, J. Devillers, Eds., Kluwer, Dordrecht (1990).

QSAR in Environmental Toxicology K. L. E. Kaiser, Ed., D. Reidel Publishing, Dordrecht (1989).

QSAR in Environmental Toxicology-II D. Reidel Publishing, Dordrecht (1987).

QSAR in Drug Design and Toxicology D. Hadzi, B. Jerman-Blažič, Eds., Elsevier, Amsterdam (1987).

QSAR and Strategies in the Design of Bioactive Compounds J. K. Seydel, Ed., VCH, Weinheim (1985).

An article listing many descriptors is

M. Cocchi, M. C. Menziani, F. Fanelli, P. G. de Benedetti, *J. Mol. Struct. (Theochem)* **331**, 79 (1995).

31 Computing NMR Chemical Shifts

Nuclear magnetic resonance (NMR) spectroscopy is a valuable technique for obtaining chemical information. This is because the spectra are very sensitive to changes in the molecular structure. This same sensitivity makes NMR a difficult case for molecular modeling.

Computationally predicting coupling constants is much easier than predicting chemical shifts. Because of this, the ability to predict coupling constants is sometimes incorporated into software packages that have little or no ability to predict chemical shifts. Computed coupling constants differ very little from one program to the next. This chapter will focus on the more difficult problem of computing NMR chemical shifts.

31.1 *AB INITIO* METHODS

NMR chemical shifts can be computed using *ab initio* methods, which actually compute the shielding tensor. Once the shielding tensors have been computed, the chemical shifts can be determined by subtracting the isotropic shielding values for the molecule of interest from the TMS values. Computing shielding tensors is difficult because of gauge problems (dependence on the coordinate system's origin). A number of techniques for correcting this are in use. It is extremely important that the shielding tensors be computed for equilibrium geometries with the same method and basis that were used to complete the geometry optimization.

It is also important that sufficiently large basis sets are used. The 6–31G(d) basis set should be considered the absolute minimum for reliable results. Some studies have used locally dense basis sets, which have a larger basis on the atom of interest and a smaller basis on the other atoms. In general, this results in only minimal improvement since the spectra are due to interaction between atoms, rather than the electron density around one atom.

One of the most popular techniques is called GIAO. This originally stood for gauge invariant atomic orbitals. More recent versions have included ways to relax this condition without loss of accuracy and subsequently the same acronym was renamed gauge including atomic orbitals. The GIAO method is based on perturbation theory. This is a means for computing shielding tensors from HF or DFT wave functions.

The individual gauge for localized orbitals (IGLO) and localized orbital local origin (LORG) methods are similar. Both are based on identities and closure relations that are rigorously correct for complete basis sets. These are reasonable approximations for finite basis sets. The two methods are equivalent in the limit of a complete basis set.

The individual gauges for atoms in molecules (IGAIM) method is based on Bader's atoms in molecules analysis scheme. This method yields results of comparable accuracy to those of the other methods. However, this technique is seldom used due to large CPU time demands.

There have also been methods designed for use with perturbation theory and MCSCF calculations. Correlation effects are necessary for certain technically difficult molecules, such as CO, N_2, HCN, F_2, and N_2O.

Density functional theory calculations have shown promise in recent studies. Gradient-corrected or hybrid functionals must be used. Usually, it is necessary to employ a moderately large basis set with polarization and diffuse functions along with these functionals.

The methods listed thus far can be used for the reliable prediction of NMR chemical shifts for small organic compounds in the gas phase, which are often reasonably close to the liquid-phase results. Heavy elements, such as transition metals and lanthanides, present a much more difficult problem. Mass defect and spin-coupling terms have been found to be significant for the description of the NMR shielding tensors for these elements. Since NMR is a nuclear effect, core potentials should not be used.

31.2 SEMIEMPIRICAL METHODS

There is one semiempirical program, called HyperNMR, that computes NMR chemical shifts. This program goes one step further than other semiempiricals by defining different parameters for the various hybridizations, such as sp^2 carbon vs. sp^3 carbon. This method is called the typed neglect of differential overlap method (TNDO/1 and TNDO/2). As with any semiempirical method, the results are better for species with functional groups similar to those in the set of molecules used to parameterize the method.

Another semiempirical method, incorporated in the VAMP program, combines a semiempirical calculation with a neural network for predicting the chemical shifts. Semiempirical calculations are useful for large molecules, but are not generally as accurate as *ab initio* calculations.

31.3 EMPIRICAL METHODS

The simplest empirical calculations use a group additivity method. These calculations can be performed very quickly on small desktop computers. They are most accurate for a small organic molecule with common functional groups. The prediction is only as good as the aspects of molecular structure being parameterized. For example, they often do not distinguish between *cis* and *trans* isomers. Due to the limited accuracy, this method is more often used as a tool to check for reasonable results, but not as a rigorous prediction method.

Another technique employs a database search. The calculation starts with a molecular structure and searches a database of known spectra to find those with the most similar molecular structure. The known spectra are then used to derive parameters for inclusion in a group additivity calculation. This can be a fairly sophisticated technique incorporating weight factors to account for how closely the known molecule conforms to typical values for the component functional groups. The use of a large database of compounds can make this a very accurate technique. It also ensures that liquid, rather than gas-phase, spectra are being predicted.

31.4 RECOMMENDATIONS

In general, the computation of absolute chemical shifts is a very difficult task. Computing shifts relative to a standard, such as TMS, can be done more accurately. With some of the more approximate methods, it is sometimes more reliable to compare the shifts relative to the other shifts in the compound, rather than relative to a standard compound. It is always advisable to verify at least one representative compound against the experimental spectra when choosing a method. The following rules of thumb can be drawn from a review of the literature:

1. Database techniques are very fast and very accurate for organic molecules with common functional groups.
2. *Ab initio* methods are accurate and can be reliably applied to unusual structures and inorganic compounds. In most cases, HF calculations are fairly good for organic molecules. Large basis sets should be used.
3. For large molecules, the choice between semiempirical calculations and empirical calculations should be based on a test case.
4. Correlated and relativistic quantum mechanical calculations give the highest possible accuracy and are necessary for heavy atoms or correlation-sensitive systems.

BIBLIOGRAPHY

Introductory descriptions are in

M. F. Schlecht, *Molecular Modeling on the PC* Wiley-VCH, New York (1998).

E. K. Wilson, *Chem & Eng. News* Sept. 28 (1998).

P. W. Atkins, R. S. Friedman, *Molecular Quantum Mechanics Third Edition* Oxford, Oxford (1997).

M. Karplus, R. N. Porter, *Atoms & Molecules: An Introduction For Students of Physical Chemistry* W. A. Benjamin, Inc., Menlo Park (1970).

Books about NMR modeling are

B. Born, H. W. Spiess, *Ab Initio Calculations of Conformational Effects on ^{13}C NMR Spectra of Amorphous Polymers* Springer-Verlag, New York (1997).

I. Ando, G. A. Webb, *Theory of NMR Parameters* Academic Press, London (1983).

The following book gives a tutorial and examples for using *ab initio* methods. Some printings have an error in the listed TMS values. An eratta is available from Gaussian, Inc.

J. B. Foresman, Æ. Frisch, *Exploring Chemistry with Electronic Structure Methods Second Edition* Gaussian, Pittsburgh (1996).

Review articles are

U. Fleischer, C. van Wüllen, w. Kutzelnigg, *Encycl. Comput. Chem.* **3**, 1827 (1998).

M. Bühl, *Encycl. Comput. Chem.* **3**, 1835 (1998).

M. Kaupp, V. G. Malkin, O. L. Malkina, *Encycl. Comput. Chem.* **3**, 1857 (1998).

C. J. Jameson, *Annu. Rev. Phys. Chem.* **47**, 135 (1996).

D. B. Chesnut, *Rev. Comput. Chem.* 8, 245 (1996).

J. R. Cheeseman, G. W. Trucks, T. A. Keith, M. J. Frisch, *J. Chem. Phys.* **104**, 5497 (1996).

D. B. Chesnut, *Annual Reports on NMR Spectroscopy* **29**, 71 (1994).

C. J. Jameson, *Chem. Rev.* **91**, 1375 (1991).

D. B. Chesnut, *Annual Reports on NMR Spectroscopy* **21**, 51 (1989).

C. Giessner-Prettre, B. Pullman, *Quarterly Reviews of Biophysics* **20**, 113 (1987).

C. J. Jameson, H. J. Osten, *Annual Reports on NMR Spectroscopy* **17**, 1 (1986).

The group additivity technique is presented in

E. Pretsch, J. Seibl, W. Simon, T. Clerc, *Tabellen zur Strukturaufklärung Organischer Verbindungen mit Spektroskopischen Methoden* Springer-Verlag, Berlin (1981).

32 Nonlinear Optical Properties

Nonlinear optical properties are of interest due to their potential usefulness for unique optical devices. Some of these applications are frequency-doubling devices, optical signal processing, and optical computers.

Most of the envisioned practical applications for nonlinear optical materials would require solid materials. Unfortunately, only gas-phase calculations have been developed to a reliable level. Most often, the relationship between gas-phase and condensed-phase behavior for a particular class of compounds is determined experimentally. Theoretical calculations for the gas phase are then scaled accordingly.

32.1 NONLINEAR OPTICAL PROPERTIES

When light is incident on a material, the optical electric field E results in a polarization P of the material. The polarization can be expressed as the sum of the linear polarization P^L and a nonlinear polarization P^{NL}:

$$P = P^L + P^{NL} \tag{32.1}$$

$$P^L = \chi^{(1)} \cdot E \tag{32.2}$$

$$P^{NL} = \chi^{(2)} \cdot EE + \chi^{(3)} \cdot EEE + \cdots \tag{32.3}$$

The susceptibility tensors $\chi^{(n)}$ give the correct relationship for the macroscopic material. For individual molecules, the polarizability α, hyperpolarizability β, and second hyperpolarizability γ, can be defined; they are also tensor quantities. The susceptibility tensors are weighted averages of the molecular values, where the weight accounts for molecular orientation. The obvious correspondence is correct, meaning that $\chi^{(1)}$ is a linear combination of α values, $\chi^{(2)}$ is a linear combination of β values, and so on.

The molecular quantities can be best understood as a Taylor series expansion. For example, the energy of the molecule E would be the sum of the energy without an electric field present, E_0, and corrections for the dipole, polarizability, hyperpolarizability, and the like:

$$E = E_0 - \mu \cdot E - \left(\frac{1}{2!}\right)\alpha \cdot E^2 - \left(\frac{1}{3!}\right)\beta \cdot E^3 - \left(\frac{1}{4!}\right)\gamma \cdot E^4 - \cdots \tag{32.4}$$

As implied by this, the polarizabilities can be formulated as derivatives of the dipole moment with respect to the incident electric field. Below these derivatives are given, with subscripts added to indicate their tensor nature:

$$\alpha_{ij} = \left(\frac{\partial \mu_i}{\partial E_j}\right)_{E\to 0} \tag{32.5}$$

$$\beta_{ijk} = \left(\frac{\partial^2 \mu_i}{\partial E_j \partial E_k}\right)_{E\to 0} \tag{32.6}$$

$$\gamma_{ijkl} = \left(\frac{\partial^3 \mu_i}{\partial E_j \partial E_k \partial E_l}\right)_{E\to 0} \tag{32.7}$$

These expressions are only correct for wave functions that obey the Hellmann–Feynman theorem. However, these expressions have been used for other methods, where they serve as a reasonable approximation. Methods that rigorously obey the Hellmann–Feynman theorem are SCF, MCSCF, and Full CI. The change in energy from nonlinear effects is due to a change in the electron density, which creates an induced dipole moment and, to a lesser extent, induced higher-order multipoles.

After examining these definitions, several conclusions can be drawn, which have been verified theoretically and experimentally. One is that a molecule with a center of inversion will have no hyperpolarizability ($\beta = 0$). Molecules with a large dipole moment and a means for electron density to shift will have large hyperpolarizabilities. For example, organic systems with electron-donating groups and electron-withdrawing groups at opposite ends of a conjugated system generally have large hyperpolarizabilities.

The definitions given above reflect static polarizabilities that are due to the presence of a static electric field. Nonlinear optical properties are the result of the oscillating electric field component of the incident light. Static hyperpolarizabilities are often computed and then employed to predict nonlinear optical properties by using an experimentally determined correction factor. Alternatively, time-dependent calculations can be used to predict experimental results directly. There are several different nonlinear optical properties due to several incoming photons of light (ν_1, ν_2, ν_3) and result in an exiting photon of the same or a different frequency (ν_σ). The list of outgoing and incoming photons is typically denoted with the notation $-\nu_\sigma; \nu_1, \nu_2, \nu_3$. The nonlinear optical properties are summarized in Table 32.1. Each of these can be computed from the appropriate frequency-dependent terms.

32.2 COMPUTATIONAL ALGORITHMS

There are several ways in which to compute polarizabilities and hyperpolarizabilities from semiempirical or *ab initio* wave functions. One option is to take

TABLE 32.1 Nonlinear Optical Properties

$-\nu_\sigma;\nu_1,\nu_2,\nu_3$	Abbreviation	Name
	Polarizability α	
$0;0$		Static polarizability
$-\nu;\nu$		Frequency-dependent polarizability
	Hyperpolarizability β	
$0;0,0$		Static hyperpolarizability
$-\nu;\nu,0$	EOPE	Electro-optics Pockels effect
$-2\nu;\nu,\nu$	SHG	Second harmonic generation
$0;\nu,-\nu$	OR	Optical rectification
$-(\nu_1+\nu_2);\nu_1,\nu_2$		Two-wave mixing
	Second Hyperpolarizability γ	
$0;0,0,0$		Static second hyperpolarizability
$-3\nu;\nu,\nu,\nu$	THG	Third harmonic generation
$-\nu;\nu,\nu,-\nu$	IDRI or DFWM	Intensity-dependent refractive index or degenerate four-wave mixing
$-\nu_1;\nu_1,\nu_2,-\nu_2$	OKE	Optical Kerr effect or AC Kerr effect
$0;0,\nu,-\nu$	DCOR	DC-induced optical rectification
$-2\nu;0,\nu,\nu$	DC-SHG or EFISH	DC-induced second harmonic generation or electric-field-induced second harmonic
$-\nu;\nu,0,0$	EOKE	Electro-optic Kerr effect
$-\nu_\sigma;\nu_1,\nu_1,\nu_2$		Three-wave mixing
$-(\nu_1+\nu_2);0,\nu_1,\nu_2$		DC-induced two-wave mixing
$-2\nu_1+\nu_2;\nu_1,\nu_1,-\nu_2$	CARS	Coherent anti-Stokes Raman scattering

the derivatives defined above either analytically or numerically. Analytic derivatives have been formulated for a few methods. This is sometimes called the derivative Hartree–Fock method or DHF (note that the acronym DHF is also used for the Dirac–Hartree–Fock method). Numerical derivatives can be used with any method but require a large amount of CPU time. The researcher should pay close attention to numerical precision when using numerical derivatives.

A second method is to use a perturbation theory expansion. This is formulated as a sum-over-states algorithm (SOS). This can be done for correlated wave functions and has only a modest CPU time requirement. The random-phase approximation is a time-dependent extension of this method.

The electric field can be incorporated in the Hamiltonian via a finite field term or approximated by a set of point charges. This allows the computation of corrections to the dipole only, which is generally the most significant contribution.

Time-dependent calculations have been completed with a number of different methods. There are three formulations giving equivalent results; TDHF,

RPA, and CPHF. Time-dependent Hartree–Fock (TDHF) is the Hartree–Fock approximation for the time-dependent Schrödinger equation. CPHF stands for coupled perturbed Hartree–Fock. The random-phase approximation (RPA) is also an equivalent formulation. There have also been time-dependent MCSCF formulations using the time-dependent gauge invariant approach (TDGI) that is equivalent to multiconfiguration RPA. All of the time-dependent methods go to the static calculation results in the $\nu = 0$ limit.

32.3 LEVEL OF THEORY

Polarizabilities and hyperpolarizabilities have been calculated with semiempirical, *ab initio*, and DFT methods. The general conclusion from these studies is that a high level of theory is necessary to correctly predict nonlinear optical properties.

Semiempirical calculations tend to be qualitative. In some cases, the correct trends have been predicted. In other cases, semiempirical methods give incorrect signs as well as unreasonable magnitudes.

Ab initio methods can yield reliable, quantitatively correct results. It is important to use basis sets with diffuse functions and high-angular-momentum polarization functions. Hyperpolarizabilities seem to be relatively insensitive to the core electron description. Good agreement has been obtained between ECP basis sets and all electron basis sets. DFT methods have not yet been used widely enough to make generalizations about their accuracy.

Explicitly correlated wave functions have been shown to give very accurate results. Unfortunately, these calculations are only tractable for very small molecules.

There have been some attempts to compute nonlinear optical properties in solution. These studies have shown that very small variations in the solvent cavity can give very large deviations in the computed hyperpolarizability. The valence bond charge transfer (VB-CT) method created by Goddard and coworkers has had some success in reproducing solvent effect trends and polymer results (the VB-CT-S and VB-CTE forms, respectively).

32.4 RECOMMENDATIONS

Unfortunately, it is necessary to use very computationally intensive methods for computing accurate nonlinear optical properties. The following list of alternatives is ordered, starting with the most accurate and likewise most computation-intensive techniques:

1. Time-dependent calculations with highly correlated methods
2. Explicitly correlated methods
3. CCSD(T)

4. CISD, CCSD, or MP4
5. TDHF, RPA, or CPHF
6. MP2 or MP3
7. SCF or DFT
8. Semiempirical methods where they have been shown to reproduce the correct trends

BIBLIOGRAPHY

Introductory descriptions are in

S. P. Karna, A. T. Yeates, *Nonlinear Optical Materials: Theory and Modeling* S. P. Karna, A. T. Yeates, Eds., 1, American Chemical Society, Washington (1996).

R. W. Boyd, *Nonlinear Optics* Academic Press, San Diego (1992).

Review articles are

H. A. Kurtz, D. S. Dudis, *Rev. Comput. Chem.* **12**, 241 (1998).

R. J. Bartlett, H. Sekino, *Nonlinear Optical Materials. Theory and Modeling* S. P. Karna, A. T. Yeates, Eds., 23, American Chemical Society, Washington (1996).

D. M. Bishop, *Adv. Quantum Chem.* **25**, 1 (1994).

D. P. Shelton, J. E. Rice, *Chem. Rev.* **94**, 3 (1994).

J. L. Brédas, C. Adant, P. Tackx, A. Persoons, *Chem. Rev.* **94**, 243 (1994).

D. R. Kanis, M. A. Ratner, T. J. Marks, *Chem. Rev.* **94**, 195 (1994).

A. A. Hasanein, *Adv. Chem. Phys.* **85**, 415 (1994).

W. T. Coffey, Y. D. Kalmykov, E. S. Massawe, *Adv. Chem. Phys.* **85**, 667 (1994).

D. M. Bishop, *Rev. Mod. Phys.* **62**, 343 (1990).

Mathematical treatments are in

W. Alexiewicz, B. Kasprowicz-Kielich, *Adv. Chem. Phys.* **85**, 1 (1994).

D. L. Andrews, *Adv. Chem. Phys.* **85**, 545 (1994).

G. C. Schatz, M. A. Ratner, *Quantum Mechanics in Chemistry* Prentice Hall, Englewood Cliffs (1993).

C. E. Dykstra, J. D. Augspurser, B. Kirtman, D. J. Malik, *Rev. Comput. Chem.* **1**, 83 (1990).

Other pertinent articles are

P. Korambath, H. A. Kurtz, *Nonlinear Optical Materials: Theory and Modeling* S. P. Karna, A. T. Yeates, Eds., 133, American Chemical Society, Washington (1996).

W. A. Goddard, III, D. Lu, G. Chen, J. W. Perry, *Computer-Aided Molecular Design* 341 C. H. Reynolds, M. K. Holloway, H. K. Cox, Ed., American Chemical Society, Washington (1995).

W. A. Parkinson, J. Oddershede, *J. Chem. Phys* **94**, 7251 (1991).

H. A. Kurtz, J. J. P. Stewart, K. M. Dieter, *J Comput. Chem.* **11**, 82 (1990).

33 Relativistic Effects

The Schrödinger equation is a nonrelativistic description of atoms and molecules. Strictly speaking, relativistic effects must be included in order to obtain completely accurate results for any *ab initio* calculation. In practice, relativistic effects are negligible for many systems, particularly those with light elements. It is necessary to include relativistic effects to correctly describe the behavior of very heavy elements. With increases in computer capability and algorithm efficiency, it will become easier to perform heavy atom calculations and thus an understanding of relativistic corrections is necessary.

This chapter provides only a brief discussion of relativistic calculations. Currently, there is a small body of references on these calculations in the computational chemistry literature, with relativistic core potentials comprising the largest percentage of that work. However, the topic is important both because it is essential for very heavy elements and such calculations can be expected to become more prevalent if the trend of increasing accuracy continues.

33.1 RELATIVISTIC TERMS IN QUANTUM MECHANICS

The fact that an electron has an intrinsic spin comes out of a relativistic formulation of quantum mechanics. Even though the Schrödinger equation does not predict it, wave functions that are antisymmetric and have two electrons per orbital are used for nonrelativistic calculations. This is necessary in order to obtain results that are in any way reasonable.

Mass defect is the phenomenon of the electrons increasing in mass as they approach a significant percentage of the speed of light. This is particularly significant for *s* orbitals near the nucleus of heavy atoms. Mass defect must only be included in calculations on the heaviest atoms, typically atomic number 55 and up. The effect of mass defect is to contract the *s* and *p* orbitals closer to the nucleus. This creates an additional shielding of the nucleus, causing the *d* and *f* orbitals to expand, making bond lengths longer. This effect is most pronounced for the group 11 elements: gold, silver, and copper.

There are many moving charges within an atom. These motions are the intrinsic electron spin, electron orbital motion, and nuclear spin. Every one of these moving charges creates a magnetic field. Spin couplings are magnetic interactions due to the interaction of these magnetic fields. Spin–orbit coupling tends to be most significant for the lightest transition metals and spin–spin couplings tend to be important for the heaviest actinides. For elements between these extremes, spin–orbit coupling is often included and other spin-coupling terms are sometimes included. The size of *p* orbitals is often relatively unchanged by relativistic effects due to the mass defect and spin–orbit effects canceling out.

Also arising from relativistic quantum mechanics is the fact that there should be both negative and positive energy states. One of these corresponds to electron energies and the other corresponds to the electron antiparticle, the positron.

33.2 EXTENSION OF NONRELATIVISTIC COMPUTATIONAL TECHNIQUES

The relativistic Schrödinger equation is very difficult to solve because it requires that electrons be described by four component vectors, called spinnors. When this equation is used, numerical solution methods must be chosen.

The most common description of relativistic quantum mechanics for Fermion systems, such as molecules, is the Dirac equation. The Dirac equation is a one-electron equation. In formulating this equation, the terms that arise are intrinsic electron spin, mass defect, spin couplings, and the Darwin term. The Darwin term can be viewed as the effect of an electron making a high-frequency oscillation around its mean position.

The Dirac equation can be readily adapted to the description of one electron in the field of the other electrons (Hartree–Fock theory). This is called a Dirac–Fock or Dirac–Hartree–Fock (DHF) calculation.

33.3 CORE POTENTIALS

The most common way of including relativistic effects in a calculation is by using relativisticly parameterized effective core potentials (RECP). These core potentials are included in the calculation as an additional term in the Hamiltonian. Core potentials must be used with the valence basis set that was created for use with that particular core potential. Core potentials are created by fitting a potential function to the electron density distribution from an accurate relativistic calculation for the atom. A calculation using core potentials does not have any relativistic terms, but the effect of relativity on the core electrons is included.

The use of RECP's is often the method of choice for computations on heavy atoms. There are several reasons for this: The core potential replaces a large number of electrons, thus making the calculation run faster. It is the least computation-intensive way to include relativistic effects in *ab initio* calculations. Furthermore, there are few semiempirical or molecular mechanics methods that are reliable for heavy atoms. Core potentials were discussed further in Chapter 10.

33.4 EXPLICIT RELATIVISTIC CALCULATIONS

There are also ways to perform relativistic calculations explicitly. Many of these methods are plagued by numerical inconsistencies, which make them applicable only to a select set of chemical systems. At the expense of time-consuming numerical integrations, it is possible to do four component calculations. These calculations take about 100 times as much CPU time as nonrelativistic Hartree–Fock calculations. Such calculations are fairly rare in the literature.

Many researchers have performed calculations that include the two large-magnitude components of the spinnors. This provides a balance between high accuracy and making the calculation tractable. Such calculations are often done on atoms in order to obtain the wave function description used to create relativistic core potentials.

There are several ways to include relativity in *ab initio* calculations more efficiently at the expense of a bit of accuracy. One popular technique is the Dirac–Hartree–Fock technique, which includes the one-electron relativistic terms. Another option is computing energy corrections to the nonrelativistic wave function without changing that wave function.

Relativistic density functional theory can be used for all electron calculations. Relativistic DFT can be formulated using the Pauli formula or the zero-order regular approximation (ZORA). ZORA calculations include only the zero-order term in a power series expansion of the Dirac equation. ZORA is generally regarded as the superior method. The Pauli method is known to be unreliable for very heavy elements, such as actinides.

Molecular mechanics and semiempirical calculations are all relativistic to the extent that they are parameterized from experimental data, which of course include relativistic effects. There have been some relativistic versions of PM3, CNDO, INDO, and extended Huckel theory. These relativistic semiempirical calculations are usually parameterized from relativistic *ab initio* results.

33.5 EFFECTS ON CHEMISTRY

As described above, relativistic effects are responsible for shifts in the bond lengths of compounds, particularly those involving group 11 elements. This is called the gold maximum. For example, the Ag_2 bond length predicted by nonrelativistic calculations will be in error by −0.1 Å. The AuH deviation of −0.2 Å and Au_2 deviation of −0.4 Å along with the $Hg_2{}^{2+}$ deviation of −0.3 Å are among the largest known.

Relativistic effects are cited for changes in energy levels, resulting in the yellow color of gold and the fact that mercury is a liquid. Relativistic effects are also cited as being responsible for about 10% of lanthanide contraction. Many more specific examples of relativistic effects are reviewed by Pyykkö (1988).

33.6 RECOMMENDATIONS

The most difficult part of relativistic calculations is that a large amount of CPU time is necessary. This makes the problem more difficult because even nonrelativistic calculations on elements with many electrons are CPU-intensive. The following lists relativistic calculations in order of increasing reliability and thus increasing CPU time requirements:

1. Relativistic semiempirical calculations
2. Relativistic effective core potentials
3. Dirac–Hartree–Fock
4. Relativistic density functional theory
5. Relativistic correlated calculations using the DHF Hamiltonian
6. Two-component calculations
7. Four-component calculations

BIBLIOGRAPHY

Introductory descriptions are

F. Jensen, *Introduction to Computational Chemistry* John Wiley & Sons, New York (1999).

M. Jacoby, *Chem. & Eng. News* **March 23**, 48 (1998).

C. B. Kellogg, *An Introduction to Relativistic Electronic Structure Theory in Quantum Chemistry* http://zopyros.ccqc.uga.edu/~kellogg/docs/rltvt/rltvt.html (1996).

I. N. Levine, *Quantum Chemistry Fourth Edition* Prentice Hall, Englewood Cliffs (1991).

W. H. E. Schwartz, *Theoretical Models of Chemical Bonding* Z. B. Maksič, Ed., Springer-Verlag, Berlin (1990).

K. S. Pitzer, *Acct. Chem. Res* **12**, 271 (1979).

P. Pyykkö, J.-P. Desclaux, *Acct. Chem. Res.* **12**, 276 (1979).

Review articles are

C. van Wullen, *J. Comp. Chem.* **20**, 51 (1999).

K. Balasubramanian, *Encycl. Comput. Chem.* **4**, 2471 (1998).

P. Schwerdtfeger, M. Seth, *Encycl. Comput Chem.* **4**, 2480 (1998).

B. A. Hess, *Encycl Comput. Chem.* **4**, 2499 (1998).

J. Almlof, O. Gropen, *Rev. Comput. Chem.* **8**, 203 (1996).

E. Engel, R. M. Dreizler, *Density Functional Theory II* Springer, Berlin (1996).

Y. Ishikawa, U. Kaldor, *Computational Chemistry – Reviews of Current Trends* J. Leszczynski, Ed., 1, World Scientific, Singapore (1996).

B. A. Hess, C. M. Marian, S. D. Peyerimhoff, *Modern Electronic Structure Theory* D. R. Yarkony, Ed., 152, World Scientific, Singapore (1995).

K. Balasubramanian, *Handbook on the Physics and Chemistry of Rare Earths* K. A. Gschneidner, Jr., L. Eyring, Eds., **18**, 29, Elsevier, Amsterdam (1994).

P. Pyykkö, *Chem. Rev.* **88**, 563 (1988).

S. Wilson, *Methods of Computational Chemistry Volume II* Plenum, New York (1988).

P. Pyykkö, *Adv. Quantum Chem.* **11**, 353 (1978).

S. R. Langhoff, C. W. Kern, *Modern Theoretical Chemistry* 381, Plenum, New York (1977).

M. Barfield, R. J. Spear, S. Sternkell, *Chem. Rev.* **76**, 593 (1976).

M. Barfield, B. Chakrabarti, *Chem. Rev.* **69**, 757 (1969).

Books on relativistic quantum theory are

K. Balasubramanian, *Relativistic Effects in Chemistry* John Wiley & Sons, New York (1997).

R. Landau, *Quantum Mechanics II* John Wiley & Sons, New York, (1996).

W. Greiner, *Relativistic Quantum Mechanics* Springer-Verlag, Berlin (1990).

A database of relativistic quantum mechanics references is at

http://www.csc.fi/lul/rtam/

34 Band Structures

In molecules, the possible electronic energies are discrete, quantized energy levels. As molecules become larger, these energy levels move closer together. In a crystal, the energy levels have merged together so closely that they are continuous bands of available energies for all practical purposes. Thus, the electronic structure of a crystal is described by its band structure.

34.1 MATHEMATICAL DESCRIPTION OF ENERGY BANDS

The electronic structure of an infinite crystal is defined by a band structure plot, which gives the energies of electron orbitals for each point in k-space, called the Brillouin zone. This corresponds to the result of an angle-resolved photo electron spectroscopy experiment.

k-space is not a physical space. It is a description of the bonding nature of orbitals. In an infinitely long string of atoms, the phases of orbitals might be anywhere from all bonding to all antibonding (these extremes are labeled as $k = 0$ and $k = \pi/a$). Somewhere in between are combinations with three atoms in a row oriented for bonding, followed by an antibonding one or any other such combination. As k-space is defined, $k = 0$ corresponds to complete bonding symmetry for some orbitals and complete antibonding for others, depending on the symmetry of the atomic orbitals.

k-space will have three dimensions (k_x, k_y, k_z) for three-dimensional crystals. Certain points in k-space are given names. The designator Γ refers to the point where $k = 0$ in all dimensions. M is the point with $k = \pi/a$ in all directions. X, Y, K, and A are points with $k = 0$ in some directions and $k = \pi/a$ in others, depending on the symmetry of the crystal. The typical band structure plot, called a spaghetti plot, maps the orbital energies along the paths between these points, as shown in Figure 34.1. These designations are discussed in more detail in the literature cited in the bibliography.

As orbitals spread into bands, orbitals oriented for σ or σ^* bonds spread into the widest bands. π orbitals form narrower bands and δ bonding orbitals form the narrowest bands.

34.2 COMPUTING BAND GAPS

In some cases, researchers only need to know the band gap for a crystal. Once a complete band structure has been computed, it is, of course, simple to find the

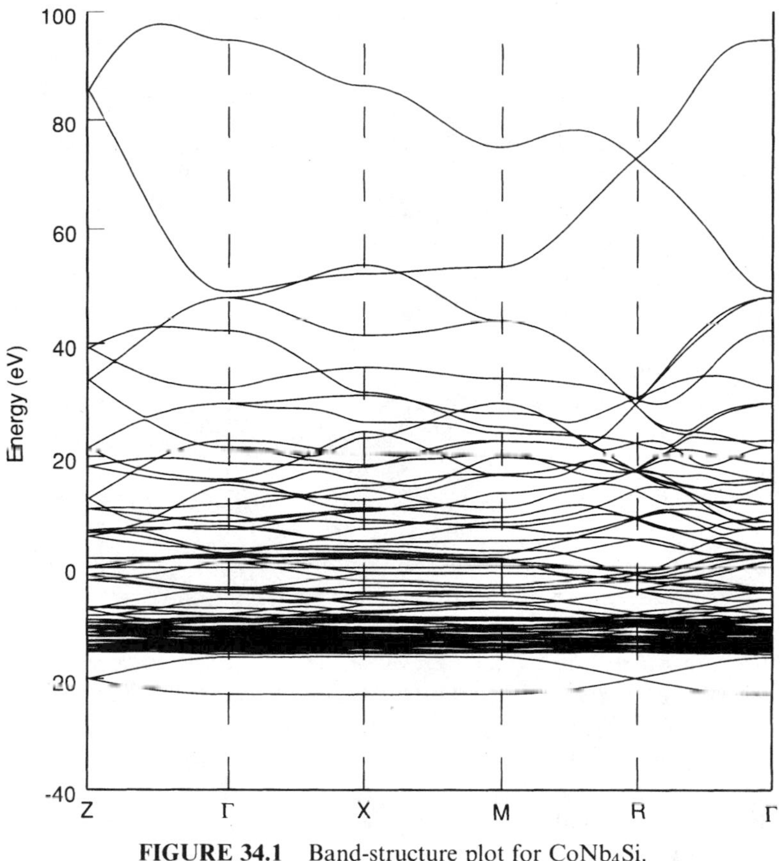

FIGURE 34.1 Band-structure plot for $CoNb_4Si$.

band gap by inspection. However, computing the entire band structure could involve an extensive amount of work to obtain a lot of unneeded information. There are ways of estimating the band gap, although these are not completely reliable.

Simply doing electronic structure computations at the *M*, *K*, *X*, and Γ points in the Brillouin zone is not necessarily sufficient to yield a band gap. This is because the minimum and maximum energies reached by any given energy band sometimes fall between these points. Such limited calculations are sometimes done when the computational method is very CPU-intensive. For example, this type of spot check might be done at a high level of theory to determine whether complete calculations are necessary at that level.

Some researchers use molecule computations to estimate the band gap from the HOMO–LUMO energy separation. This energy separation becomes smaller as the molecule grows larger. Thus, it is possible to perform quantum mechanical calculations on several molecules of increasing size and then extrapolate the energy gap to predict a band gap for the infinite system. This can be useful for polymers, which are often not crystalline. One-dimensional band structures are also used for such systems, thus assuming crystallinity or at least a high degree of order.

34.3 COMPUTING BAND STRUCTURES

Since *ab initio* and semiempirical calculations yield orbital energies, they can be applied to band-structure calculations. However, if it is time-consuming to calculate the energy for a molecule, it is even more time-consuming to calculate energies for a list of points in the Brillouin zone. Since these calculations are so computationally intensive, extended Hückel has been the method of choice unless more accurate results are needed. In the realm of band-structure calculations, extended Hückel is sometimes called the tight binding approximation. In recent years, there has been an increasing tendency to use *ab initio* or DFT methods.

Like molecular calculations, an *ab initio* method requires both a set of basis functions and a means for computing the energy. The choice of basis sets for band structure calculations is somewhat different than for molecular calculations. Large basis sets with diffuse functions can contain contractions having a large overlap, with their image in the adjacent unit cell. When this happens, it creates a linear dependency that prevents the self-consistent equations from being solved. Most often, small- to medium-size basis sets are used to avoid this problem. The linear combination of atomic orbitals (LCAO) scheme used for molecular calculations can be applied to crystal calculations, but it is not the only option. Actually, the basis functions centered on atoms are formed into Bloch functions, which obey the translational symmetry of the system, although the term LCAO is still used.

Another basis technique that is popular for modeling crystals is the use of plane wave basis functions. Plane waves were proposed because they reflect the infinite symmetry of a crystal. There have been several different plane wave techniques proposed. The earliest plane wave calculations assumed the Schrödinger equation was spherically symmetric in a region around each atom (dubbed a muffin tin potential), but suffered from an inability to conserve charge. These muffin tin calculations gave reasonable results for ionic crystals. They are no longer performed since algorithms and computer hardware improvements make more accurate and reliable calculations feasible. A technique still used is the augmented plane wave (APW) technique, which is a cellular calculation over the Vigner–Seitz cell. There are many other basis function methods that are used for certain types of problems.

Band structure calculations have been done for very complicated systems; however, most of software is not yet automated enough or sufficiently fast that anyone performs band structures casually. Setting up the input for a band structure calculation can be more complex than for most molecular programs. The molecular geometry is usually input in fractional coordinates. The unit cell lattice vectors and crystallographic angles must also be provided. It may be necessary to provide a list of k points and their degeneracies. It is safest to check the convergence of any input option affecting the calculation accuracy. Manuals accompanying the software may give some suggested values. Researchers wishing to complete band structure calculations should expect to put a lot of time into their efforts particularly during the learning stage for the software to be used.

As mentioned above, the preferred computational methods for modeling crystals have changed over the years. Below is a list of basis function schemes, with the most often used appearing first:

1. Linear combination of atomic orbitals (LCAO)
2. Augmented plane wave method (APW)
3. Green's function method of Korringa, Kohn, and Rostoker (sometimes denoted KKR)
4. Orthogonalized plane wave (OPW)
5. Pseudopotential method
6. Various approximate or empirical methods

Any orbital-based scheme can be used for crystal-structure calculations. The trend is toward more accurate methods. Some APW and Green's function methods use empirical parameters, thus edging them toward a semiempirical classification. In order of preference, the commonly used methods are:

1. Self-consistent *ab initio* or DFT methods
2. Semiempirical methods
3. Methods using an ad hoc or model potential

34.4 DESCRIBING THE ELECTRONIC STRUCTURE OF CRYSTALS

The population analysis techniques used for molecular calculations are not directly applicable to band structure calculations. A series of techniques for analyzing the band structure have been introduced. These are generally presented as graphical plots. The data for these plots come from a series of calculations at various points in k-space. Very good plots can be obtained by calculating a very large number of points. In order to reduce computer time, a more widely spaced set of points can be generated; then the plots are smoothed by some kind of interpolation algorithm. It is always prudent to perform several calculations with points increasingly close together to see if the plot changes significantly.

One important question is how many orbitals are available at any given energy level. This is shown using a density of states (DOS) diagram as in Figure 34.2. It is typical to include the Fermi level as denoted by the dotted line in this figure. A material with a half-filled energy band is a conductor, but it may be a

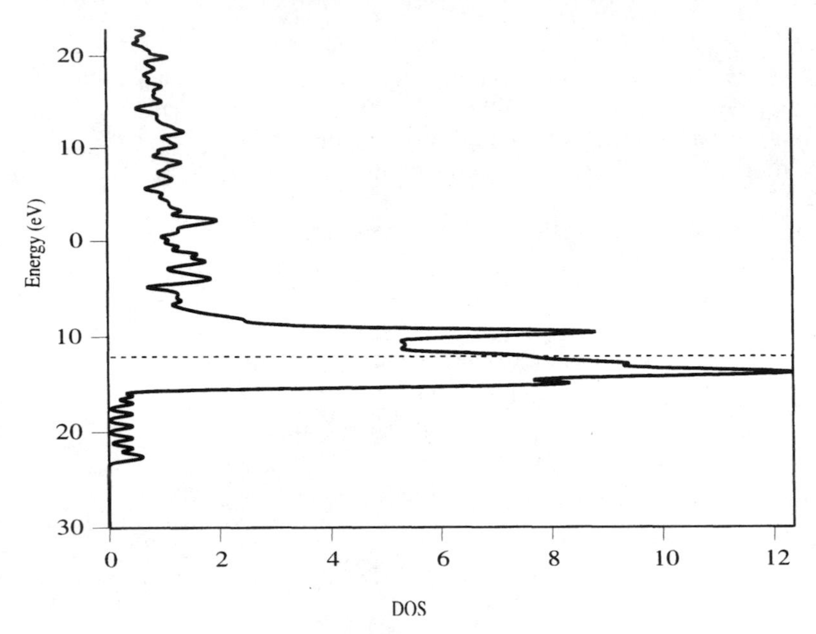

FIGURE 34.2 Density of states plot for $CoNb_4Si$.

very poor conductor if there are very few unfilled orbitals available. In some cases, the contributions of a particular orbital to the DOS are plotted on the same graph as a shaded region or dotted line.

Another question is whether the filled orbitals are of a bonding or antibonding character. This is displayed on a crystal orbital overlap population (COOP) plot as shown in Figure 34.3. Typically, the positive bonding region is plotted to the right of the zero line.

The Fermi energy is the energy of the highest-energy filled orbital, analogous to a HOMO energy. If the orbital is half-filled, its energy will be found at a collection of points in *k*-space, called the Fermi surface.

34.5 COMPUTING CRYSTAL PROPERTIES

There has not been as much progress computing the properties of crystals as for molecular calculations. One property that is often computed is the bulk modulus. It is an indication of the hardness of the material.

It may be desirable to predict which crystal structure is most stable in order to predict the products formed under thermodynamic conditions. This is a very difficult task. As of yet, no completely automated way to try all possible crystal structures formed from a particular collection of elements (analogous to a molecular conformation search) has been devised. Even if such an effort were attempted, the amount of computer power necessary would be enormous. Such studies usually test a collection of likely structures, which is by no means infal-

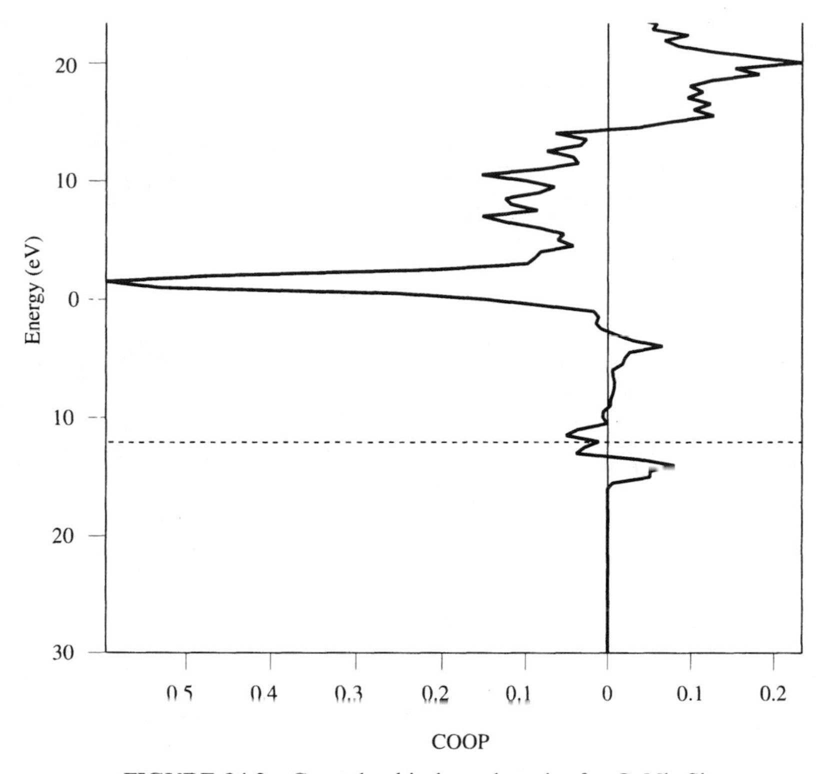

FIGURE 34.3 Crystal orbital overlap plot for $CoNb_4Si$.

lible. Energy minimizations can be performed, provided that the starting structure has the correct symmetry.

34.6 DEFECT CALCULATIONS

Sometimes, the system of interest is not the infinite crystal, but an anomaly in the crystal, such as an extra atom adsorbed in the crystal. In this case, the infinite symmetry of the crystal is not rigorously correct. The most widely used means for modeling defects is the Mott–Littleton defect method. It is a means for performing an energy minimization in a localized region of the lattice. The method incorporates a continuum description of the polarization for the remainder of the crystal.

BIBLIOGRAPHY

Text books describing band structures are

P. Y. Yu, M. Cardone, *Fundamentals of Semicondutors* Springer-Verlag, Berlin (1996).

R. Hoffmann, *Solids and Surfaces A Chemist's View of Bonding in Extended Structures* VCH, New York (1988).

I. M. Tsidilkovski, *Band Structure of Semiconductors* Pergamon, Oxford (1982).

W. A. Harrison, *Solid State Theory* Dover, New York (1979).

A. A. Levin, *Solid State Quantum Chemistry* McGraw-Hill, New York (1977).

Orbital Theories of Molecules and Solids N. H. March, Ed., Clarendon, Oxford (1974).

J. C. Slater, Quantum Theory of Molecules and Solids McGraw-Hill, New york (1963).

Review articles are

J. K. Burdett, *J. Phys. Chem.* **100**, 13274 (1996).

C. R. A. Catlow, J. D. Gale, R. W. Grimes, *J. Solid State Chem.* **106**, 13 (1993).

E. Canadell, M.-H. Whangbo, *Chem. Rev.* **91**, 965 (1991).

J. H. Harding, *Rep. Prog. Phys.* **53**, 1403 (1990).

C. R. A. Catlow, G. D. Price, *Nature* **347**, 243 (1990).

JCS Faraday Trans. II Vol. **85**, part 5 (1989).

R. Hoffmann, *Angew. Chem. Int. Ed. Engl.* **26**, 846 (1987).

C. R. A. Catlow, *Ann. Rev. Mater. Sci.* **16**, 517 (1986).

J. C. Slater, *Adv. Quantum Chem.* **1**, 35 (1964).

The program used to create the figures is YAeHMOP, which has two main executables, which are

G. A. Landrum, W. V. Glassey, Bind 3.0 (1998).

G. A. Landrum, Viewkel 2.0 (1998).

both at http://overlap.chem.cornell.edu:8080/yaehmop.html

35 Mesoscale Methods

Most of the methods described in this book are designed to model individual molecules and interactions between molecules. Engineering simulations generally treat materials as a continuum with a particular set of properties. Some molecular structures, such as suspensions, colloids, and polymer blends, are not described well by either of these techniques. These are important for the modeling of cosmetics, paper, composites, foods, detergents, and many other materials. This is because the properties of these materials are dependent on units with a length scale too large to be modeled with molecular simulations, but also not homogeneous on a microscopic level. These are typically cases in which the material is not annealed to an equilibrium configuration at the molecular level. Furthermore, the dynamics of these materials tend to be on a longer timescale than is appropriate for modeling by molecular dynamics. Thus it is useful for large timescale processes such as chrystallization. Mesoscale simulations are designed to model these materials, which have a structure with a length scale on the order of hundreds of nanometers.

Mesoscale simulations model a material as a collection of units, called beads. Each bead might represent a substructure, molecule, monomer, micelle, microcrystalline domain, solid particle, or an arbitrary region of a fluid. Multiple beads might be connected, typically by a harmonic potential, in order to model a polymer. A simulation is then conducted in which there is an interaction potential between beads and sometimes dynamical equations of motion. This is very hard to do with extremely large molecular dynamics calculations because they would have to be very accurate to correctly reflect the small free energy differences between microstates. There are algorithms for determining an appropriate bead size from molecular dynamics and Monte Carlo simulations.

Many of the mesoscale techniques have grown out of the polymer SCF mean field computation of microphase diagrams. Mesoscale calculations are able to predict microscopic features such as the formation of capsules, rods, droplets, mazes, cells, coils, shells, rod clusters, and droplet clusters. With enough work, an entire phase diagram can be mapped out. In order to predict these features, the simulation must incorporate shape, dynamics, shear, and interactions between beads.

35.1 BROWNIAN DYNAMICS

The conceptual forerunner to mesoscale dynamics is Brownian dynamics. Brownian simulations used equations of motion modified by a random force and viscous frictional force. The random force accounts for the statistical tendency of the path of a particle to be affected by collisions with the solvent molecules. This results in a simulation that reproduces diffusion, but not hydrodynamic fluid flow properties. Fluid flow is not represented correctly because the simulation does not obey the Navier–Stokes equations.

35.2 DISSIPATIVE PARTICLE DYNAMICS

Dissipative particle dynamics (DPD) is a technique for simulating the motion of mesoscale beads. The technique is superficially similar to a Brownian dynamics simulation in that it incorporates equations of motion, a dissipative (random) force, and a viscous drag between moving beads. However, the simulation uses a modified velocity Verlet algorithm to ensure that total momentum and force symmetries are conserved. This results in a simulation that obeys the Navier–Stokes equations and can thus predict flow. In order to set up these equations, there must be parameters to describe the interaction between beads, dissipative force, and drag.

There are several permutations on this general scheme. Several different components of the mixture can be simulated by permitting different-size beads with different interaction potentials. At the present stage of development, predominantly spherical beads have been used. Porous glasses have been modeled by a collection of cylinders. Often, periodic boundary conditions are incorporated in order to simulate a bulk medium with a constant number density (a canonical ensemble). It is possible to incorporate shearing "sliding brick" boundaries, thus representing a shear stress on the medium and making it possible to calculate a viscosity.

If the magnitudes of the dissipative force, random noise, or the time step are too large, the modified velocity Verlet algorithm will not correctly integrate the equations of motion and thus give incorrect results. The values that are valid depend on the particle sizes being used. A system of reduced units can be defined in which these limits remain constant.

Polymer simulations can be mapped onto the Flory–Huggins lattice model. For this purpose, DPD can be considered an off-lattice version of the Flory–Huggins simulation. It uses a Flory–Huggins χ (chi) parameter. The best way to obtain χ is from vapor pressure data. Molecular modeling can be used to determine χ, but it is less reliable. In order to run a simulation, a bead size for each bead type and a χ parameter for each pair of beads must be known.

35.3 DYNAMIC MEAN-FIELD DENSITY FUNCTIONAL METHOD

The dynamic mean-field density functional method is similar to DPD in practice, but not in its mathematical formulation. This method is built around the density functional theory of coarse-grained systems. The actual simulation is a time integration of the Langevin equations. The thermodynamic driving forces are obtained from a Gaussian-chain molecular model. The Gaussian chain model is important because it is possible to find a mathematical mapping of parameters from molecular-level descriptions to the Gaussian chain description.

35.4 NONDYNAMIC METHODS

The techniques listed above are dynamical simulations. It is also possible to use bead interaction potentials for strictly thermodynamic calculations. For example, the following steps have been used for protein-folding problems:

1. Use conventional conformation search techniques to optimize side chains.
2. Represent the side chains as cylinders with a net interaction potential.
3. Perform a conformation search of the protein backbone using the mesoscale side-chain representation.

Similar techniques have been used for simulating lipids in a membrane and other systems. The primary limitation of this method is developing interaction potentials accurately.

35.5 VALIDATION OF RESULTS

Because mesoscale methods are so new, it is very important to validate the results as much as possible. One of the best forms of validation is to compare the computational results to experimental results. Often, experimental results are not available for the system of interest, so an initial validation calculation is done for a similar system for which experimental results are available. Results may also be compared to any other applicable theoretical results. The researcher can verify that a sufficiently long simulation was run by seeing that the same end results are obtained after starting from several different initial configurations.

35.6 RECOMMENDATIONS

Of all the topics discussed in this text, mesoscale simulations are probably at the most infantile stage of development. The idea of the mesoscale calculations is very attractive and physically reasonable. However, it is not as simple as one might expect. The choice of bead sizes and parameters is crucial to obtaining physically relevant results. More complex bead shapes are expected to be incorporated in future versions of these techniques. When using one simulation technique to derive parameters for another simulation, very small errors in a low-level calculation could result in large errors in the final stages.

The current generation of mesoscale calculations has proven useful for predicting microphases of complex materials. More work is needed in order to predict additional properties, such as mechanical strength of the material or adsorption of small molecules.

In principle, mesoscale methods can provide a means for connecting one type of simulation to another. For example, a molecular simulation can be used to describe a lipid. One can then derive the parameters for a lipid–lipid potential. These parameters can then be used in a simulation that combines lipids to form a membrane, which, in turn, can be used to compute parameters describing a membrane as a flexible sheet. Such parameters could be used for a simulation with many cells in order to obtain parameters that describe an organ, which could be used for a whole-body biological simulation. Each step, in theory, could be modeled in a different way using parameters derived not from experiment but from a more low-level form of simulation. This situation has not yet been realized, but it is representative of one trend in computational technique development.

The applicability of mesoscale techniques to systems difficult to describe in any other manner makes it likely that these simulations will continue to be used. At the present time, there is very little performance data available for these simulations. Researchers are advised to carefully consider the fundamental assumptions of these techniques and validate the results as much as possible.

BIBLIOGRAPHY

Some review articles are

R. D. Groot, P. B. Warren, *J. Chem. Phys.* **107**, 4423 (1997).

B. Schmittmann, R. K. P. Zia, *Phase Transitions and Critical Phenomena* 220 C. Domb, J. Lebowitz, Eds., Academic, London (1994).

K. V. Damodaran, K. M. Merz, Jr., *Rev. Comput. Chem.* **5**, 269 (1994).

M. C. Cross, P. C. Hohenberg, *Rev. Mod. Phys.* **65**, 851 (1993).

Some DPD references are

J. C. Shillcock, *Dissipative Particle Dynamics User Guide* Molecular Simulations Ltd., Cambridge (1998).

R. D. Groot, T. J. Madden, *J. Chem. Phys.* **108**, 8713 (1998).

P. Español, P. Warren, *Europhys. Lett.* **30**, 191 (1995).

J. M. V. A. Koelman, P. J. Hoogerbrugge, *Europhys. Lett.* **21**, 363 (1993).

P. J. Hoogerbrugge, J. M. V. A. Koelman, *Europhys. Lett.* **19**, 155 (1992).

Some density functional references are

http://www.msi.com/solutions/products/cerius2/modules/MesoDyn/MesoDyn_article.html

J. G. E. M. Fraaije, B. A. C. van Vimmeren, N. M. Maurits, M. Postma, O. A. Evers, C. Hoffmann, P. Altevogt, G. Goldbeck-Wood, *J. Chem. Phys.* **106**, 4260 (1997).

36 Synthesis Route Prediction

Just as a researcher will perform a literature synthesis for a compound, there are computer programs for determining a synthesis route. These programs have a number of names, among them synthesis design systems (SDS) or computer-aided organic synthesis (CAOS) or several other names.

In principal, synthesis route prediction can be done from scratch based on molecular calculations. However, this is a very difficult task since there are so many possible side reactions and no automated method for predicting all possible products for a given set of reactants. With a large amount of work by an experienced chemist, this can be done but the difficulty involved makes it seldom justified over more traditional noncomputational methods. Ideally, known reactions should be used before attempting to develop unknown reactions. Also, the ability to suggest reasonable protective groups will make the reaction scheme more feasible.

36.1 SYNTHESIS DESIGN SYSTEMS

Creating synthesis route prediction programs has been the work of a relatively small number of research groups in the world. There are nearly as many algorithms as there are researchers in the field. However, all these can be roughly classified into three categories.

36.1.1 Formal Generalized Reaction Systems

These programs systematically determine which bonds could be broken or formed in order to obtain the desired product. This results in generating a very large number of possible synthesis paths, many of which may be impossible or impractical. Much work has been done to weed out the unwanted synthesis routes. One major strength of this technique is that it has the capacity to indicate previously unexplored reactions.

36.1.2 Database Systems

Database searches can be used to find a reference to a known compound with a matching substructure. This is a particularly good technique if a portion of the molecule has an unusual structure. It may indicate a synthesis route or simply identify a likely starting material.

A second scheme uses a database of known chemical reactions. This more often results in synthesis routes that will work. However, this occurs at the expense of not being able to suggest any new chemistry. This method can also give many possible synthesis routes, not all of which will give acceptable yield or be easily carried out. The quality of results will depend on the database of known reactions and the means for determining which possible routes are best. These are often retrosynthetic algorithms, which start with the desired product and let the researcher choose from a list of possible precursors.

36.1.3 Expert Systems

The third category avoids generating a large number of possible routes. These are systems most similar to the expert system approach to artificial intelligence. The program attempts to mimic the decision-making process of a synthetic chemist. It thus eliminates many possible synthesis strategies as unreasonable without having worked out the entire synthesis path. The quality of results depends on the reactions known to the program and the heuristics it has been given to choose between possible paths.

36.1.4 Algorithmic Concerns

Within the three broad categories listed above, there are many different approaches to designing the software algorithms, which are by no means equivalent. In many cases, these techniques have not been studied thoroughly enough to clearly show which method is better. However, understanding some of these details is necessary in order to determine the strengths or weaknesses of various programs.

The largest number of programs have been designed to model only a select type of chemistry, such as heterocyclic chemistry, phosphorous compounds, or DNA. A number of programs have been constructed to describe organic chemistry in general. There has been very little work toward full periodic table systems.

The question most widely studied is how to select from many possible reaction paths. There have been many techniques proposed, but none is clearly superior in all cases. One simple technique is to specify a minimum yield and then discontinue following a reaction route when it falls below the target yield. Another technique is to define a chemical distance, where the shortest chemical distance is the route with the fewest steps. Some programs look for critical bonds, such as those that are most likely to be successful in closing rings. Some programs attempt to classify known reactions as those that should be used or avoided. Some programs use simple rules, called heuristics, such as adding labile groups last. Others have the ability to include common techniques, such as using protective groups.

Most programs take a retrosynthetic approach. This is a means for systematically working backward from the target compound to available precursors (some programs interface to chemical company catalogs). A less often used, but effective technique is to first search for a known compound that is very similar to the target compound and then find a reaction synthesis route from one to the other. Very few studies have addressed the inverse problem of finding a valuable compound that can be created from a given starting compound or asking what will happen when a given set of compounds are mixed, heated, and so forth.

Internally, molecules can be represented several different ways. One possibility is to use a bond-order matrix representation. A second possibility is to use a list of bonds. Matrices are convenient for carrying out mathematical operations, but they waste memory due to many zero entries corresponding to pairs of atoms that are not bonded. For this reason, bond lists are the more widely used technique.

Some programs work in a batch processing mode that gives the results once all the possibilities have been examined. Other programs run in an interactive mode that allows the user to accept or reject possible paths at each step. There are programs that learn by remembering which possibilities the user chooses and giving them a higher preference in the future.

Some programs incorporate very simple energy calculations. These are very quickly calculated estimations of electronegativity and other properties. Although not accurate enough to be completely reliable, these techniques are useful in weeding out undesirable reaction routes.

Another important issue is the "care and feeding" of a synthesis program. These programs have the ability to grow by including new reactions or heuristics. The manner in which this information is updated is as important to success as the choice of algorithmic techniques. On the one hand, entering every known reaction will slow down the operation and increase the memory requirements even though many of these reactions may never be applicable to the work done in a particular lab. On the other hand, updating only the most obviously relevant reactions may fail to include reactions that would be used in intermediate steps of the problem. This is where the experience and understanding of the human operator are invaluable.

36.2 APPLICATION OF TRADITIONAL MODELING TECHNIQUES

The traditional energy-based methods, such as *ab initio*, semiempirical, and molecular mechanics techniques, are too time-consuming for exploring every possible reaction path. However, they do have advantages over reaction databases. Even though the synthesis design system may know how a particular reaction works, it is not necessarily able to identify when subsidiary factors, such as ring strain or steric hindrance, affect the reaction. A second difficulty is determining whether the reaction will be selective or yield other undesired products. Reaction barrier energies and stereochemistry are also often not considered. Asking these types of questions, particularly for the best predicted reaction paths, is best done with traditional molecular modeling techniques. Most often, this type of work is done "manually" one compound at a time, rather than attempting to automate the process.

36.3 RECOMMENDATIONS

Computer programs are by no means threatening to replace synthetic chemists. However, in some fields such as organic chemistry, they have shown enough success to become a tool that synthetic chemists should consider using. Computer-based methods particularly excel in examining a very large number of possible reaction paths without forgetting to consider any of them. Some of these programs have the ability to indicate potentially useful new areas of chemistry, although such predictions may not be feasible. There have been a number of techniques presented in the research literature, but very few software packages have actually been made publicly available. Only time will tell which of these techniques become "must have" tools and which are relegated to historical footnotes.

BIBLIOGRAPHY

A book on the subject is

Computer-Assisted Organic Synthesis W. T. Wipke, W. J. Howe, Eds., ACS, Washington (1977).

Review articles are

R. Barone, M. Chanon, *Encycl. Comput. Chem.* **4**, 2931 (1998).

E. Hladka, J. Koca, M. Kratochvil, V. Kvasnicka, L. Matyska, J. Pospichal, V. Potucek, *Computer Chemistry* Springer-Verlag, Berlin (1993).

M. Marsili, *Computer Chemistry* CRC, Boca Raton (1990).

R. Barone, M. Chanon, *Computer Aids to Chemistry* 19 G. Vernin, M. Chanon, Eds., Ellis Horwood–John Wiley, New York (1986).

R. Carlson, C. Albano, L. G. Hammarstöm, E. Johansson, A. Nilsson, R. E. Carter, *J. Molecular Science* **2**, 1 (1983).

A. K. Long, S. D. Rubenstein, L. J. Joncas, *Chem. Eng. News* **9**, 22 (1983).

B. J. Haggin, *Chem. Eng. News* **9**, 7 (1983).

M. Bersohn, A. Esack, *Chem. Rev.* **76**, 269 (1976).

Reviews of traditional technique application are

B. S. Jursic, *Theoretical Organic Chemistry* 95, C. Párkányi Ed., Elsevier, Amsterdam (1998).

K. B. Lipkowitz, M. A. Peterson, *Chem. Rev.* **93**, 2463 (1993).

PART III
Applications

37 The Computational Chemist's View of the Periodic Table

37.1 ORGANIC MOLECULES

Organic molecules are the easiest to model and the easiest for which to obtain the most accurate results. This is so for a number of reasons. Since the amount of computational resources necessary to run an orbital-based calculation depends on the number of electrons, quantum mechanical calculations run fastest for compounds with few electrons. Organic molecules are also the most heavily studied and thus have the largest number of computational techniques available.

Organic molecules are generally composed of covalent bonded atoms with several well-defined hybridization states tending to have well-understood preferred geometries. This makes them an ideal case for molecular mechanics parameterization. Likewise, organic molecules are the ideal case for semiempirical parameterization.

This section provides a brief discussion of technical issues pertaining to modeling organic molecules. The bibliography focuses on pertinent review literature. Many computational chemistry methods can be applied to organic molecules. However, there are a few caveats to note as discussed here.

37.1.1 Group Additivity Methods

One of the earliest methods for predicting the properties of organic molecules are the group additivity methods. These are systems in which a table of contributions to a particular property for each functional group is derived. The property is then estimated by adding the contribution of each functional group in the molecule. Group additivity methods are most accurate for organic systems due to the reasons cited above. Regardless of the type of molecule, group additivity techniques are only applicable when the property being predicted can be described by additive equations. This generally is most accurate for predicting the properties of monofunctional compounds. Group additivity methods are discussed in more detail in Chapter 13.

37.1.2 Molecular Mechanics

A number of molecular mechanics force fields have been parameterized for specific organic systems, such as proteins or hydrocarbons. There are also a number of good force fields for modeling organic compounds in general. These methods can be very good for predicting the geometry of molecules and relative energies of conformers. Proteins, nucleic acids, and sugars are best described by force fields designed specifically for those compounds. Other organic compounds are described well by general-purpose organic force fields. Molecular mechanics methods are discussed further in Chapter 6.

37.1.3 Semiempirical Methods

Many semiempirical methods have been created for modeling organic compounds. These methods correctly predict many aspects of electronic structure, such as aromaticity. Furthermore, these orbital-based methods give additional information about the compounds, such as population analysis. There are also good techniques for including solvation effects in some semiempirical calculations. Semiempirical methods are discussed further in Chapter 4.

37.1.4 *Ab initio* Methods

Ab initio methods are applicable to the widest variety of property calculations. Many typical organic molecules can now be modeled with *ab initio* methods, such as Hartree–Fock, density functional theory, and Møller Plesset perturbation theory. Organic molecule calculations are made easier by the fact that most organic molecules have singlet spin ground states. Organics are the systems for which sophisticated properties, such as NMR chemical shifts and nonlinear optical properties, can be calculated most accurately.

Correlated calculations, such as configuration interaction, DFT, MPn, and coupled cluster calculations, can be used to model small organic molecules with high-end workstations or supercomputers. These are some of the most accurate calculations done routinely. Correlation is not usually required for qualitative or even quantitative results for organic molecules. It is needed to obtain high-accuracy quantitative results.

Core potentials are seldom used for organic molecules because there are so few electrons in the core. Relativistic effects are seldom included since they have very little effect on the result. *Ab initio* methods are discussed further in Chapter 3.

37.1.5 Recommendations

Organic molecule calculations can be done routinely to good accuracy on workstation-class hardware. It is advisable to examine tabulations of results in order to choose a method with acceptable accuracy and computational time for the property of interest. The trend toward having microcomputer versions of computational chemistry codes is making calculations on small organic molecules even more readily accessible.

37.2 MAIN GROUP INORGANICS, NOBLE GASES, AND ALKALI METALS

Modeling the elements discussed in this section is fairly similar to modeling organic compounds. This is primarily because *d* and *f* orbitals play a minor role in their chemistry. When *d* and *f* orbitals do affect the chemistry, their effect is well defined and for the most part understood.

Molecular mechanics methods have only been used to a limited extent for these classes of compounds. However, molecular mechanics methods do fairly well in describing the geometries and relative energies of compounds with these elements. It is perhaps only for historical and economic reasons that molecular mechanics has not been used more for modeling these elements. Subsequently, there are not as many force fields available.

Semiempirical, DFT, and *ab initio* methods also work well. Correlation effects are sometimes included for the sake of increased accuracy, but are not always necessary. One particular case for which correlation is often necessary is fluorine compounds.

37.2.1 Halides

Within Hartree–Fock theory, F_2 has a reasonable bond length, but its total energy is higher than the sum of the energies of two F atoms. This is because correlation is a very significant contribution to the valence description of fluorine. Correlated calculations well describe halogenated compounds. This effect is seen to a lesser extent in modeling other halide atoms. Molecular mechanics works well if the charge computation scheme correctly reflects the electronegativity of these elements.

37.2.2 Other Main Group Inorganics

Modeling the lighter main group inorganic compounds is similar to modeling organic compounds. Thus, the choice of method and basis set is nearly identical. The second-row compounds (i.e., sulfur) do have unfilled *d* orbitals, making it often necessary to use basis sets with *d* functions.

The heavier elements are affected by relativistic effects. This is most often accounted for by using relativistic core potentials. Relativistic effects are discussed in more detail in Chapters 10 and 33.

37.2.3 Noble Gases

The noble gases are mostly unreactive. In some instances, they act mostly as a place holder to fill a cavity. For dynamical studies of the bulk gas phase or liquid-phase noble gases, hard-sphere or soft-sphere models work rather well.

Paradoxically, compounds incorporating bonds with noble gases are difficult to model. This is because a very accurate method is needed in order to correctly model what little reactivity they do have. Often, correlated *ab initio* calculations with polarized basis sets are used. The worst case is the dimers, such as the He_2 dimer that is known experimentally to exist and have one bound vibrational level. He_2 has only been modeled accurately using some of the most accurate methods known, such as quantum Monte Carlo calculations.

37.2.4 Alkali Metals

The alkali metals tend to ionize; thus, their modeling is dominated by electrostatic interactions. They can be described well by *ab initio* calculations, provided that diffuse, polarized basis sets are used. This allows the calculation to describe the very polarizable electron density distribution. Core potentials are used for *ab initio* calculations on the heavier elements.

Molecular mechanics methods may work well or poorly for compounds containing alkali metals. The crucial factor is often how the force field computes charges for electrostatic interactions.

37.2.5 Recommendations

If these elements are included in an organic molecule, the choice of computational method can be made based on the organic system with deference to the exceptions listed in this section. If completely inorganic calculations are being performed, use a method that tends to correctly model the property of interest in organic systems.

37.3 TRANSITION METALS

There is a growing interest in modeling transition metals because of its applicability to catalysts, bioinorganics, materials science, and traditional inorganic chemistry. Unfortunately, transition metals tend to be extremely difficult to model. This is so because of a number of effects that are important to correctly describing these compounds. The problem is compounded by the fact that the majority of computational methods have been created, tested, and optimized for organic molecules. Some of the techniques that work well for organics perform poorly for more technically difficult transition metal systems.

Nearly every technical difficulty known is routinely encountered in transition metal calculations. Calculations on open-shell compounds encounter problems due to spin contamination and experience more problems with SCF convergence. For the heavier transition metals, relativistic effects are significant. Many transition metals compounds require correlation even to obtain results that are qualitatively correct. Compounds with low-lying excited states are difficult to converge and require additional work to ensure that the desired states are being computed. Metals also present additional problems in parameterizing semiempirical and molecular mechanics methods.

37.3.1 Molecular Mechanics Methods

In the past, when molecular mechanics methods were used for transition metals, it was by having a set of parameters for the metal that were parameterized specifically for one class of compounds. There have been a number of full periodic table force fields created, with the most successful being the UFF force field. All the full periodic molecular mechanics methods still give completely unreasonable results for certain classes of compounds.

One reason for these difficulties is that metals have fairly soft bonding. This means that there is a nearly continuous range of values experimentally observed for any given metal-organic bond length. Likewise, inorganics more often exhibit distorted or fluxional bond angles. There is also less vibrational data available to parameterize force constants.

Not all molecular mechanics methods can be adapted to metal calculations simply by adding new parameters. For example, consider a square planar Pt atom. Unlike organic atoms, some of the bond angles are 90°, whereas others are 180°. One option is to have different parameters for describing these two cases and a program that can recognize which to use. A second option is to use an angle function with two minima at 90° and 180°. In both cases, the software package must have capabilities not needed for organic molecules. Another option is to hold the Pt rigid over the course of the calculation.

Coordination creates additional problems also. Consider the metal–Cp bond in a metallocene. One option is to have five bonds from the metal to each carbon. A second option is to have a single bond connecting to a dummy atom at the center of the Cp ring.

One way that molecular mechanics methods have been adapted to transition metal applications is by including one orbital-based term in the force field to describe the metal center. These terms are typically based on semiempirical methods or even some variation of ligand field theory.

37.3.2 Semiempirical Methods

There are a few semiempirical methods for modeling transition metals. These tend to have limited applicability. None has yet become extremely far-ranging in the type of system it can model accurately.

Extended Hückel gives a qualitative view of the valence orbitals. The formulation of extended Hückel is such that it is only applicable to the valence orbitals. The method reproduces the correct symmetry properties for the valence orbitals. Energetics, such as band gaps, are sometimes reasonable and other times reproduce trends better than absolute values. Extended Hückel tends to be more useful for examining orbital symmetry and energy than for predicting molecular geometries. It is the method of choice for many band structure calculations due to the very computation-intensive nature of those calculations.

Fenske Hall is essentially a quantification of ligand field theory. The interactions are primarily electrostatic in nature. It does a reasonable job of reproducing certain trends and reflects the very soft nature of ligand bond angles. This method must be used with caution as it sometimes oversimplifies the nature of orbital interactions.

ZINDO is an adaptation of INDO specifically for predicting electronic excitations. The proper acronym for ZINDO is INDO/S (spectroscopic INDO), but the ZINDO moniker is more commonly used. ZINDO has been fairly successful in modeling electronic excited states. Some of the codes incorporated in ZINDO include transition-dipole moment computation so that peak intensities as well as wave lengths can be computed. ZINDO generally does poorly for geometry optimization.

PM3/TM is an extension of the PM3 method to transition metals. Unlike the parameterization of PM3 for organics, PM3/TM has been parameterized only to reproduce geometries. This does, of course, require a reasonable description of energies, but the other criteria used for PM3 parameterization, such as dipole moments, are not included in the PM3/TM parameterization. PM3/TM tends to exhibit a dichotomy. It will compute reasonable geometries for some compounds and completely unreasonable geometries for other compounds. It seems to favor one coordination number or hybridization for some metals.

37.3.3 *Ab initio* Methods

Ab initio methods pose problems due a whole list of technical difficulties. Most of these stem from the large number of electrons and low-energy excited state. Core potentials are often used for heavier elements to ease the computational requirements and account for relativistic effects.

Convergence problems are very common due to the number of orbitals available and low-energy excited states. The most difficult calculations are generally those with open-shell systems and an unfilled coordination sphere. All the techniques listed in Chapter 22 may be necessary to get such calculations to converge.

Many transition metal systems are open-shell systems. Due to the presence of low-energy excited states, it is very common to experience problems with spin contamination of unrestricted wave functions. Quite often, spin projection and annihilation techniques are not sufficient to correct the large amount of spin contamination. Because of this, restricted open-shell calculations are more reliable than unrestricted calculations for metal system. Spin contamination is discussed in Chapter 27.

Electron correlation is often very important as well. The presence of multiple bonding interactions, such as pi back bonding, makes coordination compounds more sensitive to correlation than organic compounds. In some cases, the HF wave function does not provide even a qualitatively correct description of the compound. If the weight of the reference determinant in a single-reference CISD calculation is less than about 0.9, then the HF wave function is not qualitatively correct. In such cases, multiple-determinant, MSCSF, CASPT2, or MRCI calculations tend to be the most efficient methods. The alternative is to include triple and quadruple excitations in single-reference CI or CC calculations. In recent years, DFT methods, particularly B3LYP, have become widely used for large metal-containing systems, such as enzyme active sites. These calculations generally give results of good accuracy with reasonable computational requirements, although it is still necessary to use correlated *ab initio* methods at times in order to obtain more accurate results.

Relativistic effects are significant for the heavier metals. The method of choice is nearly always relativistically derived effective core potentials. Explicit spin-orbit terms can be included in *ab initio* calculations, but are seldom used because of the amount of computational effort necessary. Relativistic calculations are discussed in greater detail in Chapter 33.

Because of the existence of low-energy excited states, calculations done with the software default settings very often give results for the electronic excited states rather than the ground state. There can be a significant amount of work in just computing various states of the molecule in order to ensure that the correct ground state has been determined. Chapter 25 discusses excited-state calculations and consequently the techniques to use for determining the ground state for metal systems. An initial guess algorithm based on ligand field theory is perhaps most reliable.

37.4 LANTHANIDES AND ACTINIDES

Lanthanide and actinide compounds are difficult to model due to the very large number of electrons. However, they are somewhat easier to model than transition metals because the unpaired f electrons are closer to the nucleus than the outermost d shell. Thus, all possible spin combinations do not always have a significant effect on chemical bonding.

Relativistic effects should always be included in these calculations. Particularly common is the use of core potentials. If core potentials are not included, then another form of relativistic calculation must be used. Relativistic effects are discussed in more detail in Chapter 33.

37.4.1 Methods

Molecular mechanics force fields are sometimes parameterized to describe lanthanides and actinides. This has been effective in describing the shape of the molecule, but does not go very far toward giving systematic energies. A few semiempirical methods have been parameterized for these elements, but they have not seen widespread use.

Ab initio calculations with core potentials are usually the method of choice. The researcher must make a difficult choice between minimizing the CPU time requirements and obtaining more accurate results when deciding which core potential to use. Correlation is particularly difficult to include because of the large number of electrons even in just the valence region of these elements.

Population analysis poses a particularly difficult problem for the *f* block elements. This is because of the many possible orbital combinations when both *f* and *d* orbitals are occupied in the valence. Although programs will generate a population analysis, extracting meaningful information from it can be very difficult.

BIBLIOGRAPHY

The bibliography for this chapter is perhaps the most difficult to write. The majority of references in this entire book pertain to organic molecules. The organic references listed here are just a few of the review references pertaining specifically to organic chemistry. This list is incomplete, but attempts to include recent reviews, which will reference earlier work. The listing for other classes of molecules are more complete.

Some books relevant to organic chemistry are

Theoretical Organic Chemistry C. Párkyáni, Ed., Elsevier, Amsterdam (1998).

A. K. Rappé, C. J. Casewit, *Molecular Mechanics across Chemistry* University Science Books, Sausalito (1997).

D. Hadzi, *Theoretical Treatments of Hydrogen Bonding* John Wiley & Sons, New York (1997).

W. B. Smith, *Introduction to Theoretical Organic Chemistry and Molecular Modeling* John Wiley & Sons, New York (1996).

Modeling the Hydrogen Bond D. A. Smith, Ed., American Chemical Society, Washington (1994).

V. I. Minkin, B. Y. Simkin, R. M. Minyaev, *Quantum Chemistry of Organic Compounds; Mechanisms of Reactions* Springer-Verlag, Berlin (1990).

Applications of Molecular Orbital Theory in Organic Chemistry I. G. Csizmadia, Ed., Elsevier, Amsterdam (1977).

I. Fleming, *Frontier Orbitals and Organic Chemical Reactions* John Wiley & Sons, New York (1976).

M. J. S. Dewar, R. C. Daugherty, *The PMO Theory of Organic Chemistry* Plenum, New York (1975).

T. E. Peacock, *The Electronic Structure of Organic Molecules* Pergamon, Oxford (1972).

M. J. S. Dewar, *The Molecular Orbital Theory of Organic Chemistry* McGraw-Hill, New York (1969).

K. Higasi, H. Buba, A. Rembaum, *Quantum Organic Chemistry* John Wiley & Sons, New York (1965).

Group additivity methods are reviewed in

N. Cohen, S. W. Benson, *Chem. Rev.* **93**, 2419 (1993).

Organic anion calculations are reviewd in

L. Radom, *Applications of Electronic Structure Theory* H. F. Schafer, III, Ed., 333, Plenum, New York (1977).

Carbene calculations are reviewed in

H. F. Bettinger, P. v. R. Schleyer, P. R. Schreiner, H. F. Schaefer, III, *Encycl. Comput. Chem.* **1**, 183 (1998).

Carbocations are reviewed in

D. H. Aue, *Encycl. Comput. Chem.* **1**, 210 (1998).

Carbohydrates are reviewed in

A. D. Brench, *Encycl. Comput. Chem.* **1**, 233 (1998).

Cyclodextrins are reviewed in

C. Jaime, *Encycl. Comput Chem.* **1**, 644 (1998).

Diels-Alder reactions are reviewed in

J. Bertran, V. Branchadell, A. Oliva, M. Sodupe, *Encycl. Comput. Chem.* **3**, 2030 (1998).

J. J. Dannenberg, *Advances in Molecular Modeling* 2, 1, D. Liotta, Ed., JAI New York (1990).

Organic diradicals are reviewed in

W. T. Borden, *Encycl. Comput. Chem.* **1**, 709 (1998).

Electron transfer in organic molecules is reviewed in

K. D. Jordan, M. N. Paddon-Row, *Encycl. Comput. Chem.* **2**, 826 (1998).

Enthalpy of hydration of organic molecules is reviewed in

D. W. Rodgers, *Encycl. Comput. Chem.* **2**, 920 (1998).

Heat of formation for organic molecules is reviewed in

Y. Fan, *Encycl. Comput. Chem.* **2**, 1217, (1998).

Hydrogen bonding is reviewed in

J. E. Del Bene, *Encycl. Comput. Chem.* **2**, 1263 (1998).

J.-H. Lii, *Encycl. Comput. Chem.* **2**, 1271 (1998).

J. J. P. Stewart, *Encycl. Comput. Chem.* **2**, 1283 (1998).

P. A. Kollman, *Applications of Electronic Structure Theory* H. F. Schaefer, III, Ed., 109, Plenum, New York (1977).

Hyperconjugation is reviewed in

C. J. Cramer, *Encycl Comput. Chem.* **2**, 1294 (1998).

Photochemistry is reviewed in

M. A. Robb, M. Olivucci, F. Bernardi, *Encycl. Comput. Chem* **3**, 2057 (1998).

Proton affinity is reviewed in

S. Gronert, *Encycl. Comput. Chem.* **3**, 2283 (1998).

Organic heterocyclic reactions are reviewed in

G. L. Heard B. F. Yates, *Encycl. Comput Chem.* **4**, 2420 (1998).

Rotational barriers are reviewed in

K. B. Wiberg, *Encycl. Comput. Chem.* **4**, 2518 (1998).
L. Goodman, V. Pophristic, *Encycl. Comput. Chem.* **4**, 2525 (1998).

Carbohydrate solvation is reviewed in

J. W. Brady, *Encycl. Comput. Chem.* **4**, 2609 (1998).

Other review articles pertaining to organic chemistry are

J. Catalán, J. L. G. de Paz, *Computational Chemistry. Structure, Interactions and Reactivity Part A* S. Fraga, Ed., 434, Elsevier, Amsterdam (1992).

Sources giving discussions generally applicable to main group inorganics are

J. A. Alonso, L. C. Balbás, *Density Functional Theory III* R. F. Nalewajski, Ed., 119, Springer, Berlin (1996).
P. Comba, T. W. Hambly, *Molecular Modeling of Inorganic Compounds* VCH, Weinheim (1995).
J. K. Burdett, *Molecular Shapes· Theoretical Models of Inorganic Stereochemistry* John Wiley & Sons, New York (1980).
G. Doggett, *The Electronic Structure of Models: Theory and Applications to Inorganic Molecules* Pergamon, Oxford (1972).

Bromine containing compounds are discussed in

S. Guha, J. S. Francisco, *Computational Chemistry Reviews of Current Trends Volume 3* 75, J. Leszczynski, Ed., World Scientific, Singapore (1999).

Fluctional Processes in boranes and carboranes are reviewed in

M. L. McKee, *Encycl. Comput. Chem.* **2**, 1002 (1998).

The He_2 problem is examined in

J. B. Anderson, C. A. Traynor, B. M. Boghosian, *J. Chem. Phys.* **99**, 345 (1993).

Isolobal relationships are reviewed in

E. D. Jemmis, K. T. Giju, *Encycl. Comput. Chem.* **2**, 1149 (1998).

Molecular mechanics modeling of main group inorganics is reviewed in

A. K. Rappé, C. J. Casewit, *Molecular Mechanics across Chemistry* University Science Books, Sausalito (1997).

M. Zimmer, *Chem. Rev* **95**, 2629 (1995).

B. P. Hay, *Coord. Chem. Rev.* **126**, 177 (1993).

Organometallic modeling is reviewed in

A. Streitwieser, K. Sorger, *Encycl. Comput. Chem.* **3**, 2100 (1998).

Silylenes are reviewed in

L. Nyulászi, T. Veszprémi, *Encycl. Comput. Chem.* **4**, 2589 (1998).

Sulfur compounds are discussed in

D. C. Young, M. L. McKee, *Computational Chemistry Reviews of Current Trends Volume 3* 149, J. Leszczynski, Ed., World Scientific, Singapore (1999).

Zeolites are reviewed in

B. van de Graaf, S. C. Njo, K. S. Smirnov, *Rev. Coput. Chem.* **14**, 137 (2000).

J. Sauer, *Encycl. Comput. Chem.* **5**, 3248 (1998).

Other review articles pertinent to main group inorganics are

M. S. Gordon, *Modern Electronic Structure Theory Part 1* D. R. Yarkdony, Ed., 311, World Scientific, Singapore (1995).

Books relevant to transition metal modeling are

I. B. Bersuker, *Electronic Structure and Properties of Transition Metal Compounds* John Wiley & Sons, New York (1996).

P. Comba, T. W. Hambley, *Molecular Modeling of Inorganic Compounds* VCH, Weinheim (1995).

The Challenge of d and f Electrons D. R. Salahub, M. C. Zerner, Ed., American Chemical Society, Washington (1989).

V. K. Grigorovich, *The Metallic Bond and the Structure of Metals* Nova Science, Huntington, NY (1989).

Quantum Chemistry: The Challenge of transition Metals and Coordination Chemistry A. Veillard, Ed., D. Reidel, Dordrecht (1986).

J. K. burdett, *Molecular Shapes. Theoretical Models of Inorganic Stereochemistry* John Wiley & Sons, New York (1980).

K. Bernauer, M. S. Wrighton, A. Albini, H. Krisch, *Theoretical Inorganic Chemistry II* Springer-Verlag, Berlin (1976).

C. K. Jorgensen, H. Brunner, L. H. Pignolet, S. Veprek, *Theoretical Inorganic Chemistry* Springer-Verlag, Berlin (1975).

Review articles pertinent to transition metal modeling are

Chem. Rev **100**, number 2 (2000).

G. Frenking, T. Wagener, *Encycl. Comput. Chem.* **5**, 3073 (1998).

C. W. Bauschlicker, Jr., *Encycl. Comput. Chem.* **5**, 3084 (1998).

P. Pyykkö, *Chem. Rev.* **97**, 597 (1997).

A. Berces, T. Ziegler, *Density Functional Theory III* R. F. Nalewajski, Ed., 41, Springer-Verlag, Berlin (1996).

G. Frenking, I. Antes, M. Boehme, S. Dapprich, A. W. Ehlers, V. Jonas, A. Neuhaus, M. Otto, R. Stegmann, A. Veldkamp, S. F. Vyboisikov, *Rev. Comput. Chem.* **8**, 63 (1996).

T. R. Cundari, M. T. Benson, M. L. Lutz, S. O. Sommerer, *Rev. Comput. Chem.* **8**, 63 (1996).

I. Bytheway, M. B. Hall, *Chem. Rev.* **94**, 639 (1994).

A. Veillard, *Chem. Rev.* **91**, 743 (1991).

M. G. Cory, M. C. Zerner, *Chem. Rev.* **91**, 813 (1991).

N. Koga, K. Morokuma, *Chem. Rev.* **91**, 823 (1991).

D. E. Ellis, J. Guo, H.-P. Cheng, J. J. Low, *Adv. Quantum Chem.* **22**, 125 (1991).

S. R. Langhoff, C. W. Bauschlicker, Jr., *Ann. Rev Phys. Chem.* **39**, 181 (1988).

A. Dedieu, M.-M. Rohmer, A. Veillart, *Adv. Quantum Chem.* **16**, 43 (1982).

G. D. Broackère, *Adv. Chem. Phys.* **37**, 203 (1978).

A. Veillard, J. Demuynck, *Applications of Electronic Structure Theory* H. F. Schaefer, III, Ed., 187, Plenum, New York (1977).

G. Berthier, *Adv. Quantum Chem.* **8**, 183 (1974).

Molecular mechanics techniques for transition metals are reviewed in

A. K. Rappé, C. J. Casewit, *Molecular Mechanics across Chemistry* University Science Books, Sausalito (1997).

C. R. Landis, D. M. Root, T. Cleveland, *Rev. Comput. Chem.* **6**, 73, (1995).

M. Zimmer, *Chem. Rev.* **95**, 2629 (1995).

B. P. Hay, *Coord. Chem. Rev.* **126**, 177 (1993).

P. Comba, *Coord. Chem Rev.* **123**, 1 (1993).

Semiempirical and *ab initio* methods for transition metals are compared in

A. J. Holder, *Encycl. Comput Chem.* **4**, 2578 (1998).

J. P. Dahl, C. J. Ballhausen, *Adv. Quantum Chem.* **4**, 170 (1968).

Catalyst design is reviewed in

S. Nakamura, S. Sieber, *Encycl. Comput. Chem.* **1**, 246 (1998).

Cluster calculations are reviewed in

V. Bonacic-Koutecky, P. Fantucci, J. Koutecky, *Encycl. Comput. Chem.* **2**, 876 (1998).

Metal complex calculations are reviewed in

B. P. Hay, O. Clement, *Encycl. Comput. Chem.* **3**, 1580 (1998).

Books discussing modeling of lanthanides and actinides are

P. Comba, T. W. Hambley, *Molecular Modeling of Inorganic Compounds* VCH, Weinheim (1995).

The Challenge of d and f Electrons D. R. Salahub, M. C. Zerner, Eds., American Chemical Society, Washington (1989).

Review articles pertinent to lanthanides and actinides are

M. Dolg, *Encycl. Comput. Chem.* **2**, 1478 (1998).

M. Dolg, H. Stoll, *Handbook on the Physics and Chemistry of Rare Earths* K. A. Gschneidner, Jr., L. Eyring, Eds., **22**, 607, Elsevier, Amsterdam (1994).

M. Pepper, B. E. Bursten, *Chem. Rev.* **91**, 719 (1991).

M. S. S. Brooks, *Actinides—Chemistry and Pysical Properties* L. Manes, Ed., 261, Springer-Verlag, Berlin (1985).

Molecular mechanics methods for lanthanides and actinides are reviewed in

M. Zimmer, *Chem. Rev.* **95**, 2629 (1995).

B. P. Hay, *Coord. Chem. Rev.* **126**, 177 (1993).

38 Biomolecules

The process of designing a new drug and bringing it to market is very complex. According to a 1997 government report, it takes 12 years and 350 million dollars for the average new drug to go from the research laboratory to patient use. At several points in this process, computer-modeling techniques provide a significant cost savings. This makes biomolecule modeling a very important part of the field. The same can be said of agrochemical research and many other applications. For the sake of convenience, this chapter discusses drug design, although most of the discussion is applicable to any biomolecular application.

Due to the incredible complexity of biological systems, molecular modeling is not at all an easy task. It can be divided into two general categories: specific and general interactions. The design of a drug or pesticide aims to elicit a very specific biological reaction by interaction of the compound with a very specific biomolecule (which may be unknown). At the opposite extreme is the need to predict general interactions, which are due to a variety of processes. Some of these general interactions are biodegradation and toxicity.

38.1 METHODS FOR MODELING BIOMOLECULES

Due to the large size of most biologically relevant molecules, molecular mechanics is most often the method of choice for biochemical modeling. There are molecular mechanics force fields for both modeling specific classes of molecules and organic molecules in general. In some cases, even molecular mechanics is too time-consuming to model a very large system and mesoscale techniques can be used (Chapter 35).

At the other extreme is a trend toward the increasing use of orbital-based techniques, particularly QM/MM calculations (Chapter 23). These orbital-based techniques are needed to accurately model the actual process of chemical bond breaking and formation.

The first step in designing a new compound is to find compounds that have even a slight amount of usefulness for the intended purpose. These are called lead compounds. Once such compounds are identified, the problem becomes one of refinement. Computational techniques are a fairly minor part of finding lead compounds. The use of computer-based techniques for lead compound identification is usually limited to searching databases for compounds similar to known lead compounds or known to treat diseases with similar causes or symptoms.

Once a number of lead compounds have been found, computational and laboratory techniques are very successful in refining the molecular structures to yield greater drug activity and fewer side effects. This is done both in the laboratory and computationally by examining the molecular structures to determine which aspects are responsible for both the drug activity and the side effects. These are the QSAR techniques described in Chapter 30. Recently, 3D QSAR has become very popular for this type of application. These techniques have been very successful in the refinement of lead compounds.

A more logical approach would be to model the binding site for a target molecule and then find molecules that will dock in this site. Unfortunately, the binding site may not be known. It is fairly easy to determine the sequences of proteins and nucleotides. However, it is much more difficult to obtain structural information by X-ray crystallography. Because of this disparity, there has been an immense amount of work on solving the protein folding problem, which is to determine the three dimensional structure of a protein from its sequence. Although computing the relative energies of conformations is one of the greatest successes of computational chemistry techniques, the incredibly huge number of possible conformers of a protein make this a daunting task. Two ingenious methods for simplifying this problem are distance geometry and homology modeling. Distance geometry is a means for imposing constraints on the problem, which are obtained from two-dimensional NMR studies. Homology modeling is used to find the known structure with the most similar sequence, then using that geometry for those sections of the unknown. These techniques are discussed in more detail in Chapter 21. Once a binding site is known, a molecule to bind in that site can be determined with the techniques described in the next section.

38.2 SITE-SPECIFIC INTERACTIONS

If it is known that a drug must bind to a particular spot on a particular protein or nucleotide, then a drug can be tailor-made to bind at that site. This is often modeled computationally using any of several different techniques. Traditionally, the primary way of determining what compounds would be tested computationally was provided by the researcher's understanding of molecular interactions. A second method is the brute force testing of large numbers of compounds from a database of available structures.

More recently, a set of techniques, called rational drug design or De Novo techniques, have been used. These techniques attempt to reproduce the researcher's understanding of how to choose likely compounds. Such an understanding is built into a software package that is capable of modeling a very large number of compounds in an automated way. Many different algorithms have been used for this type of testing, many of which were adapted from artificial intelligence applications. No clear standard has yet emerged in this area so it is not possible to say which is the best technique.

38.3 GENERAL INTERACTIONS

Interestingly, QSAR is as useful for predicting general interactions as it is for the optimization of activity for very specific interactions. In this case, QSAR rather than 3D QSAR is most effective. It has been used for predicting environmental toxicity, biodegradation, and other processes. This serves as a good screening technique to determine which compounds should be examined closer. These methods are never completely reliable and should not be considered a substitute for standard testing techniques. They are best used for categorizing compounds as having a high or low likelihood of acceptability.

It is possible to obtain the sequence of a DNA strand, but that does not give an understanding of the attribute of the organism described by a particular piece of genetic code. Homology modeling can be used to shed light on this type of information, as well as for determining structure. Homology modeling is the systematic comparison of DNA sequences to determine regions of similarities and differences. This can yield information as broad as the differences between reptiles and mammals or information as narrow as the differences between individuals.

38.4 RECOMMENDATIONS

The modeling of biomolecules is a very broad and sophisticated field. The description given in this chapter is only meant to provide the connections between the topics in this book and this field. Before embarking on a computational biochemical study, it is recommended that the researcher investigate the literature pertaining to this field more closely. The references provided below should provide a good starting point for such a survey.

BIBLIOGRAPHY

Books presenting relevant computational techniques are

A. K. Rappé, C. J. Casewit, *Molecular Mechanics across Chemistry* University Science Books, Sausalito (1997).

A. R. Leach, *Molecular Modelling Principles and Applications* Longman, Essex (1996).

H.-D. Holtje, G. Folkers, T. Bierer, W. Sippl, D. Rognan, *Molecular Modeling—Basic Principles and Applications* John Wiley & Sons, New York (1996).

G. H. Grant, W. G. Richards, *Computational Chemistry* Oxford, Oxford (1995).

Books specifically addressing biomolecules are

Practical Application of Computer-Aided Drug Design P. S. Charifson, Ed., Marcel Dekker, New York (1997).

Guidebook on Molecular Modeling in Drug Design N. C. Cohen, Ed., Academic, San Diego (1996).

H. Van de Waterbeemd, *Advanced Computer-Asissted Techiques in Drug Discovery* John Wiley & Sons, New York (1995).

G. L. Patrick, *An Introduction to Medicinal Chemistry* Oxford, Oxford (1995).

Computer-Aided Molecular Design Applications in Agrochemicals, Materials and Pharmaceuticals C. H. Reynolds, M. K. Holloway, H. K. Cox, Eds., American Chemical Society, Washington (1995).

Molecular Modelling and Drug Design J. G. Vintner, M. Gardner, Eds., CRC, Boca Raton (1994).

A. Warshel, *Computer Modeling of Chemical Reactions in Enzymes and Solutions* John Wiley & Sons, New York (1991).

Computer-Aided Drug Design Methods and Applications T. J. Perun, C. L. Propst, Eds., Dekker, New York (1989).

J. A. McCammon, S. C. Harvey, *Dynamics of Proteins and Nucleic Acids* Cambridge, Cambridge (1987).

General review articles are

D. B. Boyd, *Encycl. Comput. Chem.* **1**, 795 (1998).

G. W. A. Milne, *Encycl. Comput. Chem.* **3**, 2046 (1998).

L. M. Balbes, S. W. Mascarella and D. B. Boyd, *Rev. Comput. Chem.* **5**, 337 (1994).

P. Kollman, *Chem. Rev.* **93**, 2395 (1993).

B. Pullman, *Adv. Quantum Chem.* **10**, 251 (1977).

L. Balbes, *Guide to Rational (Computer-aided) Drug Design* is online at http://www.ccl.net/cca/documents/drug.design.shtml

There are many links to online information on Soaring Bear's web page at

http://ellington.pharm.arizona.edu/%7Ebear/

Applications of artificial intelligence are reviewed in

D. P. Dolata, *Encycl. Comput. Chem.* **1**, 44 (1998).

Predicting biodegredation is reviewed in

G. Klopman, M. Tu, *Encycl. Comput. Chem.* **1**, 128 (1998).

Carcinogenicity is reviewed in

L. v. Szentpály, R. Ghosh, *Theoretical Organic Chemistry* 447 C. Párkányi, Ed., Elsevier, Amsterdam (1998).

Chemometrics are reviewed in

K. Varmuza, *Encycl. Comput. Chem.* **1**, 347 (1998).

Conformation searching of biomolecules is reviewed in

M. Vásquez, G. Némethy, H. A. Scheraga, *Chem. Rev.* **94**, 2183 (1994).

Information about De Novo techniques is in

Rational Drug Design A. Parrill, M. R. Reddy, Eds., Oxford, Oxford (1999).

A. P. Johnson, S. M. Green, *Encycl. Comput. Chem.* **1**, 650 (1998).

H. J. Böhm, S. Fischer, *Encycl. Comput. Chem.* **1**, 657 (1998).

D. E. Clark, C. W. Murray, J. Li, *Rev. Comput. Chem* **11**, 67 (1997).

S. Borman, *Chem. and Eng. News* **70**, 18 (1992).

Distance geometry techniques are reviewed in

T. F. Havel, *Encycl. Comput. Chem.* **1**, 723 (1998).

M. P. Williamson, J. P. Walto, *Chem. Soc. Rev.* **21**, 227 (1992).

Modeling DNA is reviewed in

J. Sponer, P. Hobza, *Encycl. Comput. Chem.* **1**, 777 (1998).

R. Lavery, *Encycl. Comput. Chem.* **3**, 1913 (1998).

S. Lemieux, S. Oldziej, F. Major, *Encycl. Comput. Chem.* **3**, 1930 (1998).

D. L. Beveridge, *Encycl. Comput. Chem.* **3**, 1620 (1998).

P. Auffinger, E. Westhof, *Encycl. Comput. Chem.* **3**, 1628 (1998).

G. Ravishanker, P. Auffinger, P. R. Langley, B. Jayaram, M. A. Young, *Rev. Comput. Chem.* **11**, 317 (1997).

Docking techniques are reviewed in

C. M. Oshiro, I. D. Kuntz, R. M. A. Knegtel, *Encycl. Comput. Chem.* **3**, 1606 (1998).

M. Vieth, J. D. Hirst, A. Kolinski, C. L. Brooks, III, *J. Comput. Chem.* **19**, 1612 (1998).

M. Vieth, J. D. Hirst, B. N. Dominy, H. Daigler, C. L. Brooks, III, *J. Comput. Chem* **19**, 1623 (1998).

Electron transfer is reviewed in

T. Hayashi, H. Ogoshi, *Chem. Soc. Rev.* **26**, 355 (1997).

Ligand design is reviewed in

M. A. Murcko, *Rev. Comput. Chem.* **11**, 1 (1997).

Modeling membranes is reviewed in

S. Yoneda, T. Yoneda, H. Umeyamn, *Encycl. Comput. Chem.* **1**, 135 (1998).

H. J. c. Berendsen, D. P. Tıeleman, *Encycl. Comput. Chem.* **3**, 1638 (1998).

A. Pullman, *Chem. Rev.* **91**, 793 (1991).

J. Houk, R. H. Guy, *Chem. Rev.* **88**, 455 (1988).

Modeling micelles is reviewed in

P. L. Luisi, *Adv. Chem. Phys.* **92**, 425 (1996).

Molecular dynamics of biomolecules is reviewed in

T. P. Lybrand, *Rev. Comput. Chem.* **1**, 295 (1990).

Neural network reviews are

J. A. Burns, G. M. Whitesides, *Chem. Rev.* **93**, 2583 (1993).

Oligosaccharide modeling is reviewed in

R. J. Woods, *Rev. Comput. Chem.* **9**, 129 (1996).

Pesticide modeling is reviewed in

E. L. Plumber, *Rev. Comput. Chem.* **1**, 119 (1990).

Protein & peptide reviews are

J. Skolnick, A. Kolinski, *Encycl. Comput. Chem.* **3**, 2200 (1998).

B. Rost, *Encycl. Comput. Chem.* **3**, 2243 (1998).

L. Pedersen, T. Darden, *Encycl. Comput. Chem.* **3**, 1650 (1998).

K. E. Laidig, V. Daggett, *Encycl. Comput. Chem.* **3**, 2211 (1998).

C. L. Brooks, III, D. A. Case, *Chem. Rev.* **93**, 2487 (1993).

G. E. Marlow, J. S. Perkyns, B. M. Pettitt, *Chem. Rev.* **93**, 2503 (1993).

J. Åqvist, A. Warshel, *Chem. Rev.* **93**, 2523 (1993).

H. Scheraga, *Rev. Comput. Chem.* **3**, 73 (1992).

J. M. Troyer, F. E. Cohen, *Rev. Comput. Chem.* **2**, 57 (1991).

M. Karplus, *Modelling of Molecular Structures and Properties* J. L. Rivail, Ed., 427, Elsevier, Amsterdam (1990).

J. Skolnick, A. Kolinski, *Ann. Rev. Phys. Chem.* **40**, 207 (1989).

Adv. Chem. Phys. C. L. Brooks, III, M. Karplus, B. M. Pettitt, Eds., vol. **71** (1988).

J. A. McCammon, M. Karplus, *Ann. Rev. Phys. Chem.* **31**, 29 (1980).

QSAR reviews are

H. Kubinyi, *Encycl. Comput. Chem.* **4**, 2309 (1998).

S. P. Gupta, *Chem. Rev.* **94**, 1507 (1994).

H. H. Jaffé, *Chem. Rev.* **53**, 191 (1953).

An introduction to structure-based techniques is

I. D. Kuntz, E. C. Meng, B. K. Shoichet, *Acct. Chem. Res.* **27**, 117 (1994).

Toxicity prediction is reviewed in

D. F. Lewis, *Rev. Comput. Chem.* **3**, 173 (1992).

39 Simulating Liquids

This chapter focuses on the simulation of bulk liquids. This is a different task from modeling solvation effects, which are discussed in Chapter 24. Solvation effects are changes in the properties of the solute due to the presence of a solvent. They are defined for an individual molecule or pair of molecules. This chapter discusses the modeling of bulk liquids, which implies properties that are not defined for an individual molecule, such as viscosity.

39.1 LEVEL OF THEORY

The simplest case of fluid modeling is the technique known as computational fluid dynamics. These calculations model the fluid as a continuum that has various properties of viscosity, Reynolds number, and so on. The flow of that fluid is then modeled by using numerical techniques, such as a finite element calculation, to determine the properties of the system as predicted by the Navier–Stokes equation. These techniques are generally the realm of the engineering community and will not be discussed further here.

Nearly all liquid simulations have been done using molecular mechanics force fields to describe the interactions between molecules. A few rare simulations have been completed with orbital-based methods. It is expected that it will still be a long time before orbital-based simulations represent a majority of the studies done due to the incredibly large amount of computational resources necessary for these methods.

Monte Carlo simulations are an efficient way of predicting liquid structure, including the preferred orientation of liquid molecules near a surface. This is an efficient method because it is not necessary to compute energy derivatives, thus reducing the time required for each iteration. The statistical nature of these simulations ensures that both enthalpic and entropic effects are included.

Molecular dynamics calculations are more time-consuming than Monte Carlo calculations. This is because energy derivatives must be computed and used to solve the equations of motion. Molecular dynamics simulations are capable of yielding all the same properties as are obtained from Monte Carlo calculations. The advantage of molecular dynamics is that it is capable of modeling time-dependent properties, which can not be computed with Monte Carlo simulations. This is how diffusion coefficients must be computed. It is also possible to use shearing boundaries in order to obtain a viscosity. Molecular dynamics and Monte Carlo methods are discussed in more detail in Chapter 7.

A very important aspect of both these methods is the means to obtain radial distribution functions. Radial distribution functions are the best description of liquid structure at the molecular level. This is because they reflect the statistical nature of liquids. Radial distribution functions also provide the interface between these simulations and statistical mechanics.

Another way of predicting liquid properties is using QSPR, as discussed in Chapter 30. QSPR can be used to find a mathematical relationship between the structure of the individual molecules and the behavior of the bulk liquid. This is an empirical technique, which limits the conceptual understanding obtainable. However, it is capable of predicting some properties that are very hard to model otherwise. For example, QSPR has been very successful at predicting the boiling points of liquids.

39.2 PERIODIC BOUNDARY CONDITION SIMULATIONS

A liquid is simulated by having a number of molecules (perhaps 1000) within a specific volume. This volume might be a cube, parallelepiped, or hexagonal cylinder. Even with 1000 molecules, a significant fraction would be against the wall of the box. In order to avoid such severe edge effects, periodic boundary conditions are used to make it appear as though the fluid is infinite. Actually, the molecules at the edge of the next box are a copy of the molecules at the opposite edge of the box, as shown in Figure 39.1.

The use of periodic boundary conditions allows the simulation of a bulk fluid, but creates the potential for another type of error. If the longest-range nonbonded forces included in the calculation interact with the same atom in two images of the system, then a long-range symmetry has been unnaturally incorporated into the system. This will result in an additional symmetry in the results, such as a radial distribution function, which is an artifact of the simulation. In order to avoid this problem, the long-range forces are computed only up to a cutoff distance that must be less than half of the box's side length. This is called the minimum image convention. It ensures that the system appears to be nonperiodic to any given atom. It also limits the amount of CPU time that will be required for each iteration.

Calculating nonbonded interactions only to a certain distance imparts an error in the calculation. If the cutoff radius is fairly large, this error will be very minimal due to the small amount of interaction at long distances. This is why many bulk-liquid simulations incorporate 1000 molecules or more. As the cutoff radius is decreased, the associated error increases. In some simulations, a long-range correction is included in order to compensate for this error.

A radial distribution function can be determined by setting up a histogram for various distances and then looking at all pairs of molecules to construct the diagram. Diffusion coefficients can be obtained by measuring the net distances

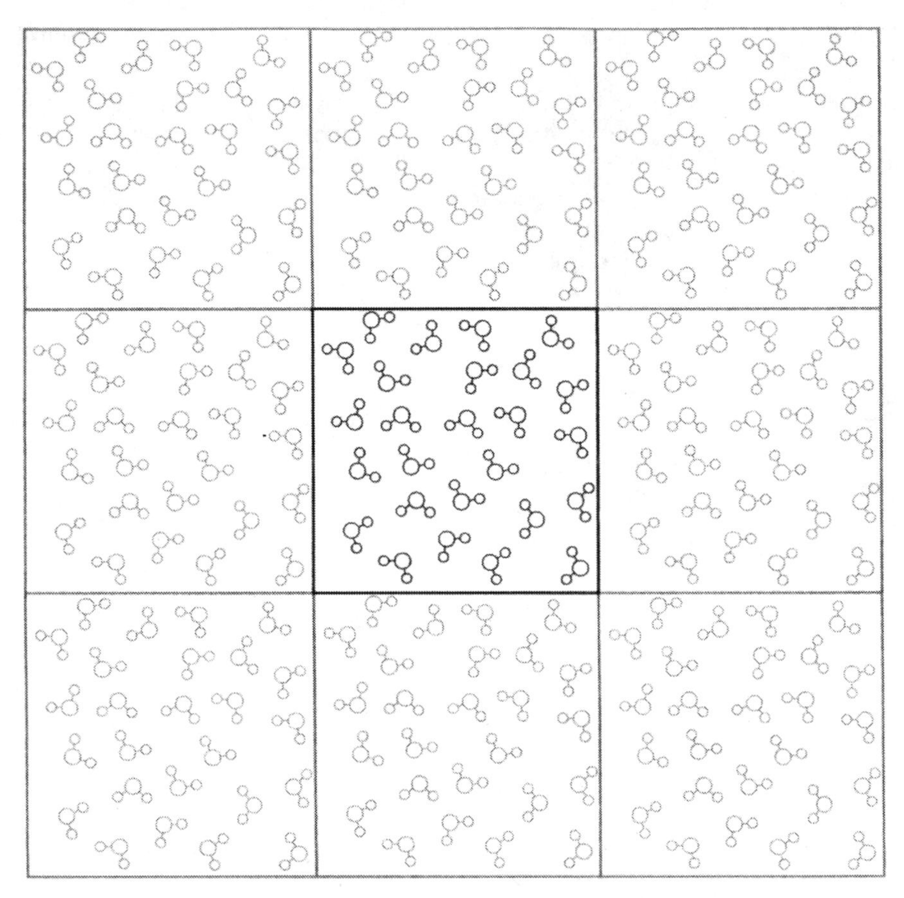

FIGURE 39.1 Periodic boundary conditions in two dimensions. The molecules that appear to be around the center box are actually copies of the center box.

moved by the solute molecules. Some statistical processes that could be the modeled in a similar way given a more sophisticated setup are chromatographic retention times, crystal growth, and adsorption of molecules on a surface.

These calculations can incorporate various types of constraints. It is most common to run simulations with a fixed number of atoms and a fixed volume. In this case, the temperature can be computed from the average kinetic energy of the atoms. It is also possible to adjust the volume to maintain a constant pressure or to scale the velocities to maintain a constant temperature.

If a sufficiently large number of iterations have been performed, the ensemble average of any given property should not change significantly with additional iterations. However, there will be fluctuations in any given property computable as a root-mean-square deviation from the ensemble average. These fluctuations can be related to thermodynamic derivatives. For example, fluctuations in energy can be used to compute a heat capacity for the fluid. Alternatively, a heat capacity can be determined from its derivative formula after running simulations at two temperatures.

It is also possible to simulate nonequilibrium systems. For example, a bulk liquid can be simulated with periodic boundary conditions that have shifting boundaries. This results in simulating a flowing liquid with laminar flow. This makes it possible to compute properties not measurable in a static fluid, such as the viscosity. Nonequilibrium simulations give rise to additional technical difficulties. Readers of this book are advised to leave nonequilibrium simulations to researchers specializing in this type of work.

39.3 RECOMMENDATIONS

Setting up liquid simulations is more complex than molecular calculations. This is because the issues mentioned in this chapter must be addressed. At least the first time, researchers should plan on devoting a significant amount of work to a liquid simulation project.

BIBLIOGRAPHY

Books dedicated to simulating liquids are

D. M. Hirst, *A Computational Approach to Chemistry* Blackwell Scientific, Oxford (1990).

M. P. Allen, D. J. Tildesley, *Computer Simulation of Liquids* Oxford, Oxford (1987).

C. G. Gray, K. E. Gubbins, *Theory of Molecular Fluids* Oxford, Oxford (1984).

J. A. Barker, *Lattice Theories of the Liquid State* Pergamon, New York (1963).

General review articles are

B. Smit, *Encycl. Comput. Chem.* **3**, 1742 (1998).

W. L. Jorgensen, *Encycl. Comput. Chem.* **4**, 2826 (1998).

K. Nakaniski, *Chem. Soc. Rev.* **22**, 177 (1993).

B. J. Alder, E. L. Pollock, *Ann. Rev. Phys. Chem.* **32**, 311 (1981).

Simulating aqueous interfaces is reviewed in

A. Pohorille, *Encycl. Comput. Chem.* **1**, 30 (1998).

Brownian dynamics simulations are reviewed in

J. D. Madura, J. M. Briggs, R. C. Wade, R. R. Gabdoulline, *Encycl. Comput. Chem.* **1**, 141 (1998).

Computing dielectric constants is reviewd in

P. Madden, D. Kivelson, *Adv. Chem. Phys.* **56**, 467 (1984).

Simulation of nonequilibrium processes is reviewed in

P. T. Cummings, A. Baranyai, *Encycl. Comput Chem.* **1**, 390 (1998).

Describing the structure of liquids is reviewed in

M. S. Wertheim, *Ann. Rev. Phys Chem.* **30**, 471 (1979).
D. Chandler, *Ann. Rev. Phys. Chem.* **29**, 441 (1978).

Simulation of supercritical fluids is reviewed in

S. C. Tucker, *Encycl. Comput. Chem.* **4**, 2826 (1998).
A. A. Chialvo, P. T. Cummings, *Encycl. Comput. Chem.* **4**, 2839 (1998).

Computing transport properties is reviewed in

J. H. Dymond, *Chem. Soc. Rev.* **14**, 317 (1985).

Computing viscosity is reviewed in

S. G. Brush, *Chem. Rev.* **62**, 513 (1962).

40 Polymers

Polymers are an extremely important area of study in chemistry. Not only are there many industrial applications, but also the study of them is a complex field of research. Polymers are in the simplest case long-chain molecules with some repeating pattern of functional groups. Most commercially important polymers are organic. The fundamental forces of bonding and intermolecular interactions are the same for polymers as for small molecules. However, many of the polymer properties are dominated by size effects (due to the length of the chain). Thus, simply applying small-molecule modeling techniques is only of limited value.

Polymers are complex systems for a number of reasons. They can be either amorphous or crystalline, or have microscopic domains of both. Most are either amorphous or amorphous with some crystalline domains. Furthermore, this is usually a nonequilibrium state since most production methods do not anneal the material slowly enough to reach an optimal conformation. Thus, the polymer properties vary with the production process (i.e., cooling rate) as well as the molecular structure. The chains of a given polymer may interact primarily by van der Waals forces, hydrogen bonding, π stacking, or charge-transfer interactions. Long-range second-order effects seem to be more important to the description of manmade polymers than is the case for proteins.

This chapter provides the connection between methods described previously and polymer simulation. It does not present the details of many of the polymer simulation methods, which can be found in the references.

40.1 LEVEL OF THEORY

One of the simplest ways to model polymers is as a continuum with various properties. These types of calculations are usually done by engineers for determining the stress and strain on an object made of that material. This is usually a numerical finite element or finite difference calculation, a subject that will not be discussed further in this book.

Polymers are difficult to model due to the large size of microcrystalline domains and the difficulties of simulating nonequilibrium systems. One approach to handling such systems is the use of mesoscale techniques as described in Chapter 35. This has been a successful approach to predicting the formation and structure of microscopic crystalline and amorphous regions.

An area of great interest in the polymer chemistry field is structure–activity relationships. In the simplest form, these can be qualitative descriptions, such as the observation that branched polymers are more biodegradable than straight-chain polymers. Computational simulations are more often directed toward the quantitative prediction of properties, such as the tensile strength of the bulk material.

One means for performing such predictions are group additivity techniques. They parameterize a table of functional groups and then add the effect of each functional group to obtain polymer properties. Group additivity methods are useful but inherently limited in the accuracy possible. They are generally less reliable when multiple functional groups appear in the repeat unit. However, group additivities are readily calculated on small computers for even the most complex repeat units. More recently, QSPR techniques have become popular (Chapter 30).

Other techniques that work well on small computers are based on the molecules' topology or indices from graph theory. These fields of mathematics classify and quantify systems of interconnected points, which correspond well to atoms and bonds between them. Indices can be defined to quantify whether the system is linear or has many cyclic groups or cross links. Properties can be empirically fitted to these indices. Topological and group theory indices are also combined with group additivity techniques or used as QSPR descriptors.

Lattice simulation techniques model a polymer chain by assuming that the possible locations of the atoms fall on the vertices of a regular grid of some sort. This technique has an ability to predict general trends well, but is fairly limited in the absolute accuracy of results. Lattice simulations can use two- or three-dimensional lattices, which might be cubic or tetrahedral (diamondlike). The shape of the polymer is simulated by growing the end of the chain with a random-walk algorithm. This means randomly choosing one of the adjacent lattice locations with the exception of doubling back to the previous spot in the chain. Simpler simulations allow the chain to recross itself by crossing the same location twice. Within the accuracy of a lattice model, this is not completely unreasonable since polymers can have twists or knots when there are nonbonded interactions between different sections of the chain. Some simulations will include an excluded volume effect preventing lattice locations from being used twice.

Another simplified model is the freely jointed or random flight chain model. It assumes all bond and conformation angles can have any value with no energy penalty, and gives a simplified statistical description of elasticity and average end-to-end distance.

The rotational isomeric state (RIS) model assumes that conformational angles can take only certain values. It can be used to generate trial conformations, for which energies can be computed using molecular mechanics. This assumption is physically reasonable while allowing statistical averages to be computed easily. This model is used to derive simple analytic equations that predict polymer properties based on a few values, such as the preferred angle between repeat units. The RIS model, with parameters to describe the specific polymer of interest, can be used to calculate the following: mean-square end-to-end distance, mean-square radius of gyration, mean-square dipole moment, mean-square optical anisotropy, optical configuration parameter, molar Kerr constant, and Cotton–Mouton parameter. RIS combined with Monte Carlo sampling techniques can be used to calculate the following: probability distributions for end-to-end lengths and radius of gyration, atom–atom pair correlation function, scattering function, and force–elongation relation for single chains. RIS structures also make a good starting point for MD simulations.

Due to the large size of polymers, most atomic level modeling is by means of molecular mechanics methods. Force fields parameterized for general organic systems work well for organic polymers. Because of this, only a few polymer force fields have been created, such as PCFF and MSXX. The united-atom approximation (implicit hydrogens) is sometimes used to reduce the computation time. However, all-atom simulations are more sensitive to chain packing and orientational correlations. Fixing bond lengths and angles while maintaining torsional and nonbond interactions improves the simulation time without any apparent loss of accuracy for many properties.

Orbital-based techniques are used for electronic properties, optical properties, and so on. It is usually necessary to make some simplifying assumptions for these calculations. Some semiempirical programs can perform calculations for the one-dimensional infinite-length chain, which gives a band gap and typical band structure electronic-state information. This assumes that interactions between adjacent polymer strands and the effects of folding are negligible. Another option is to complete simulations on successively larger oligomers and then extrapolate the results to obtain values for the infinite-length chain. A third option is to examine a typical repeat unit by using the Fock matrix elements from a repeat unit in the middle of a 100- to 200-unit oligomer.

40.2 SIMULATION CONSTRUCTION

Due to the noncrystalline, nonequilibrium nature of polymers, a statistical mechanical description is rigorously most correct. Thus, simply finding a minimum-energy conformation and computing properties is not generally sufficient. It is usually necessary to compute ensemble averages, even of molecular properties. The additional work needed on the part of both the researcher to set up the simulation and the computer to run the simulation must be considered. When possible, it is advisable to use group additivity or analytic estimation methods.

At the beginning of a project, the model system must be determined. Oligomers can be used to model properties that are a function of local regions of the chain only. Simulations of a single polymer strand can be used to determine the tendency to fold in various manners and to find mean end-to-end distances and other properties generally considered the properties of a single molecule. Multiple chains must often be included in a bulk polymer simulation, which is often necessary for modeling the physical properties of the macroscopic material. Periodic boundary conditions are used to simulate the polymer in the bulk system. Liquid-state simulations can also be performed.

The structure of a polymer, which is actually many structures generated by a sampling of conformation space, can be obtained via a number of techniques. Some of the most widely used techniques are as follows:

- Conventional molecular dynamics or Monte Carlo simulations.
- Chain growth algorithms that build up the chain one unit at a time with some randomness in the way units are added. This process can be repeated to yield multiple conformations.
- A reptation algorithm removes units from one end of the chain and adds them to the other end.
- A kink-jump algorithm displaces a few atoms at some point in the chain at each step.
- Conformation search techniques can be used to find very-low-energy conformations, which are most relevant to polymers that will be given a long annealing time.

A number of chain growth techniques have been devised. This means adding units onto an existing structure with some random choice of conformers, excluding any conformers that would result in different parts of the chain being on top of one another. This sort of model can be implemented using points on a grid with a very simple potential function or using energetics from molecular mechanics calculations. Taking into consideration the energy of various possible ways of adding the monomers will help identify when there is a tendency to develop helical domains, and so on.

Ab initio calculations of polymer properties are either simulations of oligomers or band-structure calculations. Properties often computed with *ab initio* methods are conformational energies, polarizability, hyperpolarizability, optical properties, dielectric properties, and charge distributions. *Ab initio* calculations are also used as a spot check to verify the accuracy of molecular mechanics methods for the polymer of interest. Such calculations are used to parameterize molecular mechanics force fields when existing methods are insufficient, which does not happen too often.

40.3 PROPERTIES

The property to be predicted must be considered when choosing the method for simulating a polymer. Properties can be broadly assigned into one of two categories: material properties, primarily a function of the nature of the polymer chain itself, or specimen properties, primarily due to the size, shape, and phase of the finished object. Thus, material properties are controlled by the choice of monomers, whereas specimen properties are controlled by the manufacturing process. This chapter will focus mostly on material properties, although the mesoscale method does have some ability to predict specimen properties.

Material properties can be further classified into fundamental properties and derived properties. Fundamental properties are a direct consequence of the molecular structure, such as van der Waals volume, cohesive energy, and heat capacity. Derived properties are not readily identified with a certain aspect of molecular structure. Glass transition temperature, density, solubility, and bulk modulus would be considered derived properties. The way in which fundamental properties are obtained from a simulation is often readily apparent. The way in which derived properties are computed is often an empirically determined combination of fundamental properties. Such empirical methods can give more erratic results, reliable for one class of compounds but not for another.

Once a polymer geometry has been described, it can be used to predict density, porosity, and so forth. Geometry alone is often of only minor interest. The purpose of computational modeling is often to determine whether properties of the material justify a synthesis effort. Some of the properties that can be predicted are discussed in the following sections.

Many simulations attempt to determine what motion of the polymer is possible. This can be done by modeling displacements of sections of the chain, Monte Carlo simulations, or reptation (a snakelike motion of the polymer chain as it threads past other chains). These motion studies ultimately attempt to determine a correlation between the molecular motion possible and the macroscopic flexibility, hardness, and so on.

The following sections discuss the prediction of a selection of polymer properties. This listing is by no means comprehensive. The sources listed at the end of this chapter provide a much more thorough treatment.

40.3.1 Crystallinity

Polymers can be crystalline, but may not be easy to crystallize. Computational studies can be used to predict whether a polymer is likely to crystallize readily. One reason polymers fail to crystallize is that there may be many conformers with similar energies and thus little thermodynamic driving force toward an ordered conformation. Calculations of possible conformations of a short oligomer can be used to determine the difference in energy between the most stable conformer and other low-energy conformers.

In order to reach a crystalline state, polymers must have sufficient freedom of motion. Polymer crystals nearly always consist of many strands with a parallel packing. Simply putting strands in parallel does not ensure that they will have the freedom of movement necessary to then find the low-energy conformer. The researcher can check this by examining the cross-sectional profile of the polymer (viewed end on). If the profile is roughly circular, it is likely that the chain will be able to change conformation as necessary.

The tests in the two previous paragraphs are often used because they are easy to perform. They are, however, limited due to their neglect of intermolecular interactions. Testing the effect of intermolecular interactions requires much more intensive simulations. These would be simulations of the bulk materials, which include many polymer strands and often periodic boundary conditions. Such a bulk system can then be simulated with molecular dynamics, Monte Carlo, or simulated annealing methods to examine the tendency to form crystalline phases.

40.3.2 Flexibility

It is generally recognized that the flexibility of a bulk polymer is related to the flexibility of the chains. Chain flexibility is primarily due to torsional motion (changing conformers). Two aspects of chain flexibility are typically examined. One is the barrier involved in determining the lowest-energy conformer from other conformers. The second is the range of conformational motion around the lowest-energy conformation that can be accessed with little or no barrier. There is not yet a clear consensus as to which of these aspects of conformational flexibility is most closely related to bulk flexibility. Researchers are advised to first examine some representative compounds for which the bulk flexibility is known.

40.3.3 Elasticity

Elastic polymers have long chains with many conformers of nearly identical energy. In the relaxed state, entropy is the driving force for causing the chains to take on some randomly coiled conformation with a mean-square end-to-end distance that is much less than the length of the chain in the most linear conformation, but little different in energy. Thus, the elastic restoring force is primarily entropic, although there may be a slight energetic component as well. Prediction of elasticity is based on finding a large difference in length and small difference in energy between relaxed and linear conformations. This tends to be a qualitative prediction.

Once a rubberband is stretched beyond its elastic region, it becomes much harder to stretch and soon breaks. At this point, the polymer chains are linear and more energy must be applied to slide chains past one another and break bonds. Thus, determining the energy required to break the material requires a different type of simulation.

Polymers will be elastic at temperatures that are above the glass-transition temperature and below the liquification temperature. Elasticity is generally improved by the light cross linking of chains. This increases the liquification temperature. It also keeps the material from being permanently deformed when stretched, which is due to chains sliding past one another. Computational techniques can be used to predict the glass-transition and liquification temperatures as described below.

Commercially produced elastic materials have a number of additives. Fillers, such as carbon black, increase tensile strength and elasticity by forming weak cross links between chains. This also makes a material stiffer and increases toughness. Plasticizers may be added to soften the material. Determining the effect of additives is generally done experimentally, although mesoscale methods have the potential to simulate this.

40.3.4 Glass-transition Temperature

It is generally recognized that polymers with flexible chains tend to have low glass-transition temperatures. However, there is not yet any completely reliable way of making quantitative predictions. Group additivity methods have been proposed, but are only reliable for limited classes of compounds. The best method available is a structure–property relationship, which predicts T_g based on cohesive energy, the solubility parameter, and structural parameters that quantify chain rigidity. Methods combining group additivity with results from molecular mechanics simulations have also yielded encouraging results.

40.3.5 Volumetric Properties

The van der Waals volume of a molecule is the volume actually occupied by the atoms. It is reliably computed with a group additivity technique. Connectivity indices can also be used.

The molar volume is usually larger than the van der Waals volume because two additional influences must be added. The first is the amount of empty space in the bulk material due to constraints on how tightly together the chains can pack. The second is the additional space needed to accommodate the vibrational motion of the atoms at a given temperature.

Many polymers expand with increasing temperature. This can be predicted with simple analytic equations relating the volume at a given temperature $V(T)$ to the van der Waals volume V_w and the glass transition temperature, such as

$$V(T) = V_w \left[1.42 + 0.15 \left(\frac{T}{T_g} \right) \right] \qquad (40.1)$$

However, this approach is of limited predictive usefulness due to the difficulty in predicting T_g accurately. Methods have been proposed for computing the molar volume at 298 K and thus extrapolation to other temperatures, which results in some improvement. These use connectivity indices. Note that it is necessary to employ different thermal expansion equations above and below T_g.

40.3.6 Thermodynamic Properties

Thermodynamic properties, such as enthalpy, energy, entropy, and the like, are related to one another. Thus, some information must be obtained from the polymer structure, whereas other data are obtained through thermodynamic relationships. Most often, it is heat capacity as a function of temperature $C_p(T)$ that is computed from the molecular structure.

The heat capacity can be computed by examining the vibrational motion of the atoms and rotational degrees of freedom. There is a discontinuous change in heat capacity upon melting. Thus, different algorithms are used for solid- and liquid-phase heat capacities. These algorithms assume different amounts of freedom of motion.

40.3.7 Solubility Parameters

The solubility parameter is not calculated directly. It is calculated as the square root of the cohesive energy density. There are a number of group additivity techniques for computing cohesive energy. None of these techniques is best for all polymers.

40.3.8 Optical Properties

Computed optical properties tend not to be extremely accurate for polymers. The optical absorption spectra (UV/VIS) must be computed from semiempirical or *ab initio* calculations. Vibrational spectra (IR) can be computed with some molecular mechanics or orbital-based methods. The refractive index is most often calculated from a group additivity technique, with a correction for density.

40.3.9 Mechanical Properties

For engineering applications, mechanical properties are extremely important. They are expressed as stress–strain relationships that quantify the amount of energy (stress) required to give a certain amount of deformation of the material (strain). These properties are dependent on crystallinity, orientation, and cross linking. They are also dependent on the material processing, thus making them difficult to predict with molecular modeling techniques. Mesocscale techniques are probably the best *a priori* prediction method for these. However, structure–property relationships are often used instead for practical reasons (simplicity and minimal computer time). This section will focus on the simpler cases of completely crystalline and completely amorphous phases.

Several techniques are applicable to amorphous phases. QSPR techniques give mechanical properties as a function of glass transition temperature and the repeat unit size. These techniques are not reliable near the glass transition temperature. Molecular mechanics can also be used if the structure was obtained with a molecular-mechanics-based simulation. This consists of finding an energy for a section of the bulk material (often within a periodic boundary) and then shifting the size of the box and reoptimizing to obtain a second energy. Molecular dynamics and Monte Carlo simulations can be used to predict behavior near the glass-transition temperature.

The molar sound velocity can be predicted with group additivity techniques. It, in turn, may be used to predict the mechanical properties due to high-frequency deformations.

Rubbery materials are usually lightly cross-linked. Their properties depend on the mean distance between cross links and chain rigidity. Cross linking can be quantified by the use of functions derived from graph theory, such as the Rao or molar Hartmann functions. These can be incorporated into both group additivity and QSPR equations.

For crystalline polymers, the bulk modulus can be obtained from band-structure calculations. Molecular mechanics calculations can also be used, provided that the crystal structure was optimized with the same method.

40.3.10 Thermal Stability

It is important to know whether a polymer will be stable, that is, whether it will not decompose at a given temperature. There are several measures of thermal stability, the most important of which (from an economic standpoint) is the Underwriters Laboratories (UL) temperature index.

Unfortunately, there is not at a present a computational method for predicting the UL temperature index. There is a QSPR method for predicting $T_{d,1/2}$, the temperature of half-decomposition, meaning the temperature at which a material loses half of its mass due to pyrolysis. The QSPR method uses a connectivity index and weights for the number of various functional groups.

40.4 RECOMMENDATIONS

Polymer modeling is a fast-growing field. It remains primarily the realm of experts because the preferred methods and limitations of existing methods are still changing, thus requiring the researcher to constantly stay abreast of new developments. Group additivity and QSPR methods have been the mainstay of the field due to the difficulty of alternative methods. However, mesoscale and other bulk simulations are becoming more commonplace. Researchers are advised to first consider what properties need to be computed and to then explore the methods and software packages available for those specific properties.

BIBLIOGRAPHY

Books discussing polymer modeling are

A. K. Rappé, C. J. Casewit, *Molecular Mechanics Across Chemistry* University Science Books, Sausalito (1997).

J. Bicerano, *Prediction of Polymer Properties* Marcel Dekker, New York (1996).

Polymeric Systems, Adv. Chem. Phys. vol **94** (1996).

A. Ya. Gol'dman, *Prediction of the Deformation Properties of Polymeric and Composit Materials* American Chemical Society, Washington (1994).

Computational Modeling of Polymers J. Bicerano, Ed., Dekker, new York (1992).

Computer Simulation of Polymers E. A. Coulbourne, Ed., Longman-Harlow, London (1992).

Computer Simulation of Polymers R. J. Roe, Ed., Prentice Hall, New York (1991).

H. R. Allcock, F. W. Lampe, *Contemporary Polymer Chemistry* Prentice-Hall, Englewood Cliffs (1990).

D. W. van Krevelen, *Propertis of Polymers* Elsevier, Amsterdam (1990).

P. J. Flory, *Statistical Mechanics of Chain Molecules* Hanser, New York (1989).

Review articles covering polymer modeling in general are

J. J. Ladik, *Encycl. Comput. Chem.* **1**, 591 (1998).

I. Szleifer, *Encycl. Comput. Chem.* **3**, 2114 (1998).

V. Galiatsatos, *Rev. Comput. Chem.* **6**, 149 (1995).

A. Baumgaertner, *Top. Appl. Phys.* **71**, 285 (1992).

K. A. Dill, J. Naghizadeh, J. A. Marqusee, *Ann. Rev. Phys. Chem.* **39**, 425 (1988).

Reviews of polymer dynamics are

J. Skolnick, A. Kolinski, *Adv Chem. Phys.* **78**, 223 (1990).

T. P. Lodge, N. A. Rotstein, S. Prager, *Adv. Chem. Phys.* **79**, 1 (1990).

A review of graph theory techniques for describing polymers is

S. I. Kuchanov, S. V. Korolev, S. V. Panykov, *Adv. Chem. Phys.* **72**, 115 (1988).

A review of liquid crystal modeling is

G. Marrucci, F. Greco, *Adv. Chem Phys.* **86**, 331 (1993).

A review of Monte Carlo simulations is

J. J. de Pablo, F. A. Escobedo, *Encycl. Comput. Chem.* **3**, 1763 (1998).

Reviews of optical property simulation are

G. Orlandi, F. Zerbetto, M. Z. Zgierski, *Chem. Rev.* **91**, 867 (1991).

J.-M. André, J. Delhalle, *Chem. Rev.* **91**, 843 (1991).

Phase behavior modeling is reviewed in

K. S. Schweizer, J. G. Curro, *Adv. Chem. Phys.* **98**, 1 (1997).

A. Yethiraj, *Encycl. Comput. Chem.* **3**, 2119 (1998).

Quantum mechanical modeling of polymer systems is reviewed in

M. Kertész, *Adv. Quantum Chem.* **15**, 161 (1982).

J.-M. André, *Adv. Quantum Chem.* **12**, 65 (1980).

A review of semiempirical calculation applied to polymers is

J. J. P. Stewart, *Encycl. Comput. Chem.* **3**, 2130 (1998).

Reviews of the statistical mechanical aspects of the problem are

K. W. Foreman, K. F. Freed, *Adv. Chem. Phys.* **103**, 335 (1998).
S. G. Whittington, *Adv Chem. Phys.* **51**, 1 (1982).
H. Yamakawa, *Ann. Rev. Phys. Chem.* **25**, 179 (1974).
K. F. Freed, *Adv. Chem. Phys.* **22**, 1 (1972).

41 Solids and Surfaces

Solids can be crystalline, molecular crystals, or amorphous. Molecular crystals are ordered solids with individual molecules still identifiable in the crystal. There is some disparity in chemical research. This is because experimental molecular geometries most often come from the X-ray diffraction of crystalline compounds, whereas the most well-developed computational techniques are for modeling gas-phase compounds. Meanwhile, the information many chemists are most worried about is the solution-phase behavior of a compound.

41.1 CONTINUUM MODELS

The modeling of solids as a continuum with a given shear strength, and the like is often used for predicting mechanical properties. These are modeled using finite element or finite difference techniques. This type of modeling is usually employed by engineers for structural analysis. It will not be discussed further here.

41.2 CLUSTERS

One way to model a solid is to use software designed for gas-phase molecular computations. A large enough piece of the solid can be modeled so that the region in the center for practical purposes describes the region at the center of an infinite crystal. This is called a cluster calculation.

When this calculation is done, the structure must be truncated in some fashion. If no particular truncation is used, the atoms at the outer edge of the cluster will have dangling bonds. This changes the behavior of those atoms, which in turn will affect adjacent atoms that, in turn, requires more atoms in the simulation. For covalent bonded organic compounds, truncating the structure with hydrogen atoms is very reasonable since the electronegativity of a hydrogen atom is similar to that of a carbon atom and H atoms take the least amount of computational resources. For very ionic compounds, a set of point charges, called a Madelung potential, is reasonable. For compounds in between these two extremes, the choices are not so clear and must be made on a case-by-case basis. It is often necessary to perform a small study to determine which is the best choice.

41.3 BAND STRUCTURES

As described in the chapter on band structures, these calculations reproduce the electronic structure of infinite solids. This is important for a number of types of studies, such as modeling compounds for use in solar cells, in which it is important to know whether the band gap is a direct or indirect gap. Band structure calculations are ideal for modeling an infinite regular crystal, but not for modeling surface chemistry or defect sites.

41.4 DEFECT CALCULATIONS

The chemistry of interest is often not merely the infinite crystal, but rather how some other species will interact with that crystal. As such, it is necessary to model a system that is an infinite crystal except for a particular site where something is different. The same techniques for doing this can be used, regardless of whether it refers to a defect within the crystal or something binding to the surface. The most common technique is a Mott–Littleton defect calculation. This technique embeds a defect in an infinite crystal, which can be considered a local perturbation to the band structure.

41.5 MOLECULAR DYNAMICS AND MONTE CARLO METHODS

Molecular mechanics methods have been used particularly for simulating surface–liquid interactions. Molecular mechanics calculations are called *effective potential function* calculations in the solid-state literature. Monte Carlo methods are useful for determining what orientation the solvent will take near a surface. Molecular dynamics can be used to model surface reactions and adsorption if the force field is parameterized correctly.

41.6 AMORPHOUS MATERIALS

The modeling of amorphous solids is a more difficult problem. This is because there is no rigorous way to determine the structure of an amorphous compound or even define when it has been found. There are algorithms for building up a structure that has various hybridizations and size rings according to some statistical distribution. Such calculations cannot be made more efficient by the use of symmetry.

41.7 RECOMMENDATIONS

Overall, solid-state modeling requires more time on the part of the researcher and often more CPU-intensive calculations. Researchers are advised to plan on investing a significant amount of time learning and using solid-state modeling techniques.

BIBLIOGRAPHY

Some books on modeling these systems are

Theoretical Aspects and Computer Modelling of the Molecular Solid State A. Gavezotti, Ed., John Wiley & Sons, New York (1997).

C. Pisani, *Quantum-Mechanical Ab Initio Calculation of the Properties of Crystalline Materials* Springer-Verlag, New York (1996).

R. Hoffmann, *Solids and Surfaces; A Chemist's View of bonding in Extended Structures* VCH, New York (1988).

Structure and Bonding in Noncrystalline Solids G. E. Walrafen, A. G. Revesz, Eds., Plenum, New York (1986).

D. L. Goodstein, *States of Matter* Dover, New York (1985).

W. A. Harrison, *Solid State Theory* Dover, New York (1979).

B. Donovan, *Elementary Theory of Metals* Pergamon, Oxford (1967).

Reviews of solid state modeling in general are

A. Gavezzotti, S. L. Price, *Encycl. Comput. Chem.* **1**, 641 (1998).

The application of *ab initio* methods is reviewed in

J. Sauer, *Chem. Rev* **89**, 199 (1989).

F. E. Harris, *Theoretical Chemistry Advances and Perspectives* D. Henderson (Ed.) **1**, 147, Academic Press, New York (1975).

J. Koutecký, *Adv. Chem. Phys.* **9**, 85 (1965).

Surface adsorption is reviewed in

W. Stelle, *Chem. Rev* **93**, 2355 (1993).

E. Shustorovich, *Modelling of Molecular Structures and Properties* J.-L. Rivail, Ed., 119, Elsevier, Amsterdam (1990).

M. Simonetta, A. Gavezzotti, *Adv. Quantum Chem.* **12**, 103 (1990).

P. J. Feibelman, *Ann. Rev. Phys. Chem.* **40**, 261 (1989).

M. M. Dubinin, *Chem. Rev.* **60**, 235 (1960).

Amorphous solid simulation is reviewed in

C. A. Angell, J. H. R. Clarke, L. W. Woodcock, *Adv. Chem. Phys.* **48**, 397 (1981).

Binding at surface sites is reviewed in

P. S. Bagus, F. Illas, *Encycl. Comput. Chem.* **4**, 2870 (1998).

J. Sauer, P. Ugliengo, E. Garrone, V. R. Saunders, *Chem. Rev.* **94**, 2095 (1994).

E. I. Solomon, P. M. Jones, J. A. May, *Chem. Rev.* **93**, 2623 (1993).

Cluster calculations are reviewed in

D. Michael, P. Mingos, *Chem. Soc. Rev.* **15**, 31 (1986).

Electric double layer modeling is reviewed in

R. Parsons, *Chem. Rev.* **90**, 813 (1990).

The application of molecular mechanics methods is reviewed in

A. M. Stoneham, J. H. Harding, *Ann. Rev Phys. Chem.* **37**, 53 (1986).

A review of methods for predicting properties of solids and surfaces is

E. Wimmer, *Encycl. Comput. Chem.* **3**, 1559 (1998).

Reactions and dynamics at surfaces are reviewed in

D. G. Musaev, K. Morokuma, *Adv. Chem. Phys.* **95**, 61 (1996).

W. Schmickler, *Chem. Rev.* **96**, 3177 (1996).

B. J. Garrison, P. B. Skodali, D. Srivastava, *Chem. Rev.* **96**, 1327 (1996).

B. J. Garrison, D. Srivastava, *Ann. Rev. Phys. Chem.* **46**, 373 (1995).

B. J. Garrison, *Chem. Soc. Rev.* **21**, 155 (1992).

J. W. Gadzuk, *Ann. Rev. Phys. Chem.* **39**, 395 (1988).

R. B. Gerber, *Chem. Rev.* **87**, 29 (1987).

T. F. George, K.-T. Lee, W. C. Murphy, M. Hutchinson, H.-W. Lee, *Theory of Chemical Reaction Dynamics Volume IV* M. Baer, Ed., 139, CRC, Boca Ratan (1985).

Semiempirical modeling is reviewed in

F. Ruette, A. J. Hernández, *Computational Chemistry: Structure, Interactions, Reactivity Part B* S. Fraga, Ed., 637, Elsevier, Amsterdam (1992).

Predicting the structure of crystalline solids is reviewed in

P. Verwer, *Rev. Comput. Chem.* **12**, 327 (1998).

B. P. van Eijck, *Encycl Comput. Chem.* **1**, 636 (1998).

J. K. Burdett, *Adv. Chem. Phys.* **49**, 47 (1982).

T. Kihara, A. Koide, *Adv. Chem. Phys.* **33**, 51 (1975).

Solid vibrations and dynamics are reviewed in

M. J. Klein, L. J. Lewis, *Chem. Rev.* **90**, 459 (1990).

W. J. Briels, A. P. J. Janson, A. van Der Avoird, *Adv. Quantum Chem.* **18**, 131 (1986).

O. Schnepp, N. Jacobi, *Adv. Chem. Phys.* **22**, 205 (1972).

APPENDIX A
Software Packages

Most of the computational techniques discussed in this text have been included in a number of software packages. The general techniques are uniquely defined, meaning that a HF calculation with a given basis set on a particular molecule will give the exact same results regardless of which program is used. However, the choice of software is still important. Software packages differ in cost, functionality, efficiency, ease of use, automation, and robustness. These concerns make an enormous difference in determining what computational projects can be completed successfully and how much work will be involved.

This appendix is not intended to provide a comprehensive listing of computational chemistry software packages. Some of the software packages listed here are included because they are very widely used. Others are included because they pertained to topics discussed in this book. A few relevant pieces of software were omitted because we were not able to obtain an evaluation copy prior to publication.

Program functionality and prices change rapidly. Because of this, we have not made an attempt to list all the functions of each program. Many of the software packages can be purchased at various prices, depending on the options purchased and the existence of discounts, such as for academic use. Individual companies should be contacted for current price information. The pricing information given in this appendix is in the form of general price ranges. These are listed in Table A.1.

We have arranged this chapter by classes of software. The choice of which section each software package is listed in is based on the most common use of the package, rather than every detail of functionality. We have attempted to give an indication of what types of problems the software packages generally are or are not useful for. There are expected to be exceptions to all of these generalities. The reader of this book is urged to consider each package's specific application and discuss it with experienced computational chemists and the representatives of the software companies involved.

A.1 INTEGRATED PACKAGES

These are software packages that have the ability to perform computations using several computational techniques. Most also have an integrated graphic user interface.

TABLE A.1 Price Categories

Price (U.S. dollars)	Category
0	Free
1–100	Student
101–300	Individual
301–1000	Production
1001–5000	Departmental
>5000	Institutional
Contact	Contact the vendor for pricing

A.1.1 Alchemy

Alchemy 2000 (we tested Version 2.05) is a graphic interface for running molecular mechanics and semiempirical calculations. Calculations can be done with the built-in Tripos force field or by calling the MM3 or MOPAC programs, which are included with the package. Alchemy is designed by Tripos and sold by SciVision.

Molecules can be built using a two-dimensional sketch mode and are then converted to three-dimensional geometries by the program. It is also possible to build molecules in the three-dimensional mode. Libraries of organic functional groups are available. There is also a protein builder. The user can change the stereochemistry and conformational angles. Conformations are set by allowing the program to slowly change the conformation until the user tells it to stop. The builders can be used to create any organic structure; however, the author did not find it as convenient to work with as some other graphic builders. The screen for setting up external MOPAC jobs was convenient to work with.

The program is able to do systematic or random conformation searches using the Tripos force field. Conformation searches can include up to eight single bonds and two rings.

Output data can be printed or exported to a spreadsheet. The rendering quality is very good. Structures can be rendered and labeled in several different ways. Molecular structures can be saved in several different formats or as image files. The presentation mode allows molecular structures to be combined with text.

The program allows the user to create a database of structures. Calculations can then be run on the whole set of structures. These databases may also be used by some separately sold software packages.

There are a number of separately sold programs designed to interface to Alchemy. SciQSAR is a linear-regression-based QSAR program that interfaces to Alchemy and Chem3D. SciLogP interfaces with Alchemy and Chem3D to predict water–octanol partition coefficients using linear regression and neural networks. SciPolymer interfaces to Alchemy and computes 44 different polymer properties, which include physical properties, electrical properties, optical properties, thermodynamic properties, magnetic properties, gas permeabilities, solubility, and microscopic properties.

Price category: production
Platforms: PC
Contact information: SciVision
128 Spring St.
Lexington, MA 02173
(781) 272-4949
http://www.scivision.com/
sales@scivision.com

A.1.2 Chem3D

Chem3D (we tested Version 5.0 Ultra) is a molecular modeling package for the PC and Macintosh. It can perform calculations using MM2 and extended Hückel as well as acting as a graphic interface for MOPAC (included) or Gaussian (sold separately). There are also browser plug-ins available for viewing structures and surfaces.

Chem3D can read a wide variety of popular chemical structure files, including Gaussian, MacroModel, MDL, MOPAC, PDB, and SYBYL. Two-dimensional structures imported from ChemDraw or ISIS/Draw are automatically converted to three-dimensional structures. The Chem3D native file format contains both the molecular structure and results of computations. Data can be exported in a variety of chemical-structure formats and graphics files.

Chem3D has both graphic and text-based structure-building modes. Structures can be generated graphically by sketching out the molecule. The builder creates carbon atoms, which can be edited by typing text to substitute other elements or functional groups. As the structure is built, the valence is filled with hydrogen atoms and typical bond lengths and angles are set. Several hundred predefined functional groups are available and users can define additional ones. The text-based mode allows the user to input a simple text string (similar to SMILES, but not identical). This text mode can be used to build structures entirely or to add functional groups.

A number of mechanisms are available for manually defining aspects of the molecular geometry. These include defining dummy atoms as well as setting bond lengths, angles, and dihedral angles. It is also possible to set distances between nonbonded atoms. The molecular structure is maintained internally in both Cartesian coordinates and a Z-matrix. A number of functions for defining how the Z-matrix is constructed make this one of the best GUIs available for setting up calculations that must be done by Z-matrix.

Chem3D uses a MM2 force field that has been extended to cover the full periodic table with the exception of the *f* block elements. Unknown parameters will be estimated by the program and a message generated to inform the user of this. MM2 can be used for both energy minimization and molecular dynamics calculations. The user can add custom atom types or alter the parameters used for one specific atom in the calculation. Extended Hückel may be used for the calculation of charges and molecular surfaces.

Chem3D comes with an implementation of MOPAC 97. The computation setup includes a number of screens in which to select the level of theory, type of calculation, and properties. Menu picks are available for commonly used functions, such as geometry optimization, transition structure optimization, dipole moments, population analysis, COSMO solvation, hyperfine coupling constants, and polarizability. Molecular surfaces can be displayed for electron density, spin density, electrostatic potential, and molecular orbitals. Surfaces can be generated using the shape from one property and the colorization from another. This allows property mapping of solvent-accessible surfaces such as by charges or hydrophobicity. Users can also type in additional route card options. Although menu picks are available for the most frequently used options, one notable exception is the lack of a way to graphically display normal vibrational mode motion or displacement vectors. The interface to Gaussian 98W is similar to the MOPAC interface.

While a calculation is running, the Chem3D interface must be operational. The structure for the running calculation cannot be edited. However, other structures can be built while one is calculating. If multiple MM2 jobs are executed simultaneously, they will be automatically queued and run sequentially. The Macintosh version supports Apple Events, making it possible to write scripts to automate tasks. The PC version is an OLE automation server, making it possible to call Chem3D from other programs (i.e., Visual Basic programs).

A number of properties can be computed from various chemical descriptors. These include physical properties, such as surface area, volume, molecular weight, ovality, and moments of inertia. Chemical properties available include boiling point, melting point, critical variables, Henry's law constant, heat capacity, log P, refractivity, and solubility.

Several display modes are available. Molecules can be displayed as wire frames (lines), sticks (wider lines), ball and stick models (with line or cylindrical bonds), and as space-filling models. Protein structures can be displayed as ribbons. Dot surfaces of van der Waals radii or extended Hückel charges may be added to any of these. In the PC version, a couple of the display modes rendered the molecule, with the lines depicting bonds not quite connecting to the spheres depicting atoms. When molecular surfaces from extended Hückel, MOPAC, or Gaussian calculations are displayed, a different set of rendering algorithms with improved three-dimensional shading is used. These surfaces can be displayed as solid, mesh, dots, or translucent surfaces. The graphics quality in this display mode is very good with the exception of the translucent surface algorithm, which came out looking dithered on our test platforms. Movies can be created from operations generating multiple structures, such as molecular dynamics simulations. These movies can be viewed within Chem3D, but cannot be saved in a common movie file format.

Several versions of Chem3D are available; they differ in price and functionality. These are denoted as Ltd, Std, Pro, and Ultra. Some points of difference in functionality are molecular surface generation, molecular dynamics, MOPAC support, and Gaussian support. There are only a few differences in functionality between the Macintosh and PC versions. This review was written just prior to the release of a new version of Chem3D, which is slated to have support for GAMESS and a new SAR component that includes descriptors from MM2, MOPAC, and GAMESS.

Price category: student, individual, production
Platforms: PC (Windows), Macintosh
Contact information: CambridgeSoft
100 Cambridge Park Drive
Cambridge, MA 02140
(617) 588-9300
http://www.camsoft.com/
info@camsoft.com

A.1.3 ChemSketch and the ACD Software Suite

ChemSketch (we tested Version 4.01) is a graphic interface that can be used as the front end for a host of programs sold by Advanced Chemistry Development. Both free and commercial versions are available. It is a two-dimensional structure drawing program primarily designed for organic molecules. Although the drawing mode is essentially a two-dimensional drawing routine, it is also possible to rotate the molecule in three dimensions. The program automatically keeps track of the number of hydrogens bonded to each atom. The reviewer felt that the molecule sketch mode was convenient to use. The documentation is well written and includes many examples.

ChemSketch has some special-purpose building functions. The peptide builder creates a line structure from the protein sequence defined with the typical three-letter abbreviations. The carbohydrate builder creates a structure from a text string description of the molecule. The nucleic acid builder creates a structure from the typical one-letter abbreviations. There is a function to clean up the shape of the structure (i.e., make bond lengths equivalent). There is also a three-dimensional optimization routine, which uses a proprietary modification of the CHARMM force field. It is possible to set the molecule line drawing mode to obey the conventions of several different publishers.

ChemSketch can import and export a number of molecular structure and bit mapped graphic files. It can also export HTML or VRML files. The additional computation modules are callable from ChemSketch, so it is not necessary to copy or save data to access those functions.

There is an interpretive language called ChemBasic for automating tasks in ChemSketch. It is similar to commercial versions of BASIC. Some of the features of BASIC have been omitted. ChemBasic also incorporates additional functions pertinent to the interface with ChemSketch. It can be downloaded at no charge from the website listed below.

There is a drawing mode as well as a molecule-sketching mode. The molecule-sketching mode is capable of drawing inorganic structures, but most of the computational abilities are limited to organic molecules with common heteroatoms. ChemSketch itself can compute liquid properties such as molar refractivity, molar volume, index of refraction, surface tension, density, and dielectric constant. It uses topological and group additivity methods for these calculations. There are more sophisticated modules priced separately for predicting NMR spectra, boiling points, log *P*, log *D*, *pKa*, solubility, the Hammett sigma parameter, chromatographic retention times, and systematic names using the IUPAC and CAS index rules. A Web-based version of some of these is under development.

The boiling point module predicts the boiling temperature at various pressures, vapor pressure, flash point, and enthalpy of vaporization. The elements supported are H, B, C, N, O, F, Cl, Br, I, Si, P, S, Ge, As, Se, Sn, and Pb. The prediction algorithm incorporates a database of known boiling points and mathematical relationships to adjust for the effect of molecular weight, functional groups, and so on. In our tests, the predicted boiling point was most often within 2°C of experiment for simple compounds and became less accurate with the presence of multiple functional groups.

Programs are available to predict NMR spectra for ^{1}H, ^{13}C, ^{19}F, ^{31}P, and two-dimensional NMR results. We tested Version 4.04 of CNMR, the ^{13}C NMR program. The program can display the spectrum including line broadening, the integral curve, splitting, and off-resonance peaks. The predicted shifts include estimated confidence intervals. The program uses a database of assigned-literature NMR spectra and can be trained by the user. It predicts shifts by finding the most similar structures from the database and then adjusting them for interactions between various functional groups. The literature from ACD claims that shifts are predicted to within 3 ppm of experiment most of the time. In our tests, 95% of the shifts were within 1 ppm of the experimental values. The ^{1}H module predicts both shifts and coupling constants, as well as taking into account the Karplus equation for three-dimensional optimized structures.

Price category: ChemSketch is free; the modules have a wide price range

Platforms: PC

Contact information: Advanced Chemistry Development, Inc.

133 Richmond Street West, Suite 605

Toronto, Ontario, Canada M5H 2L3

(800) 304-3988

http://www.acdlabs.com/

sales@acdlabs.com

A.1.4 HyperChem

HyperChem (we tested Version 6.0 Pro) is an integrated graphic interface, computational, and visualization package. It has seen the most use on PCs. It can also be used as a graphic interface for Q-Chem. HyperChem incorporates *ab initio*, semiempirical, and molecular mechanics programs. These can be used for computing vibrational frequencies, transition states, electronic excited states, QM/MM, molecular dynamics, and Monte Carlo simulations. Several different versions of the program with varying functionality and price are available.

The program has a drawing mode in which the backbone can be sketched out and then hydrogens added automatically. This sketcher does not set the bond lengths or angles, so the use of a molecular mechanics optimization before doing more time-consuming calculations is highly advised. Building biomolecules is made easier with a sugar builder and amino-acid sequence editor. Periodic systems can be constructed with a crystal builder. A polymer builder was added in Version 6. Overall, the builder is very easy to use.

The graphic interface incorporates a variety of rendering modes. It is possible to visualize molecular surfaces and animations of vibrational modes. Both electronic and vibrational spectra can be displayed with intensities. The program can produce good-quality graphics, including ray-traced renderings, suitable for publication. The GUI is integrated tightly with the computational modules, thus changing settings in one menu and the options available in other menus. A number of common-structure file formats can be read and written.

By default, the calculation results are displayed on screen, but they are not saved to disk. The user can specify that all results for a given session be written to a log file. While a calculation is running, no actions can be taken other than changing the molecular orientation on screen. By default, calculations are run on the PC that the GUI is running on, but it can also be configured to run calculations on a networked SGI or HP-UX workstation. The program has a scripting ability that can be used to automate tasks. The built-in scripting allows the automation of menu selections and execution of jobs. Tcl scripts can be called for more sophisticated tasks.

The molecular mechanics force fields available include MM+, OPLS, BIO+, and AMBER. Parameters missing from the force field will be automatically estimated. The user has some control over cutoff distances for various terms in the energy expression. Solvent molecules can be included along with periodic boundary conditions. The molecular mechanics calculations tested ran without difficulties. Biomolecule computational abilities are aided by functions for superimposing molecules, conformation searching, and QSAR descriptor calculation.

The semiempirical techniques available include EH, CNDO, INDO, MINDO/3, ZINDO, MNDO, AM1, and PM3. The ZINDO/S, MNDO/d, and PM3(TM) variations are also available. The semiempirical module seems to be rather robust in that it did well on some technically difficult test calculations.

There were some problems with the eigenvalue following transition-structure routine jumping from one vibrational mode to another. The semiempirical geometry optimization routines work well.

The *ab initio* module can run HF, MP2 (single point), and CIS calculations. A number of common basis sets are included. Some results, such as population analysis, are only written to the log file. One test calculation failed to achieve SCF convergence, but no messages indicating that fact were given. Thus, it is advisable to examine the iteration energies in the log file.

Price category: student, individual, production, departmental
Platforms: PC (Windows), SGI, HP-UX, Win CE
Contact information: Hypercube, Inc.
1115 NW 4th Street
Gainesville, FL 32601
(352) 371-7744
http://www.hyper.com/
info@hyper.com

A.1.5 NWChem

NWChem (we tested Version 3.2.1) is a program for *ab initio*, band-structure, molecular mechanics, and molecular dynamics calculations. The DFT band-structure capability is still under development and was not included in the Linux version tested. NWChem is unique in that it was designed from scratch for efficient parallel execution. The user agreement is more restrictive than most, apparently because the code is still under active development. At the time of this book's publication, limited support was available for users outside of the EMSL facility.

The program can be run either in a serial or parallel execution mode. Both execution modes were stable when tested on a multiprocessor Linux system. Parallel calculations can be run either on parallel computers or networked workstations. Benchmark information is available at the website listed below.

NWChem uses ASCII input and output files. The input format allows geometry to be input as Cartesian coordinates or a Z-matrix. If symmetry is specified, only the Cartesian coordinates of the symmetry-unique atoms are included. Some sections of the code require additional input files.

The *ab initio* methods available include HF, DFT, MPn, MCSCF, CI, and CC. In the version tested, the CI methods were still under development. There are a large number of basis sets available, including ECP sets. Dynamic calculations can be performed at *ab initio* levels of theory.

The program can use conventional, in-core, or direct integral evaluation. The default *ab initio* algorithm checks the disk space and memory available. It then uses an in-core method if sufficient memory is available. If memory is not available for in core evaluation, the program uses a conventional method if sufficient disk space is available. Finally, it switches to a semidirect method, with the disk files taking up to 95% of the space available. The user can specify the use of a direct integral evaluation algorithm.

The molecular mechanics force fields available are AMBER95 and CHARMM. The molecular mechanics and dynamics portion of the code is capable of performing very sophisticated calculations. This is implemented through a large number of data files used to hold different types of information along with keywords to create, use, process, and preprocess this information. This results in having a very flexible program, but it makes the input for simple calculations unnecessarily complex. QM/MM minimization and dynamics calculations are also possible.

NWChem is part of the Molecular Science Software Suite (MS^3) which has been recognized by *R&D Magazine* as one of the 100 most technologically significant new products and processes of 1999. The other elements of MS^3 are Ecce, which is a problem-solving environment, and ParSoft, which is the underlying libraries and tools for parallel communication and high-performance input/output. All of the MS^3 components are available publicly.

Price category: free

Platforms: PC (Linux only), SGI, Cray, Paragon, SP2, KSR, Sun, DEC, IBM

Contact information: W. R. Wiley Environmental Molecular Sciences Laboratory
Pacific Northwest National Laboratory
902 Battelle Blvd.
P. O. Box 999
Richland, WA 99352
http://www.emsl.pnl.gov/pub/docs/nwchem
nwchem-support@emsl.pnl.gov

A.1.6 SPARTAN

SPARTAN (we tested Version 5.1 for SGI) is a program with *ab initio*, DFT, semiempirical, and molecular mechanics methods integrated with an easily used graphic interface. This program has become a favorite of both experimental chemists and educational institutions. The primary strengths of this program are ease of use and robustness (calculations failing to complete less often than with many other programs). The price of this robustness is that calculations sometimes take longer than the exact same calculation would take using a program designed for efficiency.

Some of the other features of this program are the ability to compute transition states, coordinate driving, conformation searches, combinatorial tools, and built-in visualization. The builder includes atoms and fragments for organics, inorganics, peptides, nucleotides, chelates, high-coordination geometries, and more. The builder lets users define the bonding change involved in a reaction and then searches a database of known transition structures to find a starting geometry for a transition-structure optimization. The graphic interface is so easy to use that the primary difficulty users have is not running the program, but rather, understanding the limitations of computational techniques and expected CPU time requirements. If a program can be too easy to use, this is it.

The amount of functionality within each method is not yet as great as with some other programs. The only correlated *ab initio* method is MP2. The DFT module is limited to LDA calculations. Few properties are available from molecular mechanics calculations. Vibrational frequencies can be computed, but not intensities. Some of the PC and Macintosh versions of the program only support a subset of the functionality available in the UNIX version. PC Spartan Pro is nearly identical to the UNIX version in functionality. The Titan program is the Spartan graphic interface integrated with the Jaguar computational program. The Linux version is the Spartan graphic interface integrated with the Q-Chem computational program.

A few of the methods available are applicable to inorganic compounds. These include the PM3/TM method. However, the program is most useful for modeling organic compounds due to a lack of technical features often needed to contend with spin contamination, convergence failure, and so forth.

The program comes with its own job queue system. Jobs are submitted to this queue via a script, which can be edited to utilize third-party batch-queuing systems instead.

Price category: individual, production, departmental

Platforms: PC (Windows and Linux), Macintosh, SGI, RS/6000, Alpha, HP-UX

Contact information: Wavefunction, Inc.
18401 Von Karman Ave.
Suite 370
Irvine, CA 92612
(949) 955-2120
http://www.wavefun.com/
sales@wavefun.com

A.1.7 UniChem

UniChem (we tested Version 4.1) is a graphic interface made for running calculations on remote machines. The UniChem GUI runs on the local workstation and submits the computations to be run on a remote machine. The server software comes with MNDO, DGauss, and CADPAC. It can also be used as a graphic interface for Gaussian and Q-Chem. A toolkit can be purchased separately, which allows users to create an interface to their own programs.

The builder works very well. It has an atom-based mode in which the user chooses the element and hybridization. There are also libraries of functional groups, rings, and so on. It has some extra features that are not found in many other builders, such as being able to clean up the symmetry.

The calculation setup screens list a good selection of the options that are most widely used. However, it is not a complete list. The user also chooses which queue to use on the remote machine and can set queue resource limits. All of this is turned into a script with queue commands and the job input file. The user can edit this script manually before it is run. Once the job is submitted, the inputs are transferred to the server machine, the job is run and the results can be sent back to the local machine. The server can be configured to work with an NQS queue system. The system administrator and users have a reasonable amount of control in configuring how the jobs are run and where files are stored. The administrator should look carefully at this configuration and must consider where results will be sent in the case of a failed job or network outage.

Once the job is completed, the UniChem GUI can be used to visualize results. It can be used to visualize common three-dimensional properties, such as electron density, orbital densities, electrostatic potentials, and spin density. It supports both the visualization of three-dimensional surfaces and colorized or contoured two-dimensional planes. There is a lot of control over colors, rendering quality, and the like. The final image can be printed or saved in several file formats.

Price category: contact
Platforms: Cray, SGI, RS/6000
Contact information: Oxford Molecular Group, Inc.
2105 South Bascom Ave., Suite 200
Campbell, CA 95008
(800) 876-9994
http://www.oxmol.com/
products@oxmol.com

A.2 *AB INITIO* AND DFT SOFTWARE

Some of these software packages also have semiempirical or molecular mechanics functionality. However, the primary strength of each is *ab initio* calculation. There are also *ab initio* programs bundled with the Unichem, Spartan, and Hyperchem products discussed previously in this appendix.

A.2.1 ADF

ADF (we tested Version 1999.02) stands for Amsterdam density functional. This is a DFT program with several notable features, including the use of a STO basis set and the ability to perform relativistic DFT calculations. Both LDA and gradient-corrected functionals are supported. The program can use an ASCII input file, or be run from the Cerius^2 graphic interface sold by Molecular Simulations, Inc. There is also a separate program for band structure calculations. The documentation is well written.

The ASCII input format is a bit more complex than that of some other popular *ab initio* programs, but still usable. Atom calculations must be run before running the molecular calculation to define the reference for the molecular properties that are related to a sum of fragment energies. The geometry specification can include trigonometric functions defining the relationship between various geometric coordinates. The input for more sophisticated calculations, such as the counterpoise correction of basis set superposition error, can be somewhat lengthy. The output gives a detailed description of the results.

ADF uses a STO basis set along with STO fit functions to improve the efficiency of calculating multicenter integrals. It uses a fragment orbital approach. This is, in essence, a set of localized orbitals that have been symmetry-adapted. This approach is designed to make it possible to analyze molecular properties in terms of functional groups. Frozen core calculations can also be performed.

A number of types of calculations can be performed. These include optimization of geometry, transition structure optimization, frequency calculation, and IRC calculation. It is also possible to compute electronic excited states using the TDDFT method. Solvation effects can be included using the COSMO method. Electric fields and point charges may be included in the calculation. Relativistic density functional calculations can be run using the ZORA method or the Pauli Hamiltonian. The program authors recommend using the ZORA method.

A number of molecular properties can be computed. These include ESR and NMR simulations. Hyperpolarizabilities and Raman intensities are computed using the TDDFT method. The population analysis algorithm breaks down the wave function by molecular fragments. IR intensities can be computed along with frequency calculations.

The band-structure code, called "BAND," also uses STO basis sets with STO fit functions or numerical atomic orbitals. Periodicity can be included in one, two, or three dimensions. No geometry optimization is available for band-structure calculations. The wave function can be decomposed into Mulliken, DOS, PDOS, and COOP plots. Form factors and charge analysis may also be generated.

Price category: student and up

Platforms: Cray, SGI, PC (Pentium Pro or newer with Linux), DEC, Fujitsu, RS/6000, NEC, HP

Contact information: Scientific Computing & Modelling NV
Vrije Universiteit, Theoretical Chemistry
De Boelelaan 1083
1081 HV Amsterdam

The Netherlands
+31-20-44 47626
http://www.scm.com/
admin@scm.com

A.2.2 Crystal

Crystal (we tested Crystal 98 1.0) is a program for *ab initio* molecular and band-structure calculations. Band-structure calculations can be done for systems that are periodic in one, two, or three dimensions. A separate script, called LoptCG, is available to perform optimizations of geometry or basis sets.

Both HF and DFT calculations can be performed. Supported DFT functionals include LDA, gradient-corrected, and hybrid functionals. Spin-restricted, unrestricted, and restricted open-shell calculations can be performed. The basis functions used by Crystal are Bloch functions formed from GTO atomic basis functions. Both all-electron and core potential basis sets can be used.

The program uses two ASCII input files for the SCF and properties stages of the calculation. There is a text output file as well as a number of binary or ASCII data files that can be created. The geometry is entered in fractional coordinates for periodic dimensions and Cartesian coordinates for nonperiodic dimensions. The user must specify the symmetry of the system. The input geometry must be oriented according to the symmetry axes and only the symmetry-unique atoms are listed. Some aspects of the input are cumbersome, such as the basis set specification. However, the input format is documented in detail.

Crystal can compute a number of properties, such as Mulliken population analysis, electron density, multipoles, X-ray structure factors, electrostatic potential, band structures, Fermi contact densities, hyperfine tensors, DOS, electron momentum distribution, and Compton profiles.

Although Crystal is nongraphic, there are a number of programs available for graphic input creation and output visualization. There is a module that allows the use of the Cerius^2 interface from MSI for setting up input files and viewing the output. The molecular structure can be output in a format readable by the MOLDRAW program, which is a PC program for the display of periodic structures. The Crgra98 program is used to make postscript files of band structures and contour maps. XCrySDen is an X-window program for generating input and viewing properties.

Price category: production, departmental, institutional
Platforms: PC (Linux, Windows-98, Windows-NT), UNIX
Contact information: Theoretical Chemistry Group
Dipartimento di Chimica IFM
Via Giuria 5-I-10125 Torino, Italy
crystal@ch.unito.it
http://www.ch.unito.it/ifm/teorica/crystal.html

or

CCLRC Daresbury Laboratories
Daresbury
Warrington WA4 4AD, United Kingdom
crystal@dl.ac.uk
http://www.dl.ac.uk/TCSC/Software/CRYSTAL

A.2.3 GAMESS

GAMESS stands for general atomic and molecular electronic structure system (we reviewed a version dated Dec. 2, 1998). It is an *ab initio* and semiempirical program, and has seen the most widespread use for *ab initio* calculations. The ASCII input file format is usable but somewhat more lengthy than some other programs. The fact that GAMESS is a free, high-quality software makes it a favorite of many academic researchers.

The *ab initio* methods available are RHF, UHF, ROHF, GVB, MCSCF along with MP2 and CI corrections to those wave functions. The MNDO, AM1, and PM3 semiempirical Hamiltonians are also available. Several methods for creating localized orbitals are available.

GAMESS can compute transition structures, reaction coordinates, vibrational frequencies, and intensities. There is also a scheme for decomposing vibrational modes into separate atomic motions. The program has a number of options for computing optical and nonlinear optical properties. It can also include an approximate spin-orbit coupling correction. The program can compute the classical trajectory on an *ab initio* potential energy surface. Several solvation models are included. Distributed multipole analysis, Morokuma energy decomposition, and population analysis can also be obtained.

The macmolplt graphics package is designed for displaying the output of GAMESS calculations. It can display molecular structures, including an animation of reaction-path trajectories. It also may be used to visualize properties, such as the electron density, orbitals, and electrostatic potential in two or three dimensions.

GAMESS is designed to have robust algorithms and give the user a fairly detailed level of control over those routines. This makes it better than many other codes at modeling technically difficult systems, such as transition metals and electronic excited states.

GAMESS has been parallelized for use on multiprocessor computers and collections of networked workstations.

Price category: free
Platforms: PC, Macintosh, Linux, Unix, VMS
Contact information: Mike Schmidt
(515) 294-9796
http://www.msg.ameslab.gov/GAMESS/GAMESS.html
mike@si.fi.ameslab.gov

A.2.4 Gaussian

Gaussian (we tested G98 rev A.6) is a monolithic *ab initio* program. Gaussian probably incorporates the widest range of functionality of any *ab initio* code. It does include a few semiempirical and molecular mechanics methods that can be used alone or as part of QM/MM calculations. It uses one of the simplest ASCII input file formats. There are also many graphic interfaces available for creating Gaussian files and viewing results, such as GaussView from Gaussian Inc., Cerius^2 from Molecular Simulations Inc., UniChem from Oxford Molecular, the AMPAC GUI from Semichem, Chem3D from CambridgeSoft, Viewmol, and Spartan from Wavefunction, Inc. The documentation is fairly well written, but many users purchase additional books available from Gaussian Inc. in order to have a source of information on using this complex program.

Gaussian contains a wide range of *ab initio* functionality, such as HF, ROHF, MPn, CI, CC, QCI, MCSCF, CBS, and G2. A number of basis sets and pseudopotentials are available. It also supports a large number of DFT functionals. Semiempirical methods available include AM1, PM3, and ZINDO (single point only). Molecular mechanics methods are Amber, Dreiding, and UFF. There are a wide range of molecular properties that can be computed, such as NMR chemical shifts, nonlinear optical properties, several population analysis schemes, vibrational frequencies, and intensities and data for use in visualization programs. QM/MM calculations can be performed using the ONIOM method. Transition structures and intrinsic reaction coordinates may also be computed. It is additionally possible to manually specify which sections of code are to be called and in what order.

Gaussian has one of the ASCII input formats most convenient to use without a graphic interface. Even though graphic interfaces are avaliable, many researchers still construct input files manually due to the amount of control this gives them over the choice of computation method and molecular geometry constraints. There are a large number of options for controlling how the algorithms are executed. There are also a variety of options that allow the user to make efficient use of the hardware configuration, such as in core, direct, and semidirect integral evaluation. In addition, Gaussian can take advantage of parallel architectures.

The price for all this functionality is that the user must invest time in learning how to get the program to run to completion, to compute the desired properties, and to work with the available hardware configuration. Even users who utilize only a limited subset of this functionality can expect some amount of trial and error, and the need to read manuals and ask more experienced users for help. The program output contains a large amount of information, much of which may not be used by the average user. The error messages can also be cryptic. Due to the complexity of the code, there are frequent revisions of the program released to fix minor problems. In the revision we tested, the ONIOM method failed for some levels of theory and chemical systems.

Gaussian has seen the widest use in modeling organic molecules. However, there are also options for handling many of the difficulties that can be encountered in modeling inorganic systems. Nonetheless, inorganic modeling generally requires additional technical sophistication on the part of the user.

Gaussian is designed to execute as a batch job. It can readily be used with common batch-queueing systems. The program may be purchased as source code or executables and comes with hundreds of sample input and output files. These may be employed as examples of how to construct inputs. They may also be employed to verify that a compilation from source code was successful. In our experience, such verification is essential.

Price category: production, departmental, institutional

Platforms: PC (Windows and Linux), DEC, Cray, Fujitsu, HP-UX, RS/6000, NEC, SGI, Sun

Contact information: Gaussian, Inc.
Carnegie Office Park, Bldg. 6
Suite 230
Pittsburgh, PA 15106
(412) 279-6700
http://www.gaussian.com/
info@gaussian.com

A.2.5 Jaguar

Jaguar (we tested Version 3.5) is an *ab initio* program designed to efficiently run calculations on large molecules. This is achieved through the developers' choice of algorithms and optimization strategy. Jaguar can use a pseudospectral integration scheme, which gives time complexities of N^3 or better for HF, GVB, DFT, and MP2 calculations. Performance increases are also obtained from the ability to use non-Abelian symmetry groups. Jaguar was formerly called PS-GVB.

The HF, GVB, local MP2, and DFT methods are available, as well as local, gradient-corrected, and hybrid density functionals. The GVB-RCI (restricted configuration interaction) method is available to give correlation and correct bond dissociation with a minimum amount of CPU time. There is also a GVB-DFT calculation available, which is a GVB-SCF calculation with a post-SCF DFT calculation. In addition, GVB-MP2 calculations are possible. Geometry optimizations can be performed with constraints. Both quasi-Newton and QST transition structure finding algorithms are available, as well as the SCRF solvation method.

The properties available include electrostatic charges, multipoles, polarizabilities, hyperpolarizabilities, and several population analysis schemes. Frequency correction factors can be applied automatically to computed vibrational frequencies. IR intensities may be computed along with frequency calculations.

Jaguar comes with a graphic user interface, but it is not a molecule builder. The interface can be used to set the program options. The user must input the geometry by typing in Cartesian coordinates or a Z-matrix. The interface may then be employed to import geometry information from a large list of file formats. Both geometry and calculation setup information can be imported from GAMESS, Gaussian, and Spartan files. Jaguar may also symmetrize a molecule if the coordinates do not exactly match a given point group. The online help is fairly detailed. When the calculation is executed, a separate window displays messages showing which step of the calculation is being executed.

At the time of this review, a new graphic user interface was under development. Jaguar can also be purchased as part of the Titan program, which combines Jaguar with the Spartan graphic interface. An orbital viewer for Jaguar is available from Serena Software.

Alternatively, the user can construct ASCII input files manually. The file format includes many numerical flags to control the type of calculation. The researcher should plan on investing some time in learning to use the program in this way. Jaguar can be executed from the command line, making it possible to use batch processing or job queue systems.

For many researchers, Jaguar is the code of choice for running GVB or MP2 calculations on large molecules. For DFT calculations, there are two algorithms designed for efficiency in modeling large molecules. One is the pseudospectral method in Jaguar and the other is the fast multipole method, which has been incorporated in the Gaussian 98 and Q-Chem packages. Our reviewer ran identical calculations on several large molecules with all three packages. For some molecules, all three packages used nearly exactly the same amount of CPU time. For one test, Jaguar was 43% faster than the others. And for one test, Jaguar was 49% slower. Our reviewer was not able to propose any specific criteria for predicting which molecules would run faster or slower with each package. One published study shows Jaguar giving as much as a 25-fold speed advantage over Gaussian 92 [R. A. Friesner, R. B. Murphy, M. D. Beachy, M. N. Ringnalda, W. T. Pollard, B. D. Dunietz, and Y. Cao (1999). *J. Phys. Chem. A*103, 1913.].

Price category: production and higher

Platforms: SGI, RS/6000, HP-UX, Cray, Alpha, PC (Linux only)

Contact information: Schrödinger, Inc.

17 Sheffield Drive

West Grove, PA 19390

(800) 207-7482

http://www.schrodinger.com/

help@schrodinger.com

A.2.6 MOLPRO

MOLPRO (we tested Version 98.1) is an *ab initio* program designed for performing complex calculations. This program is often used for calculations that present technical difficulties or are very sensitive to electron correlation. A few portions of the code have been parallelized.

Calculations that can be performed are HF, CI, MRCI, FCI, CC, DFT, MCSCF, CASSCF, ACPF, CEPA, valence bond, and many variations of these. Perturbation theory calculations can be done from single- and multiple-determinant references spaces. The MCSCF and coupled-cluster algorithms have proven to be very efficient. Restricted, unrestricted, and restricted open-shell wave functions are available. The user has a large amount of detailed control over wave function construction. Many one-electron properties can be computed, including relativistic energy corrections, spin-orbit coupling, electric field gradients, and multipoles. A number of options for electronic excited states and transition-structure calculations are also available. It can use Gaussian basis sets with high-angular-momentum functions (spdfghi) and effective core potentials.

The ASCII input file includes elements of a scripting language. Thus, the input can contain variables, loops, and procedures. This is one of the aspects of the program that makes it possible to do very complex calculations. The documentation describes the input options, but does not discuss when and why they should be used. The user must have a solid understanding of *ab initio* theory in order to correctly utilize many of the functions in this program. It is very powerful, but not for beginners.

The program uses dynamic memory allocation within a memory limit that must be set manually if the default is insufficient. The program does store data in scratch files, but the size of these files has been kept to a minimum. The output is neatly formatted, but designed for wide carriage printers.

This program is excellent for high-accuracy and sophisticated *ab initio* calculations. It is ideal for technically difficult problems, such as electronic excited states, open-shell systems, transition metals, and relativistic corrections. It is a good program if the user is willing to learn to use the more sophisticated *ab initio* techniques.

Price category: production, departmental, institutional

Platforms: Linux, Alpha, Cray, Fujitsu, AIX, SGI, Sun, HP-UX, NEC

Contact information: P. J. Knowles

School of Chemistry
University of Birmingham
Edgbaston, Birmingham, B15 2TT
United Kingdom
+44-121-414-7472
http://www.tc.bham.ac.uk/molpro/
molpro-request@tc.bham.ac.uk

A.2.7 Q-Chem

Q-Chem (we tested Version 1.2) is an *ab intio* program designed for efficient calculations on large molecules. Q-Chem uses ASCII input and output files.

The HyperChem program from Hypercube Inc. and UniChem from Oxford Molecular can be used as graphic interfaces to Q-Chem. At the time we conducted our tests, it was not yet available on all the platforms listed as being supported. The current version is well designed for ground- and excited-state calculations on small or large organic molecules.

Q-Chem includes HF, ROHF, UHF, and MP2 Hamiltonians as well as a good selection of DFT functionals. Mulliken and NBO population analysis methods are available. Multiple options are available for SCF convergence, geometry optimization, and initial guess. IR and Raman intensities can also be computed. In addition, the documentation was well written.

One of the major selling points of Q-Chem is its use of a continuous fast multipole method (CFMM) for linear scaling DFT calculations. Our tests comparing Gaussian FMM and Q-Chem CFMM indicated some calculations where Gaussian used less CPU time by as much as 6% and other cases where Q-Chem ran faster by as much as 43%. Q-Chem also required more memory to run. Both direct and semidirect integral evaluation routines are available in Q-Chem.

Gaussian users will find that Q-Chem feels familiar. The ASCII input format is a bit more wordy than Gaussian; it is more similar to GAMESS input. The output is very similar to Gaussian output, but a bit cleaner. The code can easily be used with a job-queueing system.

Q-Chem also has a number of methods for electronic excited-state calculations, such as CIS, RPA, XCIS, and CIS(D). It also includes attachment–detachment analysis of excited-state wave functions. The program was robust for both single point and geometry optimized excited-state calculations that we tried.

Price category: production, departmental, institutional

Platforms: PC (Windows and Linux), DEC, Cray, Fujitsu, RS/6000, SP2, SGI, Sun

Contact information: Q-Chem, Inc.
Four Triangle Drive
Suite 160
Export, PA 15632-9255
(724) 325-9969
http://www.q-chem.com/
sales@q-chem.com

A.3 SEMIEMPIRICAL SOFTWARE

Three popular semiempirical programs, AMPAC, AMSOL, and MOPAC, are actually derivations of the same original code. AMPAC 1.0 and MOPAC 3.0 were both created from Version 2.0 of MOPAC. AMSOL was derived from

Version 2.1 of AMPAC. All three of these programs have similar input and output files, but have been developed for different purposes. New additions to AMSOL have been almost exclusively methods for including solvent effects. AMPAC was designed for efficient operation on vector computers and for finding and testing transition structures. AMPAC is also the only one incorporating the SAM1 method. MOPAC is designed to be robust and to compute a large number of molecular properties, including algorithms for large-molecule calculations.

There are also semiempirical programs bundled with the Unichem, Spartan, and Hyperchem products discussed previously in this appendix.

A.3.1 AMPAC

AMPAC (we tested Version 6.51) is a semiempirical program. It comes with a graphic user interface (we tested Version 6.0). The documentation included with the package is well written.

The graphic interface has a molecule builder that is very easy to use. This is the same GUI as the GaussView interface from Gaussian Inc., that was licensed from Semichem. The one sold with AMPAC has screens for setting up and running AMPAC calculations. It can also be used to set up and analyze Gaussian 94 calculations, but the interface is not identical to the one in GaussView 1.0. The GUI is integrated well with the computational portion. See the GaussView description for more information on the GUI.

AMPAC can also be run from a shell or queue system using an ASCII input file. The input file format is easy to use. It consists of a molecular structure defined either with Cartesian coordinates or a Z-matrix and keywords for the type of calculation. The program has a very versatile set of options for including molecular geometry and symmetry constraints.

AMPAC supports a number of semiempirical methods: AM1, SAM1, SAM1/d, MNDO, MNDO/d, MNDOC, MINDO/3, and PM3. The solvation methods available are SM1-SM3 and COSMO. Types of calculation available include single-point energies, geometry optimization, frequency calculation, IRC, and a reaction path and an annealing algorithm. It incorporates some transition structure finding algorithms that are not in other semiempirical programs, such as the CHAIN and TRUST algorithms. A simulated annealing algorithm is available for conformation searching. The code incorporates many alternative algorithms and settings to control how the calculation is performed. Property prediction functions include ESR and nonlinear optical properties.

Price category: production, departmental, institutional

Platforms: PC (Windows & Linux), SGI, RS6000, Alpha, Sun, HP-UX

Contact information: Semichem

P.O. Box 1649

Shawnee Mission, KS 66222

(913) 268-3271
http://www.semichem.com/
sales@semichem.com

A.3.2 MOPAC

MOPAC (we tested Versions 6 and 2000) stands for molecular orbital package. It is one of the most widely used semiempirical software packages and is designed for robustness and a wide range of functionality. Programs that can be used as a graphic interface for MOPAC are WinMOPAC from Fujitsu, Alchemy from SciVision, PC Model from Serena Software, Chem3D from CambridgeSoft, and HyperChem from HyperCube Inc.

The earlier versions of MOPAC were available at little or no cost. Many graphic interface programs are shipped with MOPAC, usually Version 6, which was the last free version. MOPAC 7 is a free beta version of the MOPAC 93 commercial code. MOPAC 6 is often preferred over Version 7 because of known bugs in Version 7. Older versions of MOPAC have also been incorporated into the Gaussian, Cache, and GAMESS programs. Since free versions of MOPAC have been bundled and sometimes modified by commercial companies, some variations exist. These are most often differences in the size of molecules that can be modeled. Slight differences in performance and input format can also be found. The newest versions of MOPAC were developed by Fujitsu and marketed in North America by Schrödinger, Inc.

MOPAC runs in batch mode using an ASCII input file. The input file format is easy to use. It consists of a molecular structure defined either with Cartesian coordinates or a Z-matrix and keywords for the type of calculation. The program has a very versatile set of options for including molecular geometry and symmetry constraints. Version 6 and older have limits on the size of molecule that can be computed due to the use of fixed array sizes, which can be changed by recompiling the source code. This input format allows MOPAC to be run in conjunction with a batch job-queueing system.

The MNDO, MINDO/3, AM1, and PM3 Hamiltonians are supported. Semiempirical calculations can be performed on high-spin systems and excited states using configuration interaction. Transition structures and intrinsic reaction coordinates can be computed, as well as vibrational modes including the transition dipole. The program includes the ability to perform a computation with periodic boundary conditions in one, two, and three dimensions for modeling polymers, layer systems, and solids. Hyperfine coupling constants and static and frequency-dependent hyperpolarizabilities can also be computed. In addition, a set of population analysis functions is available. Classical dynamics simulations on the semiempirical potential energy surface can also be performed.

MOPAC 2000 is the most recent commercial version of MOPAC. It includes improvements in a number of areas. The MNDO/d and AM1-d Hamiltonians are also now available. The program uses dynamic memory allocation. This results in calculations requiring less memory for small molecules and at the same time accommodates exteremely large molecules. It also has improved algorithms for computations on very large biomolecules, which include faster calculations and determination of the net charge. The program additionally includes algorithms for modeling excited states in solution using the COSMO and Tomasi methods. It is possible to find the geometry at which intersystem crossing is most likely to occur. The manual has been expanded to include a section addressing the accuracy of semiempirical calculations.

A semiempirical crystal band structure program, called BZ, is bundled with MOPAC 2000. There is also a utility, referred to as MAKPOL, for generating the input for band structure calculations with BZ. With the use of MAKPOL, the input for band-structure computations is only slightly more complicated than that for molecular calculations.

Price category: free (older versions), production, departmental, bundled with other programs

Platforms: PC, many UNIX systems

Contact information: Schrödinger, Inc.
17 Sheffield Drive
West Grove, PA 19390
(800) 207-7482
http://www.schrodinger.com/
help@schrodinger.com

A.3.3 YAeHMOP

YAeHMOP stands for yet another extended Hückel molecular orbital package. The package has two main executables and a number of associated utilities. The "bind" program does molecular and crystal band structure extended Hückel calculations. The "viewkel" program is used for displaying results. We tested Version 3.0 of bind and Version 2.0 of viewkel.

The bind program requires an ASCII input and generates one or more ASCII output files. The molecular geometry can be defined as a Z-matrix, Cartesian coordinates, or fractional coordinates. Periodic boundary condition calculations may be done in one, two, or three dimensions. The program has built-in parameters to describe most elements, although it is also possible to enter parameters manually. There is an automated function to generate the data for Walsh diagrams. Orbital occupations can be set manually in order to give electronic excited-state calculations. The program does not have an automated function for optimizing the molecular or unit cell geometry. The user must enter a list of *k* points and their weights in order to perform average property calculations. Information that can be computed includes the band structure, DOS, COOP, fragment molecular and crystal orbital analysis, MOs, Fermi energy, and Mulliken population analysis.

The viewkel program allows the graphic display of results. It is not a graphic molecular structure builder and does not require any graphic display libraries other than those included in X-windows. It uses line-drawn graphics and can shade atoms by element. The program can also be used to display molecular and crystal structures, orbital isosurfaces, Walsh diagrams, and various plots of the band structure, DOS, and so forth. The controls are not too difficult to use, although it will probably be necessary to read the manual. The display can be saved as a postscript file. Viewkel is useful for generating illustrations for publication without color, but it does not have the quality of three-dimensional shaded rendering that is now common in commercial applications. Viewkel can also be used to display some results from the ADF program. At the time this book was written, Version 2.0 of viewkel was known to have some bugs, but Version 3.0 was not yet available.

The documentation is clearly written and generally adequate, although some of the less frequently used functions and utilities were not documented. The user should have a basic understanding of band structure theory before attempting to read the documentation.

Price category: free

Platforms: PC (Linux), Macintosh, RS/6000, HP-UX, SGI

Contact information: **http://overlap.chem.cornell.edu:8080/yaehmop.html** and **yaehmop@xtended.chem.cornell.edu**

A.4 MOLECULAR MECHANICS/MOLECULAR DYNAMICS/MONTE CARLO SOFTWARE

The following are programs created specifically for force field based simulations. There are also molecular mechanics programs bundled with the Spartan, Gaussian, and Hyperchem products discussed previously in this appendix.

A.4.1 MacroModel

MacroModel (we tested Version 6.5) is a powerful molecular mechanics program. The program can be run from either its graphic interface or an ASCII command file. The command file structure allows very complex simulations to be performed. The XCluster utility permits the analysis and filtering of a large number of structures, such as Monte Carlo or dynamics trajectories. The documentation is very thorough.

The force fields available are MM2*, MM3*, AMBER*, OPLSA*, AMBER94, and MMFF. The asterisk (*) indicates force fields that use a modification of the original description in the literature. There is support for user-defined metal atoms, but not many metals are predefined. MM2* has atom types for describing transition structures. The user can designate a substructure for energy computation.

MacroModel can be used to run molecular mechanics minimizations, Monte Carlo simulations, and molecular dynamics simulations. The GB/SA solvation model is available. Several conformation-searching options are also available, including the low-mode conformation-search algorithm that is useful for ring systems. A free energy perturbation method exits for computing the ΔG between two systems when there is only a small difference between them. The user must manually construct an ASCII command file in order to run free energy calculations. Alternatively, the MINTA algorithm can be used for the direct calculation of conformational and binding free energies. MINTA is not limited by the constraints of perturbation theory. Some tasks can be distributed over a cluster of workstations.

Our reviewer felt the molecule builder was easy to use. It is set up for organic molecules. Specialized building modes are available for peptides, nucleotides, and carbohydrates. It is also possible to impose constraints on the molecular geometry. Functions are accessed via a separate window with buttons labeled with abbreviated names. This layout is convenient to use, but not completely self-explanatory. The program is capable of good-quality rendering. At the time of this book's publication, a new three-dimensional graphic user interface called Maestro was under development.

Price category: production and higher
Platforms: SGI, RS/6000
Contact information: Schrödinger, Inc.
17 Sheffield Drive
West Grove, PA 19390
(800) 207-7482
http://www.schrodinger.com/
help@schrodinger.com

A.4.2 MOE

MOE (Version 2000.02) stands for molecular operating environment. The developers of this package took the unique approach of creating a programming language for writing molecular modeling software. The package currently includes molecular mechanics, dynamics, periodic boundary conditions, QSAR, a combinatorial builder, and many functions ideal for protein modeling, including multiple-sequence alignment and homology model building. Functions are also available for computing polymer properties and diffraction patterns. Several conformation search routines are included. Other property calculations include log P and molar refractivity calculation.

The graphic interface is a multitasking environment that works well. The protein and carbohydrate builders are particularly convenient to use. The small-molecule builder has a selection of common organic functional groups as well as individual atoms for organics and common heteroatoms. There are a number of rendering modes to create fairly nice graphic images. However, the program does not have a way to save the display as an image file. MOE can be accessed through a Web browser also.

The force fields available are AMBER '89, AMBER '94, MMFF94, and PEF95SAC. The user can control scale factors, cutoff distances, and which terms are included in the force field. It can also include customized parameters. The program would automatically generate force field parameters if literature values were not available. This is convenient, but the user must be cautions about employing these parameters. No warning messages are generated when this happens, so the user must check the missing parameter report screen.

The QSAR utilities include functions for molecular diversity, similarity, and clustering. Both two- and three-dimensional pharmacophore fingerprinting techniques are available. It also includes the binary QSAR method designed for high-throughput screening. The "hole filler" function generates a new database that spans the diversity of multiple existing databases. Correlation and contingency analysis are available to determine which descriptors should be used. Over 130 descriptors are available, which include geometry, surface areas, topological indices, energy contributions, log *P*, refractivity, and charge density description. The program was not able to compute descriptors requiring semiempirical or *ab initio* computations.

QSAR, high-throughput, and combinatorial studies are aided by the program's ability to work with a large database of molecules. Substructure searches and flexible alignment checks can be run on the database. Individual entries in the database can be very large, making it possible to store detailed fingerprinting data for large molecules.

The SVL programming language (scientific vector language) is a byte-code interpretive language that can be run interactively or from an ASCII file. The language syntax is a combination of C++ and scripting language conventions. The language includes vector arithmetic, molecular structure data types, a small amount of symbolic manipulation, pattern matching, and graphic display commands. This interpretive mode works well for molecular mechanics, but it would probably be too sluggish for *ab initio* calculations unless a native compiler is included in future versions. The SVL command language and the capacity to run MOE in batch mode (without the graphic interface) provide for the ability to automate tasks. All the functions accessed from menus in the GUI are running SVL program files. This open architecture makes it easy for others to add their own functions or modify the existing routines. The platform independence; access via GUI, batch, and Web; and flexible customizable architecture make MOE ideal for corporate deployment accessible to all researchers.

The documentation is well done. It includes function references, tutorials, and a how to section.

Price category: contact
Platforms: PC (Windows and Linux), SGI, Sun, HP-UX

Contact information: Chemical Computing Group Inc.
1010 Sherbrooke Street, Suite 910
Montreal, Quebec, Canada H3A 2R7
(514) 393-1055
http://www.chemcomp.com/
info@chemcomp.com

A.4.3 PC Model

PC Model (we tested Version 7.00) is a molecular mechanics program with a graphic user interface. It can also be used as a graphic interface for AMPAC, MOPAC, and Gaussian. Separately sold modules are available for displaying orbitals and vibrational modes for these programs as well as Hondo and Jaguar. It is able to read and write a variety of chemical-structure data file formats. The native file format can hold many structures that can be used for either batch processing or displaying an animation.

The available force fields are MMX, MM3, and MMFF94. Some of the supported calculations are geometry optimization, molecular dynamics, and a simulated annealing docking algorithm. There is also an automated algorithm for computing rotational energy barriers. A π system semiempirical calculation can be incorporated for modelling aromatic molecules.

PC Model has some features that are not found in many other molecular mechanics programs. This is one of the few programs that outputs the energy given by the force field and the heat of formation and a strain energy. Atom types for describing transition structures in the MMX force field are included. There is a metal coordination option for setting up calculations with metal atoms. There are also molecular similarity and conformation search functions.

The graphic interface can be used to build structures in either a two-dimensional drawing mode or a three-dimensional building mode. There are separate builders for amino acids, sugars, and nucleic acids. The amino acid and sugar builders would connect the units to form a chain. The nucleic acid builder placed the nucleotides on the screen, but did not connect them into a strand. There are also libraries of predefined ring groups, transition-state geometries, organic functional groups, and organometallic groups. The reviewer felt that the builder was reasonably easy to use.

Molecules can be rendered as stick figures, ball and stick, CPK, and ribbons. Dot surfaces can also be included. Some regions were incorrectly shaded for small molecules on our test system running at 800 × 600 resolution with 24-bit color. The display uses a black background, but graphics are saved or printed with a white background. Overall, the rendering is adequate.

The user manual includes both a function reference and some short tutorials. There is also a tutorial in the following reference: M. F. Schlecht (1998). *Molecular Modeling on the PC.* New York: John Wiley & Sons.

Price category: production
Platforms: PC (Windows and Linux), SGI
Contact information: Serena Software
Box 3076
Bloomington, IN 47402-3076
(812) 333-0823
http://www.serenasoft.com/
gilbert@serenasoft.com

A.4.4 TINKER

TINKER (we tested Version 3.7) is a collection of programs for performing molecular mechanics and dynamics calculations. All the executables read the same ASCII input files. TINKER does not come with a graphic interface, although it can be used with RasMol, ChemDraw, Chem3D, gOpenMol, MOLDEN, and ReView. Protein database files can be converted to TINKER files, making it possible to import molecular structures from many sources. Output can be generated in a format displayed with Sybyl, Insight II, or Xmol. Source code is available.

The force fields currently available are AMBER-95, CHARMM22, MM2(91), MM3(99), OPLS-AA, OPLS-UA, and TINKER. TINKER can be used for geometry optimization, molecular dynamics, simulated annealing, normal vibrational mode analysis, conformation searching, distance-geometry, free energy perturbation, and finding conformational transition states. Fractional coordinates may then be used for importing molecular structures and hence the parameterization of force field terms. Molecular structures can also be generated from a peptide sequence. In addition, molecular properties, such as energy analysis and molecular volumes and surface areas, can be computed.

The molecular structure input requires atom types to be assigned, which are not the same from one force field to the next. The input also includes a list of bonds in the molecule. There is not a module to automatically assign atom types. Most of the modules use a Cartesian coordinate molecular structure, except for a few that work with torsional space. The same keyword file is read by all the executables. A little bit of input is obtained by the program either interactively or from an ASCII file piped to standard input, which makes for a somewhat cryptic input file. This system of common input files and the user choosing which executables to run give TINKER the ability to run very sophisticated simulations while keeping the input required for simple calculations fairly minimal within the limitations mentioned here.

The TINKER documentation provides a description of the input, but not a tutorial. Documentation is available as html, Acrobat, or postscript. A set of example input files is provided. The researcher can expect to invest some time in learning to use this system of programs. Most of the executables seem to be fairly robust and as tolerant as possible of variations in the input format. When possible, default values are assumed to minimize the amount of input information that must be provided. TINKER is well suited for researchers wishing to call it from their own front-end programs.

Price category: free

Platforms: PC (Windows and Linux), SGI, Macintosh, RS/6000, Alpha, HP-UX, Sun

Contact information: Jay W. Ponder
Biochemistry, Box 8231
Washington University Medical School
660 South Euclid Ave.
St. Lewis, MO 63110
(314) 362-4195
http://dasher.wustl.edu/tinker/
ponder@dasher.wustl.edu

A.5 GRAPHICS PACKAGES

The programs discussed in this section are designed not for running calculations, but as graphic interfaces for constructing input files and/or viewing results. Some more general-purpose programs were omitted in favor of software designed specifically for chemistry. Many other chemistry visualizaton programs were omitted because they had narrow applicability or are less widely used. These capabilities are also integrated with computational programs in many of the packages discussed previously in this appendix.

A.5.1 GaussView

GaussView (we used Version 2.08) is a graphic interface for use with the Gaussian *ab initio* program. It can be used to build molecules, set up the options in the input file, run a calculation, and display results. GaussView uses the molecule builder that was written by SemiChem, but has screens for setting up calculations that are different from those in the AMPAC GUI sold by SemiChem.

The program has several building modes. Compounds can be built one atom at a time by selecting the element and hybridization. There are also libraries of ring systems, amino acids, nucleosides, and common organic functional groups. The user can manually set bond lengths, angles, and dihedral angles. There is a clean function that gives an initial optimization of the structure using a rule-based VSEPR algorithm. There is a Z-matrix editor that gives some control over how the Z-matrix is constructed, but does not go as far as giving the user the ability to enforce symmetry constraints. GaussView can also be used to set up ONIOM QM/MM calculations. Our reviewer felt that the graphic molecule-building functions were very easy to use.

There is a screen to set up the calculation that has menus for the most widely used functions. Many users will still need to know many of the keywords, which can be typed in. There was no default comment statement, so the input file created would not be valid if the user forgot to include a comment. A calculation can be started from the graphic interface, which will be run interactively by default. The script that launches the calculation was not too difficult to modify for use with a job-queueing system.

The molecular structures were rendered with good-quality shading on a blue background. Isosurfaces produced from cube files or checkpoint files also looked nice. Molecular vibrations can be animated on screen and vibrational displacement vectors displayed. The vibrational line spectrum may be displayed too, but the user has no control over the axes. There is no way to set the background color. The display can be saved using several image file formats.

Price category: production, departmental
Platforms: RS/6000, SGI, Alpha, PC (Windows)
Contact information: Gaussian, Inc.
Carnegie Office Park, Bldg. 6
Suite 230
Pittsburgh, PA 15106
(412) 279-6700
http://www.gaussian.com/
info@gaussian.com

A.5.2 Molden

Molden (we tested Version 3.6) is a molecular display program. It can display molecular geometries read from a number of molecular file formats. Various views of the wave function can be displayed from the output of the Gaussian and GAMESS programs. Some functionality is available from MOPAC and AMPAC files. Conversion programs are available to import wave functions from ADF, MOLPRO, ACES II, MOLCAS, DALTON, Jaguar, and HONDO.

Molden can display molecular geometries in a variety of formats, including lines, tubes, ball and stick, ribbons, and CPK. The user has some control over colors and sizes. Molden also has features designed for the display of proteins and crystal structures. The display can be exported as postscript, VRML, Povray, and image files. It can also be configured as a chemical mime viewer.

The program has a Z-matrix editor, which is not the same as a graphic molecule builder. This allows the user to display the Z-matrix and then define which parameters are fixed or equivalent to other parameters. Molden can be configured to optimize the geometry by passing data to the TINKER program. Molecular structures can be exported in formats compatible with many popular software programs.

Wave functions can be visualized as the total electron density, orbital densities, electrostatic potential, atomic densities, or the Laplacian of the electron density. The program computes the data from the basis functions and molecular orbital coefficients. Thus, it does not need a large amount of disk space to store data, but the computation can be time-consuming. Molden can also compute electrostatic charges from the wave function. Several visualization modes are available, including contour plots, three-dimensional isosurfaces, and data slices.

Molden now has the capacity to export a series of GIF files, creating an additional file with each screen update. Third-party utilities can be used to combine these images in animations with various file formats, including fli and animated GIF files. This is one of the easiest ways to create animations of chemical systems.

Program documentation is available as postscript, text, or Web pages. None of these seemed to give a comprehensive description of the program functionality and controls.

Price category: free, production

Platforms: PC (Windows, Linux, free BSD), SGI, RS/6000, Alpha, Cray, HP-UX, Sun, open VMS, OS2

Contact information: Gijs Schaftenaar
CMBI
University of Nijmegen
Toernooiveld 1
6525 ED NIJMEGEN, The Netherlands
+31 24 365 33 69
http://www.cmbi.kun.nl/~schaft/molden/molden.html
schaft@cmbi.kun.nl

A.5.3 WebLab Viewer

WebLab Viewer is a molecular graphics program. We tested two versions: WebLab® ViewerLite™ (we tested Version 3.2), a free molecular display program, and WebLab® ViewerPro™ (we tested Version 3.5) for molecule building and display. ViewerPro can also be used as a graphic interface to MedChem Explorer (for drug refinement), Diversity Explorer (in combinatorial chemistry), and Gene Explorer (in bioinformatics).

Both versions of the WebLab Viewer use a native file format that is capable of describing molecular or crystal structures along with surfaces and labels. They also read a dozen different common molecular file formats. Two-dimensional structures can be automatically converted to three-dimensions when the file is read. Work can be saved in common molecule file formats, VRML, SMILES, and bit-mapped graphic files.

WebLab Viewer gives a very-high-quality display suitable for publication and presentation. Molecules can be displayed as lines, sticks, ball and stick, CPK, and polyhedrons. In addition, different atoms within the same structure may be displayed in different ways. Text can be added to the display as well as labeling parts of the structure in a variety of ways. The user has control over colors, radii, and display quality. The program can also replicate a unit cell to display a crystal structure. Several types of molecular surfaces can be displayed.

WebLab ViewerPro has all the functionality of ViewerLite plus a number of additional features. ViewerPro can be used to build molecular structures. It has building modes for adding individual atoms, creating chains, and creating rings. All these create structures of carbon atoms. Once the backbone has been created, atoms can be changed to other elements and hydrogens added. This builder worked reasonably well for organic molecules, but seemed somewhat inconvenient for building inorganic structures. There is a function to clean up the shape of the molecule, which does a basic molecular mechanics minimization using a simple Dreiding-type force field. ViewerPro can be used to create animations and has a scripting language to automate tasks.

Price category: free, individual, production

Platforms: PC, Mac

Contact information: Molecular Simulations, Inc.
9685 Scranton Road
San Diego, CA 92121-3752
(888) 249-2292
http://www.msi.com/viewer

A.6 SPECIAL-PURPOSE PROGRAMS

A.6.1 Babel

Babel (we tested Version 1.6) is a utility for converting computational chemistry input files from one format to another. It is able to interconvert about 50 different file formats, including conversions between SMILES, Cartesian coordinate, and Z-matrix input. The algorithm that generates a Z-matrix from Cartesian coordinates is fairly simplistic, so the Z-matrix will correctly represent the geometry, but will not include symmetry, dummy atoms, and the like. Babel can be run with command line options or in a menu-driven mode. There have been some third-party graphic interfaces created for Babel.

Price category: free

Platforms: PC (Windows and Linux), Unix, Macintosh

Contact information: **http://www.ccl.net/pub/chemistry/software/UNIX/babel/**

A.6.2 CHEOPS

CHEOPS (we tested Version 3.0.1) is a program for predicting polymer properties. It consists of two programs: The analysis program allows the user to draw the repeat unit structure and will then compute a whole list of properties; the synthesis program allows the user to specify a class of polymers and desired properties and will then try the various permutations of the functional groups to find ones that fit the requirements. On a Pentium Pro 200 system, the analysis computations were essentially instantaneous and the synthesis computations could take up to a few minutes. There was no automated way to transfer information between the two programs.

CHEOPS is based on the method of atomic constants, which uses atom contributions and an anharmonic oscillator model. Unlike other similar programs, this allows the prediction of polymer network and copolymer properties. A list of 39 properties could be computed. These include permeability, solubility, thermodynamic, microscopic, physical and optical properties. It also predicts the temperature dependence of some of the properties. The program supports common organic functionality as well as halides, As, B, P, Pb, S, Si, and Sn. Files can be saved with individual structures or a database of structures.

The program is very easy to use. The help screens give step-by-step directions for various operations, which are complete but somewhat difficult to read because of poor English grammar. Additional information on the website is more readable. The synthesis program works well, although it is limited to seven classes of polymers.

Price category: departmental, institutional (initial purchase and annual license fee)

Platforms: PC

Contact information: MillionZillion Software

3306 Decatur Lane

Minneapolis, MN 55426

(612) 932-9048

http://www.millionzillion.com/cheops

ward@millionzillion.com

A.6.3 CODESSA

CODESSA (we tested Version 2.6) stands for comprehensive descriptors for structural and statistical analysis. It is a conventional QSAR/QSPR program.

CODESSA reads molecular structure files or output files created by other software packages as the starting point for QSAR analysis. It can import computational results from AMPAC, MOPAC, and Gaussian as well as structures in a number of common formats.

CODESSA can compute or import over 500 molecular descriptors. These can be categorized into constitutional, topological, geometric, electrostatic, quantum chemical, and thermodynamic descriptors. There are automated procedures that will omit missing or bad descriptors. Alternatively, the user can manually define any subset of structures or descriptors to be used.

The program incorporates several very automated procedures for choosing and testing possible QSAR equations. These procedures incorporate correlation and intercorrelation coefficients to find an equation with the best fit using a minimal number of descriptors. These automated procedures performed very well when creating an equation to predict normal boiling points using a test set that was constructed by our reviewer. There are both statistical and graphic tools, which also makes this package an excellent choice for experts desiring manual control over the process. The QSAR equations obtained are multilinear.

Price category: production, departmental, institutional

Platforms: PC (Windows and Linux), SGI, RS6000, Alpha, Sun, HP-UX

Contact information: Semichem

P.O. Box 1649

Shawnee Mission, KS 66222

(913) 268-3271

http://www.semichem.com/

sales@semichem.com

A.6.4 gNMR

gNMR (we tested Version 4.0.1) is a program for NMR spectral prediction and simulation. The simulation portion of the program draws the spectrum once the user has input the chemical shifts and coupling constants. gNMR can simulate spectra for any active nuclei, but can predict chemical shifts only for ^{1}H and ^{13}C. The computed spectrum can be compared to experimental data. Our review will only cover the prediction features pertinent to the discussion in Chapter 31.

gNMR can predict ^{1}H and ^{13}C chemical shifts and coupling constants for up to 23 active nuclei (increasing to 49 nuclei in Version 4.1). It uses additivity rules to predict chemical shifts. The computation time is negligible even on low-end microcomputers. The computed shifts are put in a tabular format. The user can click on atoms in the structure display in order to jump to the corresponding row of numerical data.

The program was made somewhat less convenient to use by the fact that it does not have a molecule builder. In order to predict chemical shifts, the molecular structure must be built with some other software package and then imported into gNMR. Structures can be imported in formats produced by a number of popular chemical drawing and modeling programs.

Price category: individual, production
Platforms: PC
Contact information: Cherwell Scientific Publishing
The Magdalen Centre
Oxford Science Park
Oxford, OX4 4GA UK
+44 (0)1865 784800
http://gNMR.cherwell.com/
gNMR@cherwell.com

A.6.5 MedChem Explorer

WebLab® MedChem Explorer™ (we tested Version 1.6) is a drug refinement package designed for researchers who do not specialize in computational chemistry. It works as a client–server system so that all functionality is available from a PC. WebLab® ViewerPro™ integrates with MedChem Explorer for molecule building and display. A Web-enabled system is then used to submit the calculations to a server.

The functionality available in MedChem Explorer is broken down into a list of available computational experiments, including activity prediction, align/pharmacophore, overlay molecules, conformer generation, property calculation, and database access. Within each experiment, the Web system walks the user through a series of questions that must be answered sequentially. The task is then submitted to a remote server, where it is performed. The user can view the progress of the work in their Web browser at any time. Once complete, the results of the calculation are stored on the server. The user can then run subsequent experiments starting with those results. The Web interface includes links to help pages at every step of the process.

Activity prediction is based on a list of models (i.e., QSAR models, pharmacophore models, etc.) that are maintained on the server. There is a second level of access so that only authorized users may be allowed to add or delete model entries.

The align/pharmacophore experiment orients the molecules to obtain maximum similarity in chemical features. This application can then generate a pharmacophore model consistent with all the molecules.

The molecular overlay experiment orients the molecules to find the best RMS or field fit. The field fit is based on electrostatic and steric interactions. The application can find either the best total alignment of all molecules or the best match of all molecules to a specified target molecule. Alignment can include a database search for conformers that show the best alignment based on the molecules under study.

Conformer generation is used to obtain a list of likely conformers of the molecule. This list can include a set number of the lowest-energy conformers or a number of conformers that give the most diversity of possible shapes.

The property calculation experiment offers a list of 34 molecular properties, including thermodynamic, electrostatic, graph theory, geometric properties, and Lipinski properties. These properties are useful for traditional QSAR activity prediction. Some are computed with MOPAC; others are displayed in the browser without units. A table of computed properties can be exported to a Microsoft Excel spreadsheet.

Database access is used to search external databases for molecules most similar to a specified target molecule. MedChem Explorer offers the following databases: ACD, BioByte, National Cancer Institute, Derwent Drug Index, Maybridge, MDL ISIS™, and Daylight, as well as any in-house or corporate databases that the user may have in any of the ISIS, Catalyst, or Daylight formats. The software is configured to link to the in-house informatics environment during installation. The user can search with queries based on shape, topology, substructure, or property information.

Overall, WebLab MedChem Explorer is very easy to use. The stepwise job setup works well, assuming that all users will be following a conventional drug refinement process. It is not a program that can be used for complex simulations requiring the researcher to manually control many details of the simulation.

Price category: contact
Client platforms: PC
Contact information: Molecular Simulations, Inc.
9685 Scranton Road
San Diego, CA 92121-3752
(888) 249-2292
http://www.msi.com/

A.6.6 POLYRATE

POLYRATE (we tested Version 8.0) is a program for computing chemical reaction rates. The MORATE, GAUSSRATE, and AMSOLRATE programs are derived from POLYRATE and designed to work with the MOPAC, GAUSSIAN, and AMSOL programs, respectively.

POLYRATE can be used for computing reaction rates from either the output of electronic structure calculations or using an analytic potential energy surface. If an analytic potential energy surface is used, the user must create subroutines to evaluate the potential energy and its derivatives then relink the program. POLYRATE can be used for unimolecular gas-phase reactions, bimolecular gas-phase reactions, or the reaction of a gas-phase molecule or adsorbed molecule on a solid surface.

The input to POLYRATE is a free-format ASCII file. There are a large number of input keywords, which the user must be familiar with to use all the features, but almost every keyword has a default value that is recommended by the authors. So, if the user is willing to accept the default, it is not necessary to understand all the options. The program output is a well-formatted ASCII file.

The reaction-rate calculations available include TST, CVT, ICVT, and μVT. A number of options to account for anharmonicity are available. Semiclassical corrections for tunneling and nonclassical reflection can also be included, and in fact the small-curvature and large-curvature multidimensional tunneling corrections available in this program are one of its key features. Dual-level calculations allow for additional corrections to energetics or both energetics and frequencies using levels of theory too time-consuming to apply to the entire potential energy surface.

The documentation is very thorough, although it does assume some familiarity with transition-state theory. New users can expect to spend some time with the manual, which is nearly 500 pages long! A collection of example files is also included.

Price category: free
Client platforms: Unix, Linux
Contact information: Benjamin Lynch or Donald G. Truhlar
Department of Chemistry
University of Minnesota
207 Pleasant Street SE
Minneapolis, MN 55455
http://comp.chem.umn.edu/polyrate/
lynch@chem.umn.edu

A.6.7 QCPE

QCPE (the Quantum Chemistry Program Exchange) is a repository for programs that have been contributed by many authors. Hundreds of programs are available with source code. There is no acceptance criteria for including a program, so programs range from those that are simplistic or difficult to use to ones that are very well written and powerful pieces of software. The catalogue is on the Web page listed below and can be searched interactively by opening a telnet session to **qcpe6.chem.indiana.edu** (using the login "anonymous" and then typing "./Catsrch"). For a small fee, updates listing new software submissions can be received.

Price category: student, individual
Platforms: varies from one program to the next
Contact information: QCPE

Creative Arts Bldg. 181
Indiana University
Bloomington, IN 47405
(812) 855-5539
http://server.ccl.net/cca/html_pages/qcpe/index.shtml
qcpe@indiana.edu

A.6.8 SynTree

SynTree (we tested Version 3.0) is a program for finding organic synthesis routes. It uses a retrosynthetic algorithm and a database of known reactions. The database of reactions includes 450 reactions typically included in an undergraduate organic curriculum. The algorithm includes the ability to recognize when protective groups are needed. There are utility programs to add additional reactions.

The program is used by first building the target molecule. It then generates a list of possible precursors. The user can choose which precursor to use and then obtain a list of precursors to it. The reaction name and conditions can also be displayed. Once a satisfactory synthesis route is found, it can be printed without all the other possible precursors included. The drawing mode worked well and the documentation was well written.

This program is marketed as an exploratory tool for undergraduate organic chemistry students. As an educational tool, it is well designed. The program, as is, might also serve as a reminder of possible options for synthetic chemists. It could also be useful to the research community if more reactions are included in future versions.

Price category: student, individual
Platforms: PC, Macintosh
Contact information: Trinity Software, Inc.
607 Tenney Mountain Hwy.
Suite 215
Plymouth, NH 03264
(800) 352-1282
http://www.trinitysoftware.com/
trsoft@lr.net

BIBLIOGRAPHY

Other listings of chemistry software packages are

D. B. Boyd, *Rev. Comput. Chem.* **11**, 373 (1997).

Computational Thermochemistry K. K. Irikura, D. J. Frurip, Eds., Appendix A., American Chemical Society, Washington (1998).

Encyclopedia of Computational Chemistry John Wiley & Sons, New York (1998). Some software packages are mentioned in sequence in this encyclopedia and others are collected at the end of volume 5.

http://server.ccl.net/

http://www.chamotlabs.com/cl/Freebies/Software.html

http://nhse.npac.syr.edu:8015/rib/repositories/csir/catalog/index.html

J. P. Bays, *J. Chem. Ed.* **69**, 209 (1992).

Reviews of individual packages are sometimes published in Journal of Computational Chemistry.

Software can be purchased directly from the company that makes it or through catalogues, such as the following. Some of these have paper catalogues as well as web stores.

http://genamics.com/software/

http://www.ChemStore.com/

http://www.chemsw.com/

http://www.falconsoftware.com/

http://www.biosoft.com/

http://www.trinitysoftware.com/

http://chemweb.com/

http://www.claessen.net/chemistry/soft_en.html

http://www.sciquest.com/

Glossary

The following are definitions of terms relevant to computational chemistry. These definitions are based on common usage in this field. They do not necessarily reflect the dictionary definitions or those in other branches of science.

μVT (microcanonical variational theory) a variational transition state theory technique

ab initio a calculation that may use mathematical approximations, but does not utilize any experimental chemical data either in the calculation or the original creation of the method

accuracy how close a computed value is to the experimental value

adiabatic process a chemical process in which the system does not make a transition from one electronic state to another

Ahhrenius equation mathematical equation for predicting reaction rate constants

AI (artificial intelligence) computer algorithms that mimic some aspects of how people think

AIM (atoms in molecules) a population analysis technique

AM1 (Austin model 1) a semiempirical method

AMBER (assisted model building with energy refinement) a molecular mechanics force field

amu (atomic mass unit) atomic unit of mass

ANO (atomic natural orbital) a way of deriving basis functions

antisymmetric function a function that only changes sign when the identities of two electrons are switched

approximation a numerical estimation of a solution to a mathematical problem

APW (augmented plane wave) a band structure computation method

atomic mass unit (amu) atomic unit of mass

atomic units a system of units convenient for formulating theoretical derivations with a minimum number of constants in the equations

B3LYP (Becke 3 term, Lee Yang, Parr) a hybrid DFT method

basis set a set of functions used to describe a wave function

B96 (Becke 1996) a gradient corrected DFT method

band structure the electronic structure of a crystalline solid

beads individual units in a mesoscale simulation

BLYP (Becke, Lee, Yang, Parr) a gradient corrected DFT method

Bohr atomic unit of length

Boltzmann distribution statistical distribution of how many systems will be in various energy states when the system is at a given temperature

Born–Oppenheimer approximation assumption that the motion of electrons is independent of the motion of nuclei

boson a fundamental particle with an integer spin

BSSE (basis set superposition error) an error introduced when using an incomplete basis set

CAOS (computer aided organic synthesis) a program for predicting a synthesis route

Cartesian coordinates system for locating points in space based on three coordinates, which are usually given the symbols x, y, z or i, j, k

CBS (complete basis set) an *ab initio* method

CC (coupled cluster) a correlated *ab initio* method

CFF (consistent force field) a class of molecular mechanics force fields

CFMM (continuous fast multipole method) a method for fast DFT calculations on large molecules

CHAIN a relaxation method for obtaining reaction paths from semiempirical calculations

charge density (electron density, number density) number of electrons per unit volume at a point in space

CHARMM (chemistry at Harvard macromolecular mechanics) a molecular mechanics force field

CHEAT (carbohydrate hydroxyls represented by external atoms) a molecular mechanics force field

CHelp an electrostatic charge calculation method

CHelpG an electrostatic charge calculation method

CI (configuration interaction) a correlated *ab initio* method

CNDO (complete neglect of differential overlap) a semiempirical method

computational chemistry computer-automated means for predicting chemistry

configuration interaction (CI) a correlated *ab initio* method

conventional integral evaluation algorithm that stores integrals in a file

convergence criteria for completion of a self-consistent field calculation

convex hull a molecular surface that is determined by running a planar probe over a molecule

COOP (crystal orbital overlap population) a plot analogous to population analysis for band-structure calculations

correlation name for the statement that there is a higher probability of finding electrons far apart than close to one another, which is reflected by some but not all *ab initio* calculations

COSMO (conductor-like screening model) a method for including solvation effects in orbital-based calculations

Coulomb's law the statement that like charges repel and unlike charges attract along with the equations for predicting the magnitude of those interactions

coupled cluster (CC) a correlated *ab initio* method

CPHF (coupled perturbed Hartree–Fock) *ab initio* method used for computing nonlinear optical properties

CPU (central processing unit) the part of a computer that does mathematical and logical operations.

CVT (canonical variational theory) a variational transition state theory technique

Davidson–Fletcher–Powell (DFP) a geometry optimization algorithm

De Novo algorithms algorithms that apply artificial intelligence or rational techniques to solving chemical problems

density functional theory (DFT) a computational method based on the total electron density

determinant a mathematical procedure for converting a matrix into a function or number

DFP (Davidson–Fletcher–Powell) a geometry optimization algorithm

DFT (density functional theory) a computational method based on the total electron density

DHF (Dirac–Hartree–Fock) relativistic *ab initio* method

DHF (derivative Hartree–Fock) a means for calculating nonlinear optical properties

diabatic process (nonadiabatic) a process in which the lowest-energy path is followed, even if it is necessary to change from one electronic state to another

diffuse functions basis functions that describe the wave function far from the nucleus

DIIS (direct inversion of the iterative subspace) algorithm used to improve SCF convergence

DIM (diatomics-in-molecules) a semiempirical method used for representing potential energy surfaces

Dirac equation one-electron relativistic quantum mechanics formulation

direct integral evaluation algorithm that recomputes integrals when needed

distance geometry an optimization algorithm in which some distances are held fixed

DM (direct minimization) an algorithm for forcing SCF calculations to converge

DPD (dissipative particle dynamics) a mesoscale algorithm

DREIDING a molecular mechanics force field

dummy atom an atom type, usually given the symbol X, used in specifying a molecular to specify a point in space at which no atom is located

ECEPP (empirical conformational energy program for peptides) a molecular mechanics force field

ECP (effective core potential) a potential function for representing the core electrons in an *ab initio* calculation

EF (eigenvector following) a geometry optimization algorithm

EFF (empirical force field) a molecular mechanics force field

eigenvector following (EF) a geometry optimization algorithm

electron density (charge density, number density) number of electrons per unit volume at a point in space

electronic structure the arrangement of electrons in a molecule

electrostatics results that are implications of Coulomb's law

electrostatic potential (ϕ) a function that gives the energy of interaction with an infinitesimal charge at any position in space (if we assume polarizability is negligible)

empirical a procedure not based purely on mathematical theory

ensemble a conceptual collection of identical chemical systems

ESP (electrostatic potential) normally used to denote charges derived from the electrostatic potential

Fenske–Hall a semiempirical method

Fermi contact density the electron density at the nucleus of an atom (if we assume that the nucleus is an infinitesimal point with a given mass and charge)

fermion a fundamental particle with a half-integer spin

Fletcher–Powell (FP) a geometry optimization algorithm

FMM (fast multipole method) a method for fast DFT calculations on large molecules

force field a set of functions and associated constants that defines the energy expression for molecular mechanics calculations

FP (Fletcher–Powell) a geometry optimization algorithm

freely jointed chain (or random flight) a polymer simulation technique

G1, G2, G3 (Gaussian theory) a method for extrapolating from *ab initio* results to an estimation of the exact energy

G96 (Gill 1996) a DFT method

Gaussian theory (G1, G2, G3) a method for extrapolating from *ab initio* results to an estimation of the exact energy

Gaussian-type orbital (GTO) mathematical function for describing the wave function of an electron in an atom

GAPT (generalized atomic polar tensor) a charge calculation method

GB/SA (generalized Born/surface area) method for computing solvation effects

generalized valence bond (GVB) an *ab initio* method

genetic algorithm an optimization algorithm based on a collection (population) of solutions that combine, mutate, and die to produce subsequent populations by a survival-of-the-fittest process

GIAO (gauge-independent atomic orbitals) technique for removing dependence on the coordinate system when computing NMR chemical shifts or optical activity

GROMOS (Gronigen molecular simulation) a molecular mechanics force field, also the name of a computer program

group additivity an empirical method for computing chemical properties

GTO (Gaussian type orbital) mathematical function for describing the wave function of an electron in an atom

GVB (generalized valence bond) an *ab initio* method

half-electron approximation an algorithm for open-shell semiempirical calculations

Hamiltonian quantum mechanical operator for energy.

hard sphere assumption that atoms are like hard billiard balls, which is implemented by having an infinite potential inside the sphere radius and zero potential outside the radius

Hartree atomic unit of energy

Hartree–Fock (HF) an *ab initio* method based on averaged electron–electron interactions

Hessian matrix the matrix of second derivatives of energy with respect to nuclear motion

HF (Hartree–Fock) an *ab initio* method based on averaged electron–electron interactions

HFS (Hartree–Fock–Slater) a DFT method

homology an algorithm that looks for similar molecules, particularly sequences of peptides or nucleotides

Hückel one of the simplest semiempirical methods

ICVT (improved canonical variational theory) a variational transition state theory technique

IGAIM (individual gauges for atoms in molecules) technique for removing dependence on the coordinate system when computing NMR chemical shifts

IGLO (individual gauge for localized orbitals) technique for removing dependence on the coordinate system when computing NMR chemical shifts

in-core integral evaluation algorithm that stores integrals in memory

INDO (intermediate neglect of differential overlap) a semiempirical method

initial guess an approximate wave function used as the starting point for an SCF calculation

intrinsic reaction coordinate (IRC, MEP, minimum-energy path) the lowest-energy route from reactants to products in a chemical process

IPCM (isosurface polarized continuum method) an *ab initio* solvation method

IRC (intrinsic reaction coordinate, MEP, minimum-energy path) the lowest-energy route from reactants to products in a chemical process

kinetic energy energy that a particle has due to its motion

Klein–Gordon equation for describing relativistic behavior of spin zero particles

Kohn–Sham orbitals functions for describing the electron density in density functional theory calculations

Koopman's theorem a means for obtaining the ionization potential from a Hartree–Fock calculation

LCAO (linear combination of atomic orbitals) refers to construction of a wave function from atomic basis functions

LDA (local density approximation) approximation used in some of the more approximate DFT methods

level shifting algorithm used to improve SCF convergence

LMP2 (local second-order Møller–Plesset) an *ab initio* perturbation theory technique

LORG (localized orbital-local origin) technique for removing dependence on the coordinate system when computing NMR chemical shifts

LSDA (local spin-density approximation) approximation used in more approximate DFT methods for open-shell systems

LSER (linear solvent energy relationships) method for computing solvation energy

MCSCF (multiconfigurational self-consistent field) a correlated *ab initio* method

MEP (IRC, intrinsic reaction coordinate, minimum-energy path) the lowest-energy route from reactants to products in a chemical process

MIM (molecules-in-molecules) a semiempirical method used for representing potential energy surfaces

MINDO (modified intermediate neglect of differential overlap) a semi-empirical method

minimum-energy path (IRC, MEP, intrinsic reaction coordinate) the lowest-energy route from reactants to products in a chemical process

MK (Mertz–Singh–Kollman) an electrostatic charge calculation method

MMFF (Merck molecular force field) a molecular mechanics force field

MMn (MM1, MM2, MM3, MM4, MMX, MM+) names of a family of similar molecular mechanics force fields

MNDO (modified neglect of diatomic overlap) a semiempirical method

model a simple way of describing something that is actually more complex than the model

molecular dynamics a time-dependent calculation in which a molecular mechanics force field is combined with classical equations of motion

molecular mechanics an empirical method for predicting molecular shape and interactions

Møller–Plesset (MPn) correlated *ab initio* method based on perturbation theory

MOMEC a molecular mechanics force field with a semiempirical term for describing transition metals

Monte Carlo a simulation technique that incorporates a random movement of atoms or molecules

Morse potential a function used to describe the energy change due to bond stretching

MPn (Møller–Plesset *n*th-order) correlated *ab initio* method based on perturbation theory

MRCI (multireference configuration interaction) a correlated *ab initio* method

multiconfigurational self-consistent field (MCSCF) a correlated *ab initio* method

multireference configuration interaction (MRCI) a correlated *ab initio* method

NBO (natural bond order) the name of a set of population analysis techniques

NDO (neglect of differential overlap) the fundamental assumption behind many semiempirical methods

neural networks computer algorithms that simulate how the brain works by having many simple units, analogous to neurons in the brain

Newton–Raphson a geometry optimization algorithm

NMR (nuclear magnetic resonance) an analytical chemistry technique

NPA (natural population analysis) one of the NBO population analysis techniques

OPLS (optimized potentials for liquid simulation) a molecular mechanics force field

OPW (orthogonalized plane wave) a band-structure computation method

P89 (Perdew 1986) a gradient corrected DFT method

parallel computer a computer with more than one CPU

Pariser–Parr–Pople (PPP) a simple semiempirical method

PCM (polarized continuum method) method for including solvation effects in *ab initio* calculations

perturbation theory an approximation method based on corrections to a solution for a portion of a mathematical problem

PES (potential energy surface) space of energies corresponding to locations of nuclei ignoring vibrational motion

PLS (partial least-squares) algorithm used for 3D QSAR calculations

PM3 (parameterization method three) a semiempirical method

PMF (potential of mean force) a solvation method for molecular dynamics calculations

potential energy energy that a particle has due to its position, particularly because of Coulombic interactions with other particles

population analysis a method of partitioning the wave function in order to give an understanding of where the electrons are in the molecule

PPP (Pariser–Parr–Pople) a simple semiempirical method

PRDDO (partial retention of diatomic differential overlap) a semiempirical method

PRISM (polymer reference interaction-site model) method for modeling homopolymer melts

PW91 (Perdew, Wang 1991) a gradient corrected DFT method

QCI (quadratic configuration interaction) a correlated *ab initio* method

QMC (quantum Monte Carlo) an explicitly correlated *ab initio* method

QM/MM a technique in which orbital-based calculations and molecular mechanics calculations are combined into one calculation

QSAR (quantitative structure–activity relationship) a technique for computing chemical properties, particularly as applied to biological activity

QSPR (quantitative structure–property relationship) a technique for computing chemical properties

quadratic configuration interaction (QCI) a correlated *ab initio* method

quantum mechanics a mathematical method for predicting the behavior of fundamental particles, which is considered to be rigorously correct when applicable (where the effects of relativity are negligible)

quantum Monte Carlo (QMC) an explicitly correlated *ab initio* method

radial distribution function a function that gives the probability of finding a particle at a given distance from another particle

RAM (random access memory) volatile computer memory

random flight (or freely jointed chain) a polymer simulation technique

RECP (relativistic effective core potential) a potential function for representing the core electrons in an *ab initio* calculation

relativity mathematical theory for describing behavior of particles near the speed of light

restricted (spin-restricted) assumption that particles of different spins can be described by the exact same spatial function, rigorously correct for singlet systems

RIS (rotational isomeric state) a polymer simulation technique

RHF (restricted Hartree–Fock) *ab initio* method for singlet systems

ROHF (restricted open-shell Hartree–Fock) *ab initio* method for open-shell systems

RPA (random-phase approximation) *ab initio* method used for computing nonlinear optical properties

SAC (symmetry-adapted cluster) a variation on the coupled cluster *ab initio* method

SACM (statistical adiabatic channel model) method for computing reaction rates

SAM1 (semi-*ab initio* method one) a semiempirical method

SASA (solvent-accessible surface area) algorithm for computing solvation effects

SCF (self-consistent field) procedure for solving the Hartree–Fock equations

SCI-PCM (self-consistent isosurface-polarized continuum method) an *ab initio* solvation method

SCR (structurally conserved regions) sections of a biopolymer sequence that are identical to that of another sequence, for which there is a known three-dimensional structure

SCRF (self-consistent reaction field) method for including solvation effects in *ab initio* calculations

SDS (synthesis design system) a program for predicting a synthesis route

self-consistent field (SCF) procedure for solving the Hartree–Fock equations

semiempirical methods that are based on quantum mechanics, but also include values obtained through an empirical parameterization

simulated annealing algorithm consisting of a molecular dynamics simulation with a gradually decreasing temperature

SINDO (symmetrically orthogonalized intermediate neglect of differential overlap) a semiempirical method

size-consistent a method is size-consistent if the energy obtained for two molecular fragments at large separation will be equal to the sum of the energies of those fragments computed separately

size-extensive a method is size-extensive if the energy is a linear function of the number of electrons

Slater type orbital (STO) mathematical function for describing the wave function of an electron in an atom, which is rigorously correct for atoms with one electron

SM1–SM5 solvation methods for use with semiempirical and *ab initio* calculations

SMILES (simplified molecular-input line-entry specification) a way of specifying a molecular formula and connectivity, but not the three-dimensional geometry

solvation effects changes in the behavior of a solute due to the presence of the solvent

SOS (sum over states) an algorithm that averages the contributions of various states of the molecule

spin contamination an error sometimes occurring in unrestricted calculations

spin-restricted (restricted) assumption that particles of different spins can be described by the exact same spatial function, rigorously correct for singlet systems

spin-unrestricted (unrestricted) calculation in which particles of different spins are described by different spatial functions

statistical mechanics mathematical theory for computing thermodynamic properties from atomic-scale properties

STO (Slater type orbital) mathematical function for describing the wave function of an electron in an atom, which is rigorously correct for atoms with one electron

TDGI (time-dependent gauge-invariant) *ab initio* method used for computing nonlinear optical properties

TDHF (time-dependent Hartree–Fock) *ab initio* method used for computing nonlinear optical properties

thermodynamics mathematical system for describing energy and entropy in macroscopic chemical systems

theoretical chemistry mathematical means for predicting chemistry

time complexity a way of denoting how much additional computational resources, particularly CPU time, will be used as the size of the system being modeled is increased

TNDO (typed neglect of differential overlap) a semiempirical method for computing NMR chemical shifts

trajectory a sequence of geometries produced by a molecular dynamics simulation

transition structure geometry of a molecular system corresponding to the energy maximum (saddle point) that must be traversed in going from reactants to products

Tripos a molecular mechanics force field, also the name of a company that sells computational chemistry software

TST (transition state theory) method for computing rate constants

UHF (unrestricted Hartree–Fock)

UFF (universal force field) a molecular mechanics force field

unrestricted (spin unrestricted) calculation in which particles of different spins are described by different spatial functions

VTST (variational transition state theory) method for predicting rate constants

VWN (Vosko, Wilks, and Nusair) a DFT method

wave function a function used to describe the electron distribution in a quantum mechanical scheme; the wave function is also called the probability amplitude because the square of the wave function gives the probability of finding an electron

Xα (X alpha) a DFT method

YETI a molecular mechanics force field

zero point energy the energy difference between the minimum on a potential energy surface and the first vibrational energy level

ZINDO (Zerner's intermediate neglect of differential overlap, synonymous with INDO/S) a semiempirical method

Z-matrix a way of writing a molecular geometry

BIBLIOGRAPHY

Other sources that have computatonal chemistry definitions are

M. F. Schlecht, *Molecular Modeling on the PC* Wiley-VCH, New York (1998).

P. W. Atkins, R. S. Friedman, *Molecular Quantum Mechanics* 315, Oxford, Oxford (1997).

P. W. Atkins, *Quanta* Oxford, Oxford (1991).

http://www.mathub.com/glossary/index.html

http://www.iupac.org/recommendations/1996/6802brown/

Index

μVT, 167, 360. *See also* transition state theory
ab initio, 19–31, 252, 284, 288, 332–339, 360. *See also* Quantum mechanics
  accuracy, 138–141
  basis sets, 78–91
  core potentials, 84–85
  time complexity, 130
  vibrational spectrum, 92–98
Absolute hardness, 246
Acceptance ratio, 63
Accuracy, 135, 137–141, 360
ACD, 326
Actinides, 289
ADF, 88, 332
Adiabatic process, 173, 360
Ahhrenius equation, 164, 360. *See also* Reaction rate prediction
Ahlrichs basis, 82, 87
AI, *see* Artificial intelligence
AIM (atoms in molecules), 101, 360. *See also* Population analysis
Alchemy, 323
Alkali metals, 285
Almlöf, Taylor ANO, 88
AM1 (Austin model 1), 36, 360. *See also* Semiempirical
AMBER (assisted model building with energy refinement), 53, 360. *See also* Molecular mechanics
Amorphous solids, 319
AMPAC, 210, 341
AMSOL, 210
amu, *see* Atomic mass unit
Anharmonic frequencies, 94
ANO (atomic natural orbital), 85, 360
Antisymmetric function, 360. *See also* Wave function
Approximation, 2, 136, 360
APW, *see* augmented plane wave
Artificial intelligence (AI), 109, 278, 360
Assisted model building with energy refinement, *see* AMBER
Atomic charges, 102
Atomic mass unit (amu), 360. *See also* Atomic units
Atomic natural orbital, *see* ANO
Atomic units, 9, 360
Atoms in molecules, *see* AIM
Augmented plane wave (APW), 268, 360
Austin model 1, *see* AM1

B3LYP (Becke 3 term, Lee Yang, Parr), 44, 360. *See also* Density functional theory
B96 (Becke 1996), 44, 360. *See also* Density functional theory
Babel, 352
Basis set, 19, 78–91, 360
  accuracy, 89–90
  contraction, 78–81
  core potentials, 84–85
  customization, 231–238
  DFT and, 46
  effects, 80
  excited states, 218
  notation, 81–84
  superposition error, 361
Balaban index, 245
Band gap, 266–268
Band structure, 266–272, 319, 361
Bange, Burrientos, Bunge, Corordan STO, 88
Basch, 87
Bauschlicker ANO, 88
Beads, 273, 361. *See also* Mesoscale
Beeman's algorithm, 61
Berny, 70, 152
Binning/Curtis, 87
Biological activity, 113, 296–301
Biomolecules, 296–301
BFGS, 131
BLYP (Becke, Lee, Yang, Parr), 44, 361. *See also* Density functional theory

Bohr, 361. *See also* Atomic units
Boiling point, 114
Bond order, 245
Boltzmann distribution, 13, 361. *See also* Statistical mechanics
Born model, 210
Born–Oppenheimer approximation, 11, 28, 237, 361. *See also* ab initio
Boson, 361
Brownian dynamics, 273
Brueckner correction, 225
BSSE (basis set superposition error), 361. *See also* Basis set

Canonical variational theory, *see* CVT
CAOS (computer aided organic synthesis), 277, 361. *See also* Synthesis route prediction
Carbohydrate hydroxyls represented by external atoms, *see* CHEAT
CARS, 258
Cartesian coordinates, 67–68, 361. *See also* Molecular geometry
Cartesian *d* & *f* functions, 80
CASSCF, 25
Castro & Jorge universal, 88
CBS (complete basis set), 83, 88–89, 361. *See also* Basis sets
  accuracy, 141
CC, *see* Coupled cluster
CCD, 24
cc-pVnZ, 88
cc-pCVnZ, 88
CCSD, 25
CCSDT, CCSD(T), 25
Central processing unit, *see* CPU
CFF (consistent force field), 54, 361. *See also* Molecular mechanics
CFMM (continuous fast multipole method), 361. *See also* Density functional theory
CHAIN, 361. *See also* Transition structure, Reaction coordinate
Chain-Growth, 186
Chaos theory, 193
Charge, *see* Atomic charges
Charge density, 361. *See also* Electron density
CHARMM (chemistry at Harvard macromolecular mechanics), 53, 361. *See also* Molecular mechanics
CHEAT (carbohydrate hydroxyls represented by external atoms), 54, 361. *See also* Molecular mechanics
CHelp, 102, 361. *See also* Population analysis
CHelpG, 102, 361. *See also* Population analysis
Chem 3D, 324
Chemical accuracy, 3
ChemSketch, 326
CHEOPS, 353
Chemistry at Harvard macromolecular mechanics, *see* CHARMM
Chipman, 87
CI, *see* Configuration interaction
CID, 25
Circular dichrosim (CD), 113
CIS, 216
CISD, 24
CISDT, CISD(T), 24
CISDTQ, 24
Clausius–Mossotti equation, 112
Clementi and Roetti STO, 89
Cluster, 318
CNDO (complete neglect of differential overlap), 34, 361. *See also* Semiempirical
CODESSA, 353
COLUMBUS, 218
Comparative QSAR, 249
Complete basis set, *see* CBS
Complete neglect of differential overlap, *see* CNDO
Computational chemistry, 1, 361
Conductor-like screening model, *see* COSMO
Configuration interaction (CI), 23–24, 216, 218, 361. *See also* Correlation, ab initio
  accuracy, 27
Conformation search, 179–192. *See also* Molecular geometry
  chain-growth, 186
  distance-geometry, 185
  fragment approach, 186
  genetic algorithm, 189
  grid search, 180
  homology, 187–189
  Monte Carlo, 182
  ring systems, 189
  rule-based, 186
  simulated annealing, 183
Conjugate gradient, 70, 131
Connectivity index, 245
Connolly surface, 111
Consistent force field, *see* CFF
Continuous fast multipole method, *see* CFMM

Continuum methods
  liquids, 302
  solids, 318
  solvation, 208–212
Conventional integral evaluation, 79, 361. *See also* Integral evaluation
Convergence, 193–197, 361. *See also* Hartree–Fock
Convex hull, 111, 361.
COOP (crystal orbital overlap population), 270, 362. *See also* Band structure
Core potential(s), 84. *See also* ab initio, Basis sets
Correlation, 21–22, 362. *See also* ab initio
Correlation-consistent basis, 82, 88
Correlation function, 26
Cosine function, 50–53
COSMO (COnductor-like Screening MOdel), 212, 362. *See also* Solvation effects
Coulomb's law, 8, 362
Coupled cluster (CC), 25, 362. *See also* ab initio, Correlation
  accuracy, 27, 140–141
Coupled perturbed Hartree–Fock, *see* CPHF
CPHF (coupled perturbed Hartree–Fock), 259, 362. *See also* Hartree–Fock
CPU (central processing unit), 61, 79, 83, 85, 93, 128–132, 362
CREN, 85
CRENBL, 85
Crystal, 334
Crystal orbital overlap population, *see* COOP
Crystallinity, 311
Cubic potential, 50–53
CVFF, 54
CVT (canonical variational theory), 167, 362. *See also* Transition state theory

3D QSAR, 247
D95, 86
D95V, 86
Database, 108–109
Davidson correction, 224–225
Davidson–Fletcher–Powell (DFP), 70, 362. *See also* Geometry optimization
DCOR, 258
DC-SHG, 258
Defects in crystals, 319
De Novo algorithms, 109, 362. *See also* Artificial intelligence
Density functional theory (DFT), 42–48, 218, 332–339, 362
  accuracy, 47, 137–138
  functionals, 44
  time complexity, 130
Density of states, 269
Derivative Hartree–Fock, *see* DHF
Descriptors, 224
DET, 86
Determinant, 20, 362. *See also* Wave function
DFP, *see* Davidson–Fletcher–Powell
DFT, *see* Density functional theory
DFWM, 258
DHF (Dirac–Hartree–Fock), 262, 362. *See also* Relativity
DHF (Derivative Hartree–Fock), 258, 362
Diabatic process (non-adiabatic), 173, 362
Diatomics-in-molecules, *see* DIM
Dielectric constant, 112
Diffuse functions, 82, 362. *See also* basis set
Diffusivity, 115
DIIS (direct inversion of the iterative subspace), 195, 362. *See also* Hartree–Fock
DIM (diatomics-in-molecules), 177, 362
Dipole moment, 110
Dirac equation, 262, 362. *See also* Relativity
Dirac–Hartree–Fock, *see* DHF
Direct integral evaluation, 79, 362. *See also* Integral evaluation
Direct inversion of the iterative subspace, *see* DIIS
Direct minimization (DM), 362. *See also* Hartree–Fock
Disk space, 79
Dissipative particle dynamics (DPD), 274, 363. *See also* Mesoscale
Distance-geometry, 185, 362. *See also* Conformation search
DM, *see* Direct minimization
DN, 88
Dolg, 85, 89
DOS, 269
DPD, *see* Dissipative particle dynamics
DREIDING, 54, 363. *See also* Molecular mechanics
Duijneveldt basis, 82, 86
Dummy atom, 75, 363. *See also* Molecular geometry
Dunning basis, 82
Dunning–Hay basis, 82, 86

Dynamic mean-field density functional, 273
DZC-SET, 87
DZVP, 88

ECEPP (empirical conformational energy program for peptides), 54, 363. *See also* Molecular mechanics
ECP (effective core potential), 84, 363. *See also* ab initio, Basis sets
EF, *see* Eigenvector following
EFF (empirical force field), 54, 363. *See also* Molecular mechanics
Effective core potential, *see* ECP
Effective potential functions, 319
EFISH, 258
Eigenvector following (EF), 70, 154, 363. *See also* Geometry optimization, Transition structure
Elasticity, 312
Electron affinity, 111, 245
Electron density, 108, 363. *See also* Wave function
  charge density, 361
Electronic excited states, 216–222
  basis functions, 218
  CI, 218
  CIS, 216
  DFT, 218
  Hamiltonian, 218
  initial guess, 217
  path integral, 219
  QMC, 219
  semiempirical, 220
  spectrum intensities, 220
  spin states, 216
  state averaging, 220
  Time-dependent, 219
Electronic spacial extent, 111
Electronic spectrum, 216–222
Electronic-state crossing, 169
Electronic structure, 363
Electronically adiabatic, 173
Electrostatic(s), 8, 363
Electrostatic charges, 102
Electrostatic potential, 102, 363
Empirical, 363
Empirical conformational energy program for peptides, *see* ECEPP
Empirical force field, *see* EFF
Energy, 7–8, 107
Ensemble, 15, 363. *See also* Statistical mechanics
EOKE, 258
EOPE, 258
EPR, 111
ESP, *see* Electrostatic potential
Excited states, *see* Electronic excited states
Expectation value, 11
Expert systems, 278
Explicit solvent calculations, 207
Extended Hückel, 33

Fenske–Hall, 363. *See also* Semiempirical
Fermi contact density, 110, 363
Fermi energy, 270
Fermion, 363
Fletcher–Powell (FP), 70, 131, 363. *See also* Geometry optimization
Fletcher–Reeves, 131
Flory–Huggins, 274
Flexibility, 312
FMM (fast multipole method), 43, 363. *See also* Density functional theory
Force field(s), 363. *See also* Molecular mechanics
  customization, 239–242
  existing, 53–56
  mathematics of, 49–53
FP, *see* Fletcher–Powell
Fragment approach, 186. *See also* Conformation searching
Freely jointed chain (random flight), 363
Frose–Fischer, 88
Full CI, 24
Fuzzy logic, 109

3–21G, 86, 89
6–31G, 86, 89
6–311G, 86, 89
G1, G2, G3, *see* Gaussian theory
G96 (Gill 1996), 363. *See also* Density functional theory
GAMESS, 87, 335
Gasteiger, 103
Gauge-independent atomic orbitals, *see* GIAO
Gaussian, 336
Gaussian theory (G1, G2, G3), 38–39, 83, 89, 363
  accuracy, 141
Gaussian type orbital (GTO), 19, 79–80, 363–364. *See also* Basis set

GaussView, 349
GAPT (generalized atomic polar tensor), 364
GB/SA (generalized Born/surface area), 211, 364. *See also* Solvation effects
GDIIS, 70
Gear predictor-corrector, 61
General contraction, 79
Generalized atomic polar tensor, *see* GAPT
Generalized Born/surface area, *see* GB/SA
Generalized valence bond (GVB), 25, 364. *See also* ab initio, Correlation
Generally contracted basis, 78
Genetic algorithm, 184, 364. *See also* Conformation search
Geometry, *see* Molecular geometry
Geometry optimization, 68–71. *See also* Conformation search, Molecular geometry
GIAO (gauge-independent atomic orbitals), 113, 252, 364.
Glass-transition temperature, 313
gNMR, 354
Gradient-corrected DFT functional, 43
Graphics software, 349–351
Grid search, 180
GROMOS (Gronigen molecular simulation), 54, 61, 364. *See also* Molecular mechanics
Group additivity, 108, 208, 283, 364
GTO, *see* Gaussian type orbital
GVB, *see* Generalized valence bond

Half-electron approximation, 229, 364. *See also* Semiempirical
Hamiltonian, 10–11, 218
Hammond postulate, 153
Hard sphere, 165, 364
Harmonic oscillator, 50–53, 92–94
Hartree, 364. *See also* Atomic units
Hartree–Fock (HF), 19–20, 364. *See also* ab initio
  accuracy, 27
  convergence, 193–197, 361
Hartree–Fock–Slater (HFS), 44, 364. *See also* Density functional theory
Hay, 87, 89
Hay–Wadt, 84–85, 89
Hellmann–Feynman theorem, 12
Hessian matrix, 70, 151, 364
HF, *see* Hartree–Fock
HFS, *see* Hartree–Fock Slater
Homology, 187, 364. *See also* Conformation search
Hückel, 33, 364. *See also* Semiempirical
Huzinaga basis, 82, 87
Hybrid DFT functionals, 43
Hyperbolic map functions, 177
Hyperchem, 328
Hyperfine coupling, 112
Hyperpolarizability, 256

ICVT (improved canonical variational theory), 167, 364. *See also* Transition state theory
IDRI, 258
IGAIM (individual gauges for atoms in molecules), 253, 364. *See also* NMR spectrum prediction
IGLO (individual gauge for localized orbitals), 252, 364. *See also* NMR spectrum prediction
IMOMM, 201
IMOMO, 201
Improved canonical variational theory, *see* ICVT
In core integral evaluation, 79, 364. *See also* Integral evaluation
Individual gauges for atoms in molecules, *see* IGAIM
INDO (intermediate neglect of differential overlap), 35, 364. *See also* Semiempirical
INDO/S, *see* ZINDO
Information content, 245
Initial guess, 20, 194, 217, 365. *See also* Hartree–Fock
Integral evaluation, 44, 79
Intermediate neglect of differential overlap, *see* INDO
Intrinsic reaction coordinate (IRC), 154, 365. *See also* Reaction coordinate
Ionization potential, 111, 245
IPCM (isosurface polarized continuum method), 212, 365. *See also* Solvation effects
IRC, *see* Intrinsic reaction coordinate
Isosurface polarized continuum method, *see* IPCM

Jaguar, 337
Jahn–Teller distortion, 175

Kier and Hall indices, 245
KKR, 269
Kinetic energy, 365
Kirkwood equation, 112
Klein–Gordon, 365. *See also* Relativity
Koga, Saito, Hoffmeyer, Thakkar, 87
Koga, Tatewaki, Thakkar STO, 89
Koga, Watanabe, Kunayama, Yasuda, Thakkar STO, 89
Kohn–Sham orbitals, 42, 365. *See also* Density functional theory
Koopman's theorem, 112, 365. *See also* Ionization potential

Labor cost, 132
LANL2DZ, 85, 89
Lanthanides, 289
LCAO (linear combination of atomic orbitals), 268, 365
LDA (local density approximation), 43, 365. *See also* Density functional theory
Least motion path, 161
Leonard–Jones, 50–53
LEPS, 177
Level shifting, 194, 365. *See also* Hartree–Fock
Linear combination of atomic orbitals, *see* LCAO
Linear scaling, 43–45
Linear solvent energy relationships, *see* LSER
Linear synchronous transit (LST), 153
Liquids, 64, 302–306
LMP2 (local second order Møller–Plesset), 23, 365. *See also* Møller–Plesset
Local density approximation, *see* LDA
Local second order Møller–Plesset, *see* LMP2
Local spin density approximation, *see* LSDA
Localized orbital-local origin, *see* LORG
Log P, 115
LORG (localized orbital-local origin), 253, 365. *See also* NMR spectrum prediction
Löwdin population analysis, 100
LSDA (local spin density approximation), 43, 365. *See also* Density functional theory
LSER (linear solvent energy relationships), 207, 365. *See also* Solvation effects

MacroModel, 344
Main group inorganics, 285
MAXI, 86
McLean/Chandler, 87
MCSCF (multi-configurational self consistent field), 24–25, 365, 366. *See also* ab initio, Correlation
MedChem Explorer, 355
Melting point, 114
MEP (minimum energy path), 159, 365. *See also* Reaction coordinate
Merck molecular force field, *see* MMFF
Mertz–Singh–Kollman, *see* MK
Mesoscale, 273–276
  Brownian dynamics, 273
  DPD, 274
  dynamic mean-field density functional, 274
  validation, 275
Microcanonical variational theory, *see* μVT
MIDI, 86
MIM (molecules-in-molecules), 177, 365
MINDO (modified intermediate neglect of differential overlap), 34, 365. *See also* Semiempirical
MINI, 86
Minimum energy path, *see* MEP
MK (Mertz–Singh–Kollman), 102, 365. *See also* Population analysis
MMFF (Merck molecular force field), 55, 365. *See also* Molecular mechanics
MMn (MM1, MM2, MM3, MM4, MMX, MM+), 55, 365. *See also* Molecular mechanics
MNDO (modified neglect of diatomic overlap), 34, 366. *See also* Semiempirical
Model, 2, 366
Modified intermediate neglect of differential overlap, *see* MINDO
Modified neglect of diatomic overlap, *see* MNDO
MOE, 345
Molden, 350
Molecular descriptors, 244
Molecular dynamics, 60–62, 319, 344–348, 366
  algorithm, 60–61

time complexity, 130
Molecular geometry, 67–72, 107
algorithms, 70, 151, 152
coordinate space, 68–70
level of theory, 70
optimization, *see* Conformation search, Geometry optimization
Molecular mechanics, 49–59, 148, 283, 287, 344–348, 366
accuracy, 57, 137
force field customization, 239–242
time complexity, 130
Molecular vibrations, 92–98, 246, 314
anharmonic, 94
correction factors, 93
harmonic, 92–94
peak intensities, 95,
thermodynamic corrections, 96
zero point energy, 96
Molecular volume, 111, 245
Molecules-in-molecules, *see* MIM
Møller–Plesset (MPn), 22–23, 366. *See also* Perturbation theory
accuracy, 27, 139–140
MOLPRO, 338
MOMEC, 55, 366
Monte Carlo, 62, 182, 319, 344–348, 366
algorithm, 62–63
MOPAC, 342
Morse potential, 50–53, 366. *See also* Force Field
Mott–Littleton, 271, 319
MPn (Møller–Plesset nth order, MP2, MP3, etc.), *see* Møller–Plesset
MRCI (multi-reference configuration interaction), 25, 366. *See also* Configuration interaction
MSXX, 309
Mulliken population analysis, 99
Multi-configurational self consistent field, *see* MCSCF
Multipole moments, 110, 245
Multi-reference configuration interaction, *see* MRCI

NASA Ames ANO, 88
Natural bond order, *see* NBO
Natural orbitals, 27
Natural population analysis, *see* NPA
NBO (natural bond order), 100, 366. *See also* Population analysis
NDO, *see* Neglect of differential overlap
Neglect of differential overlap (NDO), 34, 366. *See also* Semiempirical
Neural networks, 109, 253, 366. *See also* Artificial intelligence
Newton–Raphson, 70, 366. *See also* Geometry optimization
NMR (nuclear magnetic resonance), 366
ab initio, 252
empirical, 253
semiempirical, 253
spectrum prediction, 111, 252–255
Noble gases, 285
Non-adiabatic, *see* Diabatic process
Non-linear optical properties, 256–260
NPA (natural population analysis), 100, 366. *See also* Population analysis
Nuclear magnetic resonance, *see* NMR
Nuclear Overhauser effect (NOE), 185
Number density, *see* Electron density
NWChem, 329

Octupole moment, 110
OKE, 258
ONIOM, 201
OPLS (optimized potentials for liquid simulation), 55, 61, 366. *See also* Molecular mechanics
Optical activity, 113
Optical rotary dispersion (ORD), 113
Optimized potentials for liquid simulation, *see* OPLS
OPW (orthogonalized plane wave), 269, 366
OR, 258
Orbital, 19
basis sets for, 78–91
Organic molecules, 283–285
Orthogonality, 218–219
Orthogonalized plane wave, *see* OPW

Parallel computer, 132, 366
Parameterization method three, *see* PM3
Pariser–Parr–Pople, *see* PPP
Partial least squares, *see* PLS
Partial retention of diatomic differential overlap, *see* PRDDO
Partridge uncontracted sets, 87
Path integral, 219
PCA, 248
PCFF, 309

PCM (polarized continuum method), 212, 366. *See also* Solvation effects
PC Model, 347
Periodic boundary conditions, 303–305
Perturbation theory, 366. *See also* Møller–Plesset
PES (potential energy surface), 173–178, 367
  analytic, 176
  properties, 173
  semiempirical, 177
Phase space, 12
PLS (partial least squares), 248, 367
PM3 (parameterization method three), 37, 367. *See also* Semiempirical
PM3/TM, 37, 288
PMF (potential of mean force), 367
PMP2, 229
Poisson–Boltzmann, 209
Poisson equation, 9, 209
Polak–Ribiere, 70, 131
Polarizability, 245, 256
Polymers, 307–317
POLYRATE, 169, 356
Potential energy, 367
Potential energy surface, *see* PES
Potential of mean force, *see* PMF
Pople basis sets, 82
Population analysis, 99–106, 367
  AIM, 101
  charges, 102
  Löwdin, 100
  Mulliken, 99
  NBO, 100
Powell, 131
PPP (Pariser-Parr-Pople), 33, 366, 367. *See also* Semiempirical
PRDDO (partial retention of diatomic differential overlap), 36, 367. *See also* Semiempirical
PRISM (polymer reference interaction site model), 367
Pseudo reaction coordinate, 154
PUHF, 229
Pullman, 103
Pure *d* & *f* functions, 80
PW91 (Perdew, Wang 1991), 44, 367. *See also* Density functional theory

Q-Chem, 339
QCI (quadratic configuration interaction), 26, 367. *See also* Configuration Interaction
QCPE, 357
Q-equilibrate, 103
QMC, *see* Quantum Monte Carlo
QMFF, 54
QM/MM, 198–205, 367
QSAR (quantitative structure activity relationship), 108, 114, 367. *See also* Structure-activity relationships
QSPR (quantitative structure property relationship), 108, 308, 314, 367. *See also* Structure-activity relationships
Quadratic configuration interaction, *see* QCI
Quadratic synchronous transit (QST), 153
Quadrupole moment, 110
Quantitative structure activity relationship, *see* QSAR
Quantitative structure property relationship, *see* QSPR
Quantum mechanics, 10–12, 367. *See also* ab initio, Semiempirical
Quantum Monte Carlo (QMC), 26–27, 219, 367. *See also* Correlation, ab initio time complexity, 130
Quasiclassical calculation, 168
Quasi-Newton, 70, 131, 152

Radial distribution function, 15–16, 367. *See also* Statistical mechanics
RAM (random access memory), 79, 367
Randic index, 245
Random access memory, *see* RAM
Random flight, *see* Freely jointed chain
Random phase approximation, *see* RPA
Reaction coordinate, 154, 159–163
  least motion path, 161
  minimum energy path, 159
  reaction dynamics, 162
  relaxation method, 161
Reaction dynamics, 162
Reaction rate prediction, 104–172
  Arrhenius equation, 164
  electronic-state crossing, 169
  hard-sphere theory, 165
  relative rates, 165
  statistical, 168
  trajectory calculation, 167
  transition state theory, 166
  variational transition state theory, 166
RECP (relativistic effective core potential), 84, 262, 367. *See also* Relativity, Basis sets
Redundant internal coordinates, 69

Refractivity, 245
Relative permitivity, 112
Relative reaction rates, 165
Relativistic effective core potential, *see* RECP
Relativity, 261–265, 367
Relaxation method, 155, 161
Research project, 135–142
Restricted, *see* RHF
Restricted Hartree–Fock, *see* RHF
Restricted open shell Hartree–Fock, *see* ROHF
Ring systems, 189
RIS (rotational isomeric state), 308, 367
RHF (restricted Hartree–Fock), 20, 368. *See also* Hartree–Fock
ROHF (restricted open shell Hartree–Fock), 228, 368. *See also* Hartree–Fock
Roos augmented, 88
Ross and Seigbahn, 87
Roos, Veillard, Vinot, 88
Rotational isomeric state, *see* RIS
RPA (random phase approximation), 259, 368
Rule-based systems, 186
Rydberg functions, 82

SAC (symmetry adapted cluster), 26, 368
SACM (statistical adiabatic channel model), 168, 368
Sadlej, 87
SAM1 (semi ab initio method one), 38, 368
SASA (solvent accessible surface area), 368
SBKJC, 84, 89
SCF, *see* Self-consistent field
Schrodinger equation, 10
SCI-PCM (Self consistent isosurface polarized continuum method), 212, 368. *See also* Solvation effects
SCR (structurally conserved regions), 368
SCRF (self consistent reaction field), 211, 368. *See also* Solvation effects
SDS (synthesis design system), 277, 368. *See also* Synthesis route prediction
Segmented basis, 78
Self-consistent field (SCF), 368. *See also* Hartree–Fock
  convergence, 193–197
Self consistent isosurface polarized continuum method, *see* SCI-PCM
Self consistent reaction field, *see* SCRF
Semi ab initio method one, *see* SAM1
Semidirect integral evalutaion, 79
Semiempirical, 32–41, 177, 220, 253, 284, 287, 340–343, 368. *See also* Quantum mechanics
  accuracy, 137
  AM1, 36
  CNDO, 34
  Extended Hückel, 33
  Fenske-Hall, 37
  Hückel, 33
  INDO, 35
  MINDO, 34
  MNDO, 34
  PM3, 37
  PM3/TM, 37
  PPP, 33
  PRDDO, 36
  SAM1, 38
  SINDO1, 35
  TNDO, 37
  ZINDO, 35
Shadow indices, 245
Shape factor, 245
SHG, 258
Siegbahm correction, 225
Simplex, 70, 131, 152
Simplified molecular input line entry specification, *see* SMILES
Simulated annealing, 183, 368. *See also* Conformation search
SINDO (symmetrically orthogonalized intermediate neglect of differential overlap), 368. *See also* Semiempirical
Size-consistent, 223–226, 368
Size-extensive, 223–226, 368
Slater type orbital, *see* STO
SM1–SM5, 210–211, 368. *See also* Solvation effects
SMILES (simplified molecular input line entry specification), 67, 368. *See also* Molecular geometry
Software, 322–359
Solids, 318–321
Solubility, 115, 314
Solvation effects, 155, 206–215, 369
  analytic equations, 207
  continuum, 208–212
  explicit solvent, 207
  group additivity, 208
Solvent accessible surface area, *see* SASA
Solvent-excluded volume, 111

SOS (sum over states), 258, 369
Spaghetti plot, 266
Spartan, 330
Spin contamination, 227–230, 369
  ROHF, 228
  semiempirical, 229
  spin projection, 229
Spin Eigenfunctions, 228
Spin projection, 229
Spin restricted wave function, 367, 369. *See also* RHF
Spin states, 216
Spin unrestricted, *see* UHF
SSMILES, 62
State averaging, 220
Statistical adiabatic channel model, *see* SACM
Statistical mechanics, 12–16, 369
STD-SET, 87
Steepest descent, 70
STO (Slater type orbital), 80, 368, 369
STO-nG, 86, 89
Stromberg, 87
Structurally conserved regions, *see* SCR
Structure-activity relationships, 243–251
Structure-property relationships, *see* structure-activity relationships
Stuttgart, 85
SUHF, 229
Sum over states, *see* SOS
Surface tension, 114
Surfaces, 318, 321
Symmetrically orthogonalized intermediate neglect of differential overlap, *see* SINDO
Symmetry, 125–127, 151
Symmetry adapted cluster, *see* SAC
Synthesis design system, *see* SDS
Synthesis route prediction, 277–280
SynTree, 358

Taft steric constant, 245
TCSCF, 25
TDGI (time dependent gauge-invariant), 259, 369
TDHF (time dependent Hartree–Fock), 259, 369
Thermal stability, 315
Thermodynamics, 9, 369
Theoretical chemistry, 1, 369
THG, 258
Time complexity, 128–132, 369
Time-dependent calculation, 219
Time step, 61
TINKER, 348
TNDO (typed neglect of differential overlap), 37, 253, 369. *See also* Semiempirical, NMR spectrum prediction
Trajectory, 167, 369. *See also* Molecular dynamics
Transition metals, 286–288
Transition state, *see* Transition structure(s)
Transition structure(s), 147–158, 369
  algorithms, 151
  level of theory, 149
  molecular mechanics, 148
  potential surface scans, 155
  reaction coordinates and, 154
  relaxation methods, 155
  solvent effects, 155
  symmetry use, 151
  verifying, 155
Transition state theory (TST), 166, 369. *See also* Reaction rate prediction
Tripos, 55, 369. *See also* Molecular mechanics
Typed neglect of differential overlap, *see* TNDO
TZVP, 88

UBCFF, 54
UHF (unrestricted Hartree–Fock), 21, 369. *See also* Hartree–Fock
UFF (universal force field), 56, 369. *See also* Molecular mechanics
Ultraviolet spectrum, 216–222, 314
UniChem, 331
Universal force field, *see* UFF
Unrestricted, *see* UHF
UV spectrum, *see* Ultraviolet spectrum

VAMP, 253
Van der Waals, 50–53, 111
Vapor pressure, 115
Variational transition state theory, *see* VTST
VB-CT, 259
Veillard, 87
Velocity Verlet, 61
Verlet, 61
Vibration of molecules, *see* Molecular vibration

Vibrational circular dichroism (VCD), 113
Vibronic coupling, 175
Visible spectrum, 216–222, 314
Visualization, 115–121
Volume, molecular, *see* Molecular volume
VTST (variational transition state theory), 166, 369. *See also* Transition state theory
VWN (Vosko, Wilks and Nusair), 370. *See also* Density functional theory

Wachters, 87, 89
Wave function, 10, 108, 370. *See also* Quantum mechanics
  symmetry, 127
WebLab Viewer, 351
Weiner index, 245

Xα (X alpha), 43, 370. *See also* Density functional theory
XCIS, 217

YAeHMOP, 343
YETI, 56, 370. *See also* Molecular mechanics

Zerner's intermediate neglect of differential overlap, *see* ZINDO
Zero point energy, 96, 370
ZINDO (Zerner's intermediate neglect of differential overlap, INDO/S), 35, 220, 288, 370. *See also* Semiempirical
Z-matrix, 66–67, 73–77, 370. *See also* Molecular geometry
ZORA, 263